S0-CKI-787

Physics of
Laser Driven Plasmas

Physics of Laser Driven Plasmas

HEINRICH HORA

University of New South Wales
Kensington-Sydney, Australia

A WILEY-INTERSCIENCE PUBLICATION

JOHN WILEY & SONS New York • Chichester • Brisbane • Toronto

Library of Congress Cataloging in Publication Data:

Hora, Heinrich, 1931-
Physics of laser driven plasmas.

"A Wiley-Interscience publication."
Includes bibliographical references and indexes.
1. Laser plasmas. I. Title.
QC718.5.L3H67 530.4′4 80-39792
ISBN 0-471-07880-8

Printed in the United States of America

10 9 8 7 6 5 4 3 2 1

To Sir Mark Oliphant
The Founder of Fusion Energy

Preface

The advent of the laser was a substantial confirmation of Einstein's derivation of Planck's radiation law, the discovery of quantum physics. It opened a huge new industry of quantum optics influencing our daily life, technology, and advanced defense. The interaction of laser radiation with matter opened new basic physics. When the response of dielectric materials was extended to numerous nonlinearities known before with all kind of applications in opto-electronics and communication, a much stranger physics appeared at the very high intensities where all materials are vaporized and ionized, and the response of the generated plasmas exceeded all of the previously known curiosities.

Just after generating plasmas of about 50,000° K, emitting ions of few eV energy, slightly higher intensities produced keV ions. Nuclear fusion reactions were ignited, but highly nonthermal electron energy distributions were detected by x-rays and anomalous fast groups of ions appeared. The relativistic change of the electron mass when quivering in the strong laser field results in relativistic optical constants causing quick shrinking of the laser beam to one wavelength diameter. After this theoretical prediction, ions with energies exceeding 10 MeV were observed flying against the laser beam in agreement with the theoretical expectations. The high-density plasma always realizes what could not be done by the best optical lenses: to squeeze a beam to wavelength diameter. Laser intensities of 10^{20} W/cm^2 (10^{21} times brighter than the sun) and of fields of 3×10^{11} V/cm^2 and more have been achieved, and even the excitation of nuclei by lasers (C. Yamanaka *et al.*) was done before 1980.

Though the problems to be overcome are yet very complex, there is a definite possibility of producing fusion energy by lasers, which would be cleaner and safer than the energy from coal and require negligible cost for fuel and its transportation: the basis for the golden age. Most advanced technologies for lasers, diagnostics, analysis, and theories have been developed for this purpose. Compressions of matter to densities of more than 100 times that of the solid state have been achieved as well as genuine temperatures above 100,000,000° K.

One relatively new principle was derived from these very complex and very special applications: the obvious assumption that any nonlinear approximation permits further simplifications and approximations in the initial assumptions, is wrong. Totally contradictory results may then be produced. It was found that nonlinear theory needs more accurate initial models (exact solutions of Maxwellian equations, etc.) than linear theory. This result is the *postulate of accuracy in nonlinearity* and may also be of value for similar nonlinearities in the unsolved theories of nuclear forces and elementary particles.

An introduction to the physics of these processes of the driving of plasmas by lasers has been much desired. Despite the very intensive research, however, only few steps have been taken to present a monograph for introduction. The difficulty is that the differences of the view of various research groups are too great. This book was written to present the basic facts of plasmas without necessitating cross-referencing; however, an extensive bibliography of the original literature is included for the interested reader.

This introduction is formulated for advanced students and for graduates who have heard nothing of plasmas before. After a short overview of the various phenomena of laser produced plasmas, the quantities as plasma frequency, Debye length, collision frequency, and the Boltzmann equation are derived *ab initio* on which the macroscopic hydrodynamics of plasmas is based. The optical constants are then nonlinear and relativistic. The Maxwellian theory is developed in a straightforward manner for the complicated waves in inhomogeneous plasmas. For the plasma dynamics at laser irradiation, the equation of motion with the nonlinear terms of electromagnetic interaction needs more attention to explain the role of the different terms and its relation to the Maxwellian stress tensor. We then find ponderomotive and nonponderomotive terms of the general nonlinear force.

Analytic evaluations of these results of physics arrives at momentum transfer, ion energies of laser-nonlinear force driven plasma, the Abraham–Minkowski problem, and the parametric instabilities. The numerical treatment showed the generation of density minima and profile steepening (cavitons), as measured later. Direct (highly efficient) transfer of optical energy into fast moving "cold" thick blocks of plasmas by nonlinear forces and soliton processes are seen. Continuing the general derivations for oblique incidence of radiation leads to striated motion and resonance absorption. Laser beams include a basic problem for the radial forces, leading to self-focusing by ponderomotive forces or by relativistic effects. Consequences of and applications for fast ion generation and nuclear fusion by lasers are given.

Thanks are due to Dr. M. N. Nicholson–Florence for assistance with corrections, to my secretary, Mrs. Ingrid Varley for typing, Mr. B. Thews of Datronics Sydney North for help with a Videc word processor, to Novalux Pty. Limited for continuous support, and last but not least to Rosemarie Hora, my wife, for her assistance, and to her and our children for their understanding when I was overabsorbed in the preparation of this book.

HEINRICH HORA

Nagoya, Japan
April 1980

Contents

Physics of
Laser Driven Plasmas

ONE

Aim and Scope

The study of laser produced plasmas is one of the fastest growing fields of present-day physics. It has brought about numerous innovations in materials treatment, such as quality change, welding, drilling, and related high-power beam weapons; the most exciting goal is the safe production of clean nuclear fusion energy with inexhaustive and low-cost fuel. Research projects costing several hundred million dollars have been established, some of them involving hundreds of physicists.

Despite these facts, no books have been published as monographs before 1980, apart from a short introduction [1] or a digest of the voluminous literature [2]. The first serious monograph in the classical sense was published by Hans Motz [3], while an introduction to the more basic problems of nonlinear phenomena found a preliminary formulation as lecture notes [4]. The hectic development can be seen from the fact that conference digests mostly contain short abstracts only. The Boston Conference of the Plasma Physics Division of the American Physical Society in November 1979 presented 317 papers about laser produced plasmas, while the topic conference of the Optical Society of America in San Diego, February 1980, presented—after strong refereeing—papers of 520 authors. Conferences with the intention of reviewing significant highlights using quickly printed proceedings [5] provided a way to present a synopsis of the level achieved within some set period of time.

1.1 Basic Aspects

The difficulty in finding the right motivation for presenting the new field with its obvious attractions is not so much a common denominator of the phenomena involved; it is more the question of the consciousness that really new physics has been opened up. This cannot simply be seen within the overwhelming pluralism of phenomena which individually look very trivial and without significance. There is not so much the task of advertising the fascinating applications but rather the view that one can be attracted by reasons of general physics. Singularities in this direction as for example, relativistic self-focusing, generation of high-Z GeV ions of very high density, or the way to laser induced pair production, and so on, are not the only points of significance. There is much more to report than the whole synopsis of a new physics which is on the rise.

At the beginning of this century there was one single phenomenon that was contrary to all preceding knowledge and that had been confirmed step by step in all the known world: the atomistic structure of action (quantization). This is now classical knowledge that nobody would have a doubt about.

What is dominant in the laser produced plasmas is not just one single phenomenon. It is the breakthrough of the dominance of nonlinear phenomena which, at least to some extent, have been known in part for a long time. What then, is new? It is not only the concentration of old and new nonlinear phenomena. As we shall see in detail in the simple problem of a *correct description of a laser beam and its mechanical interaction with a surrounding plasma* [6], *the nonlinear description is satisfying only if the basic physics (Maxwellian theory) is used exactly without approximations.* This fact explains the dangerous confusion that can occur with any approximate extension to nonlinearity. The more complex a nonlinear extension is, the more precise the basis must be. The desperate question on the philosophy is, then, how can we trust the basic concepts not to be embedded into a more general nonlinear background of interrelations or response?

It is the aim of this monograph to address this very basic question, which might be called *nonclassical* from the present point of knowledge. This point should appeal to the broader community of physicists who may be attracted by studying this complex and very special and apparently applied field, where nothing should be taken away from the importance of its applications for changing our basic technology, energy sources, and world-wide security.

The very complex view should not be astonishing if one remembers the painful steps in exploring the world of the high-temperature plasma. This can be seen from the description of one of the fathers of the plasma theory, Hannes Alfvén, when he wrote shortly after having received his Nobel Prize:

> The study of plasma physics developed along two parallel lines. The first one, originating about a century ago, comprised investigations into electrical dis-

> charges in gases. This approach was, to a great extent, experimental and phenomenological; only very slowly did it reach some degree of theoretical sophistication. Most theoretical physicists looked down on this field, which was complicated and awkward. The plasma exhibited striations and double layers; the electron distribution was non-Maxwellian; there were all sorts of oscillations and instabilities. In short, it was a field not at all suited for mathematically elegant theories.
>
> The other approach to plasma physics came from the highly developed kinetic theory of ordinary gases. It was thought that with a limited amount of work this field could be extended to include ionized gases as well. The theories were mathematically elegant, and the consequences of them showed that it should be possible to produce a very hot plasma and confine it magnetically. This was the starting point of thermonuclear research.
>
> However, the theories had initially very little contact with experimental plasma physics, and all the awkward and complicated phenomena that had been treated in the study of discharges in gases were simply neglected. The result of this was what has been called the 'thermonuclear crisis,' some ten years ago. It taught us that plasma physics is a very difficult field, which can only be developed by a close cooperation between theory and experiments [7].

The nonlinearities in the response of plasmas to laser irradiation will give a further magnitude of difficulties. This monograph, indeed, can only open one gate to describe this new world of physics (including applications in astrophysics or laboratory studies of matter of more than one thousand times solid-state density). This is a beginning, not a solution. It may be that we are as Winston Churchill put it at the end of the beginning.

1.2 Limitations

The scope of the following consideration is limited against low laser intensities and low laser powers. As soon as the laser radiation *produces an irreversible process* in an irradiated material, it should be of interest in our view. No reversible processes, for example, self-focusing of laser beams in a liquid without plasma production or the frequency doubling due to the nonlinear dielectric constant in a solid, will be discussed.

The phenomena of irreversible processes with the lowest possible laser intensities are, for example, the generation or the annealing of crystal defects. This has been a very wide-open field since 1977 [8], though it could have been opened up much earlier as the necessary laser techniques were available long before 1977.

Low-intensity effects of materials processing create defects of such high density that mechanical destruction of the material occurs. These processes are well known for irradiation with electron beams, where the whole crystal lattice is deformed before breaking into parts [8].

The process of laser induced gas breakdown will be treated only marginally. This field received considerable attention in the earlier years of laser development. Besides its intrinsic importance, it was also the forerunner of the field of laser-solid interactions and stimulated many important diagnostic techniques. The first laser produced gas breakdown was achieved by Terhune et al. [9] at Ford research laboratories, followed by Meyerand et al. [10] at United Aircraft. That this field is far from a reasonable understanding has been shown by Papoular [11], who observed many complex phenomena over several orders of magnitude of laser intensity and gas parameters. This very complex situation is demonstrated by some examples: the observation of luminosity before breakdown; the generation of free electrons without breakdown; the appearance of breakdown field strengths corresponding to a multiphoton breakdown process where the ionization energy was only 5 eV, while the gases under investigation had ionization energies greater than 10 eV. A review of the field of laser induced gas breakdown has been given by Zaidel et al. [12], an early pioneer.

The processes of the laser induced gas breakdown will be touched upon in the following discussion on the self-focusing of laser radiation in plasmas, and also when the influence of low-pressure gas surrounding the target is considered. Some very significant experiments are also mentioned for example, the measurement of the polarization dependent emission of electrons from a laser produced gas breakdown experiment by Yablontovich et al. [13] and the study of the nonlinear radiation forces in the laser breakdown of extremely low-density gases [14].

Another limitation in scope is the upper limit to the available laser intensities. This is quite open due to continuing improvements in technology. The growth has been very rapid since the discovery of the laser in 1960. The available laser power of 10 kW in 1961 has grown to beyond 10 TW in 1978. Laser intensities achieved in 1978 by focusing laser beams in vacuum have reached more than 10^{18} W/cm^2, while the nonlinear interaction of such single terawatt laser beams with plasmas of sufficiently high density can self-focus in plasma to intensities exceeding 10^{21} W/cm^2 [15], as concluded indirectly from the observation of MeV ions accelerated by the fields generated when the laser beam underwent relativistic self-focusing [16].

1.3 Lasers

The more advanced lasers should be mentioned without dwelling too much on the problems of their physics and development. Presently the most commonly used laser for high-power research is the neodymium glass laser. Its wavelength is 1.06 μm and its pulse duration can be anywhere between 170 fsec (=0.17 psec) [17] and cw (continuous-wave) operation. The power in beams of 25 cm diameter, using disc laser amplifiers in the last stages, can be more than 1 TW with pulse durations of more than 10 psec (usually between 0.1 and 3 nsec [18]). One system (SHIVA) uses 20 such beams and was specially developed for laser-fusion experiments. It produces 20 TW laser pulses of 0.1 nsec duration [19]. The extension of single-beam glass amplifiers to 2.5 m diameter is under design [18], and one beam will be capable of producing 100 TW in 0.1 nsec. Another large laser system is DELFIN [20], which has 216 output beams, each of 45 mm diameter. The beams produce 50 J pulses in 0.1 nsec. A laser system using glass slabs as multipass amplifiers with a final beam cross section of 32×100 cm^2, called UMI-35, is in the design stage and should produce 10 TW in pulses between 10 psec and 1 nsec [21]. Independent designs of SHIVA-like systems are underway at LLE (University of Rochester), at the Institute of Laser Engineering (University of Osaka), at Limeil (France), and at the Shanghai Institute (Chinese Academy of Science). Other projects of similar size to the ones mentioned are at the Naval Research Labs, Washington D.C., and at the Rutherford Labs in England.

The advantage of the neodymium glass laser is the highly developed technology, which makes its use preferable to other systems despite the well-known disadvantages: nonlinear refractive index, thermal birefringence of the laser glass and the low efficiency of the laser system, which does not transfer more than 1% of electrical energy into laser energy. Exceptions with respect to the efficiency are the quasi-cw tungsten filament pumped glass laser of 3% [22], or the attempts to pump by GaAs-type semiconductor lasers [23] reaching 6%. It has been shown that an efficiency of 20% can be reached [24], if the pumping is made by laser diodes emitting at a wavelength near 900 nm, where the diodes are assumed to operate with 100% efficiency. Using erbium instead of neodymium in glass, a laser of 1.54 μm wavelength results in an efficiency of 70% if pumped by 1 μm radiation [25].

Lasers, which avoid the solid-state problems of nonlinear refractive index and thermal birefringence, and provide high repetition rates up to 1 kHz and more at high-power outputs, are based on gaseous media. Attempts have been made to keep the ideal properties of the Nd^{3+} laser action by using vapor containing molecules of neodymium compounds [26]. This development has not yet reached a state necessary for high-power laser applications.

The classical high-power gas laser is the carbon dioxide laser. The highest powers achieved with transverse electron beam discharge pumping and final stage beam diameters up to 30 cm are 3 TW in 0.5 nsec. A combination of eight beams in one laser system, at Los Alamos, reached outputs of

20 TW [27] or 40 TW [28]. Pulse lengths as short as a few psec have been reached with carbon dioxide lasers [29]. The development of high-pressure laser systems has been followed through many years [30] and has been developed independently [31] in connection with a very compact amplifier system. The only problem for several applications, including possibly laser fusion, is the long laser wavelength of 10.6 μm. Over this question, however, the argument is not finally settled. For several other applications, as for material processing, the wavelength is not of primary importance *but is of secondary importance*, as can be seen from its use in surgery for example [32], where light of a wavelength 0.5 to 1 μm is diffused by very strong scattering at the cells. This leads to much higher power thresholds for cutting tissues and results in broader cuts than with CO_2 lasers.

The future development of carbon dioxide lasers can be seen in the ANTARES project at Los Alamos [27] and similar ones elsewhere, where laser pulses of several 100 kJ in 1 nsec or less should be achieved. It should be noted, from the point of view of laser fusion, as well as material processing, that one line of development of the carbon dioxide laser could be the nuclear reactor pumped laser, where pulses of several tens of megajoules within several hundreds of nanoseconds can be expected [33].

Another important high-intensity gas laser is the photochemical iodine laser with a wavelength of 1.3 μm. This is particularly useful since its wavelength is close to that of neodymium. The high-power iodine laser was developed by Hohla [34], following the outlines given by Kompa [35]. The present design, using a final amplifier of 20 cm diameter and 10 m length, produces terawatt pulses in 0.5 nsec [36]. The most significant result is an ideal optical beam quality devoid of all the complicated lateral intensity variations due to birefringence and Fresnel diffraction. It has been shown [37], that the focused beam diameter of the completely uncorrected beam was only two times larger than the diffraction limit. One current disadvantage from the point of view of laser fusion is the relatively low efficiency of the laser, which is presently at or below 1%. The use of a specially developed UV source for dissociating the iodine molecules has resulted in higher gains [38]. Another way to achieve higher gains may be the use of exploding wires in the center of the cylindrical amplifiers, where terawatt pulses have been reached [39].

There are several further candidates for high-power lasers to be considered at present. Reed Jenssen [40] succeeded in building a working HF (hydrogen fluoride) laser, where a mixture of both H_2 and F_2 at 1 atmosphere is preexcited by an electron beam to generate more than 10^{25} free atoms/sec/cm^3. The laser consisted of a cylindrical volume of 40 cm diameter and was relatively short (1.5 m long). It produced laser pulses of 3 kJ energy and of 30 nsec duration. Another interesting laser would be one in the category of the excimer laser, which was one of the schemes originally proposed for lasers by Houtermans [41], but which was realized only after high-intensity electron beams became available [42]. With these lasers, not only wavelengths in the UV and far UV, but also laser powers above 10 GW in picosecond pulses are available (refer to Bradley [17]). By generation of high-order harmonics [42] laser wavelengths below 500 Å have been obtained at useful intensities. Lasers worked with fife times amplification at 182 Å [43] and are studied at 117 Å and less [44]. The possibility of producing short wavelengths by gamma ray lasers [45] does not seem to be outside the realms of possibility; the laser medium has to be preexcited by very intense laser beams in the optical and infrared wavelength range [46]. As a first important step, Okamoto's model [46, 47] was the basis of the first laser excitation of nuclei: Yamanaka et al. [48] succeeded in exciting ^{235}U nuclei into their isomeric state by laser excitation of uranium electrons causing a resonance with nuclear levels. A further consequence of gamma-ray lasers should be preferable for extremely high-intensity emission. Another new development of high-power lasers is the free electron laser. The first working system uses electron beams of about 40 MeV energy [49] moving through a rippled magnetic field. This mechanism of laser emission had been realized for microwaves in 1952 by Hans Motz [50]. An extensive digest of the different systems of this kind has shown [50] that the emission of coherent radiation is essentially based on second-order cyclotron radiation effects [51]. The proposed idea of using cyclotrons of 45 m radius may result in a relatively extensive apparatus for laser fusion; however, the high beam quality and the possibility of producing any magnitude of beam power may be so attractive, that the development of the cyclotron-type free electron laser could be of great importance for the future [52].

There is another free electron laser system under consideration that is basically different from the cyclotron type. It does not need any additional magnetic field to produce cyclotron-type effects and requires only a high-intensity laser beam (which has to be produced in the conventional way) that interacts with appropriate electron beams of specific energies and spatial configurations [53]. The essential mechanism is the application of the nonlinear radiation force [54]. Examples have been given [53], where a carbon dioxide laser pulse of 1 TW can be amplified to a power of 10^{15} W.

The advantage is that no materials are involved that can be damaged or ionized, because the interaction process of the electrons occurs in vacuum. The wavelength is continuously variable and, in theory, applicable to x-ray lasers. A disadvantage is that the amplification follows the square of the wavelength. However, the energetic conversion efficiency for transfer of electron beam energy into optical energy is theoretically up to 100%. This is an important point in the design of power stations or similar equipment, as the exchange of cooling energy can then be reduced.

A further basically different type of a free electron laser has been proposed by Schwarz [55], where the superposition of electron beams to produce an interference field with a quantum-modulated electron current is used. Modulation is possible by laser [56] or by the Aharonov–Bohm effect [55]. The long beating electron beam emits coherent radiation. This modulation-type laser has a higher efficiency at higher frequency and is therefore of interest for very short wavelengths (Schwarz-Hora effect type laser).

1.4 Review of Phenomena and Results

This subsection will review several significant phenomena that have been observed in experiments of laser interaction with solid targets, gases, and with plasmas usually produced by the laser itself. With the discovery of the first laser by Maiman [57] in 1960, an obvious step was to use its high-intensity radiation to study the interaction of light with solid targets. At that time the use of electron beams for grinding, drilling, welding, and other kind of material treatment was well developed, where power densities of 10^8 W/cm^2 and more had been reached [58]. It was remarkable that the first spiking ruby lasers with maximum power of 10 to 100 kW could be focused down to less than 0.05 mm diameter resulting from the beginning in comparable power densities to the electron beam. In some of the first experiments, solid targets were irradiated in vacuum and the time dependence of the emitted ions reaching a Faraday probe arranged in front of the target was measured. Time-of-flight measurements on the ions showed velocities corresponding to a few electron volts. This was in full agreement with the expected temperatures of a few ten thousand degrees centigrade for the plasma generated at the target surface [59].

These measurements were confirmed later [60], though the number of such measurements reported in the broad stream of laser plasma interaction literature has been minimal. A change occurred in the years 1977 to 1978, when solid-state physicists and semiconductor technologists began to use lasers for melting, recrystallizing, generation of crystal defects, or annealing of these defects [61]. A clear distinction of these interactions from those occurring at higher intensities, leading to evaporation and plasma generation, has to be made. It should be mentioned that the use of lasers for evaporation techniques for thin films was studied earlier [62] with the surprising result that such a complicated molecule as strontium titanate was redeposited in the initial molecular state, even after laser vaporization (or plasma generation).

The use of higher laser power than that of the spiking laser became a reality when Hellwarth [63] discovered the Q-switched laser, where the ruby laser emission was very reproducibly concentrated in pulses of 10 to 40 nsec duration. This meant that peak powers of 10 to 100 MW were attainable. Linlor was able to use such a laser to irradiate targets such as carbon, tungsten, and others in vacuum [64] and arrived at the very surprising result that the aforementioned ion energies of few eV for 100 kW irradiation were increased to more than kilovolts. It was very significant that the measurement of the ion energy as a function of the laser power P or laser peak height increased in a superlinear way. Isenor [65] measured a nearly linear increase of the ion velocity v_i on the laser power P (Fig. 1.1) corresponding to a nearly quadratic increase of the ion energy ε_i on P.

$$\varepsilon_i = \text{const } P^m \qquad (m = 1.8 \text{ to } 2) \tag{1.1}$$

This type of increase of the ion energy was measured by Schwarz [66] and by Namba and Schwarz [67]. It was evident that at higher powers the superlinear increase of ion energy has to undergo saturation, resulting finally in a sublinear increase [68], Fig. 1.2.

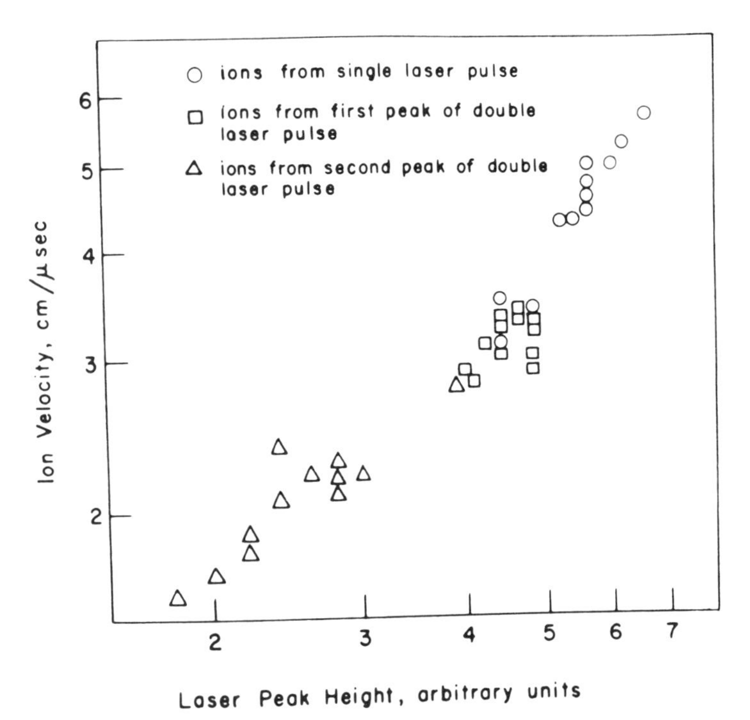

Figure 1.1 Nearly linear increase of the ion velocity with the laser power P at laser irradiation with about 10 MW (after Isenor [65]).

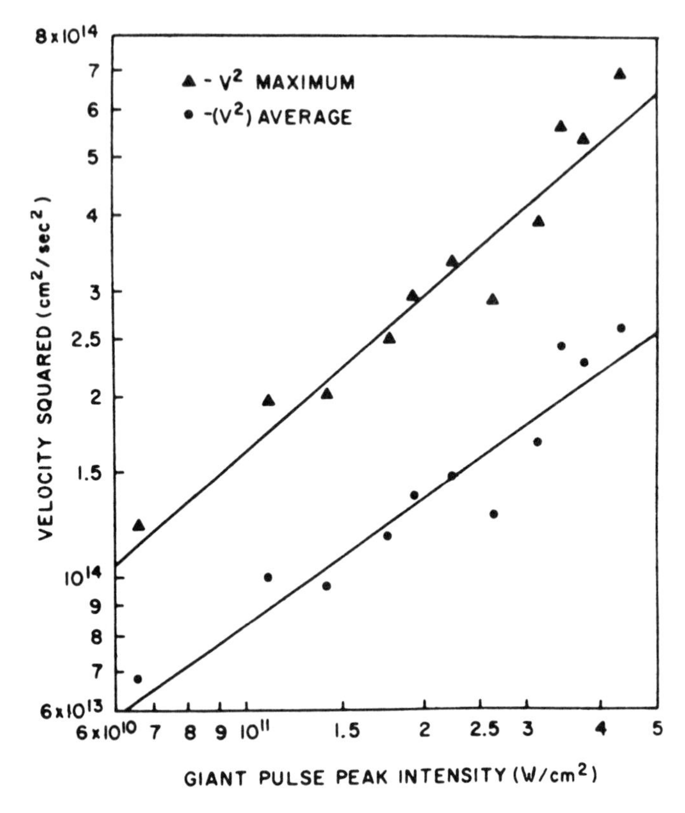

Figure 1.2 Sublinear increase of the ion energy (square of the ion velocity at 100 MW and higher laser power (after Gregg and Thomas [68]).

Another example of highly superlinear behavior with laser power using the ruby or neodymium glass wavelength around 10 MW is the measurement of the recoil exerted by the incident laser radiation on the target [69] as shown in Fig. 1.3. At higher laser powers, this increase is again saturated and merges into a sublinear slope [68, 70].

A similar mechanism between the laser powers of 1 to 10 MW for ruby or neodymium glass lasers happens for the emission of electrons. From the beginning it was an aim of workers in this field to use the interaction of laser radiation with targets for the generation of very high electron emission currents with the hope of developing "super cathodes." It was very controversial when, at the beginning of these investigations, Ready observed a fully classical behavior of electron emission with maximum currents of several hundred milliamps, completely in agreement with the space charge limitation laws of Langmuir [59]. Ready used laser powers again around,

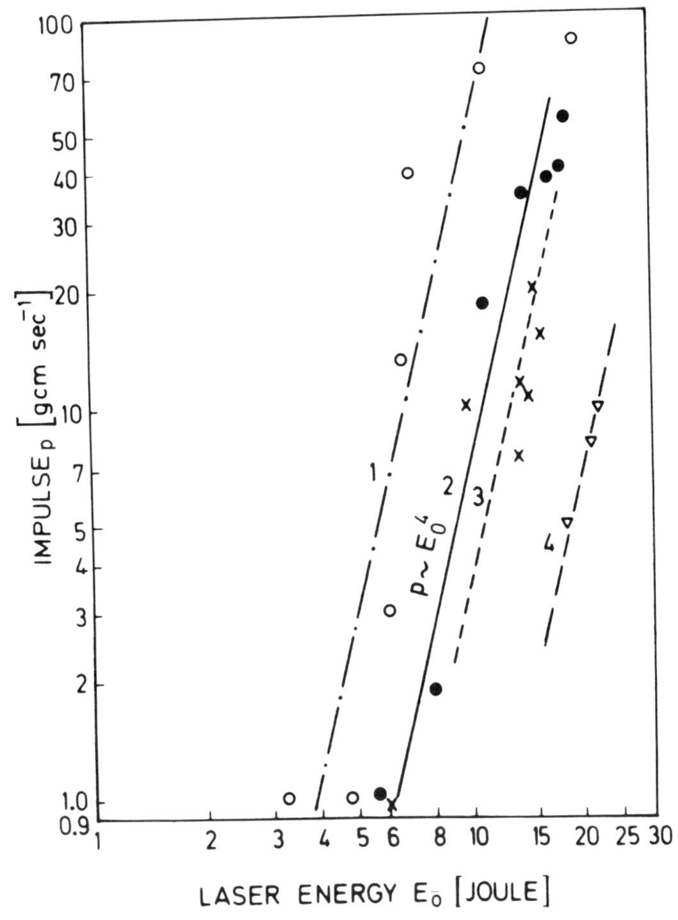

Figure 1.3 Superlinear increase of the momentum (impulse) transferred to the laser irradiated target at neodymium glass laser powers near 10 MW measured by Metz [69].

or less than, 1 MW. In contrast to this, Honig [59] measured emission currents of 100 A, which was in contradiction to any knowledge of space charge restrictions (Child–Langmuir law). This unusual result was fully reproduced [71] when using the more advanced laser techniques; emission currents of more than 1 kA were measured. It is now evident that a special mechanism takes over above powers of around 1 to 10 MW.

Another example of the complexity of the laser interaction with solids can be seen by the following examples when free spherical aluminum targets were irradiated by a laser pulse (Fig. 1.4) [72]. The framing camera

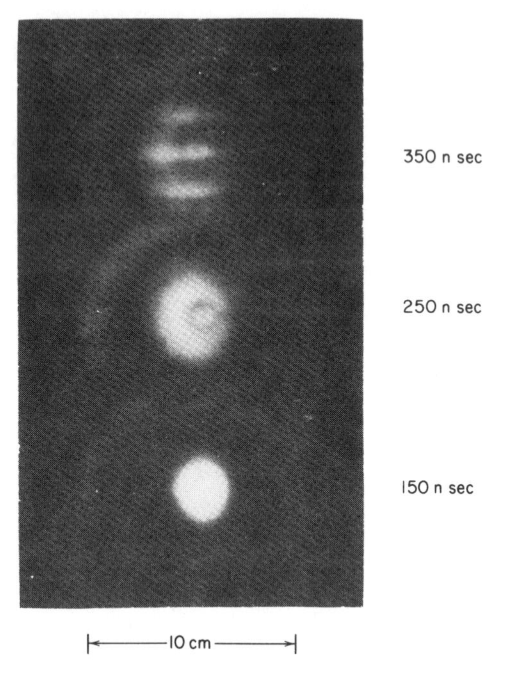

Figure 1.4 Side-on framing camera picture of a plasma produced from an aluminum sphere of 80 μm radius at the time marked, after irradiation with a 30 nsec ruby laser pulse focused to 0.4 mm diameter. The second frame shows the outer part of fast expanding plasma and an inner spherical thermally expanding part measured by Engelhardt et al. [79].

picture at 250 nsec after irradiation shows two groups of plasma: a spherical center containing 95% of the transferred laser energy expanding with a speed corresponding roughly to 10 eV temperature and a fast expanding asymmetric outer plasma with a maximum energy of 3 to 5 keV for ions moving against the laser beam direction. The fully linear or thermal behavior of the inner part was evident, while the outer part demonstrated a highly nonlinear property [72].

It appears that a special process occurs at the already mentioned ruby laser power of 1 MW, as can be seen schematically from a graph comprising the various results of measured ion energy as a function of the laser power (Fig. 1.5). This is called the Linlor effects, though Linlor has always insisted that his observations were of a fully linear nature. In the present view, including the mentioned cases of momentum transfer and of electron emission, as well as other results (e.g., [72]), the initiation of some nonlinear

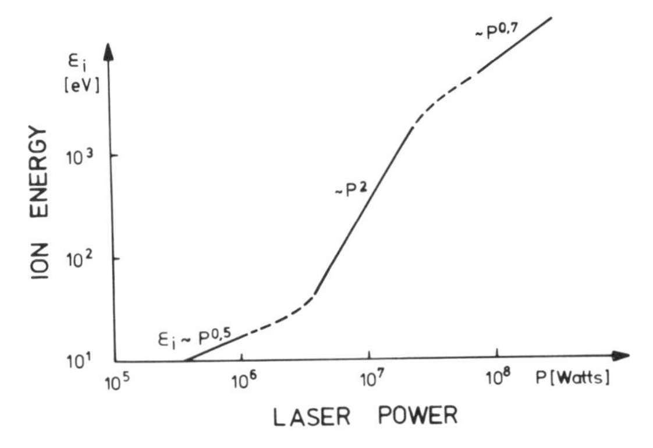

Figure 1.5 Schematic composition of measurement of the ion energy against the laser power for ruby and neodymium glass lasers with a thermal part below 1 MW, a highly superlinear part (Linlor effect) up to 10 MW, and a sublinear part above.

mechanism above a threshold somewhere in the 10 MW range of ruby laser intensities is evident.

The highly complex properties of a laser focus are another complication [73]. Figure 1.6 shows the measurement of the intensity in a laser focus where, within one small area only, the approximately Gaussian profile can be seen,

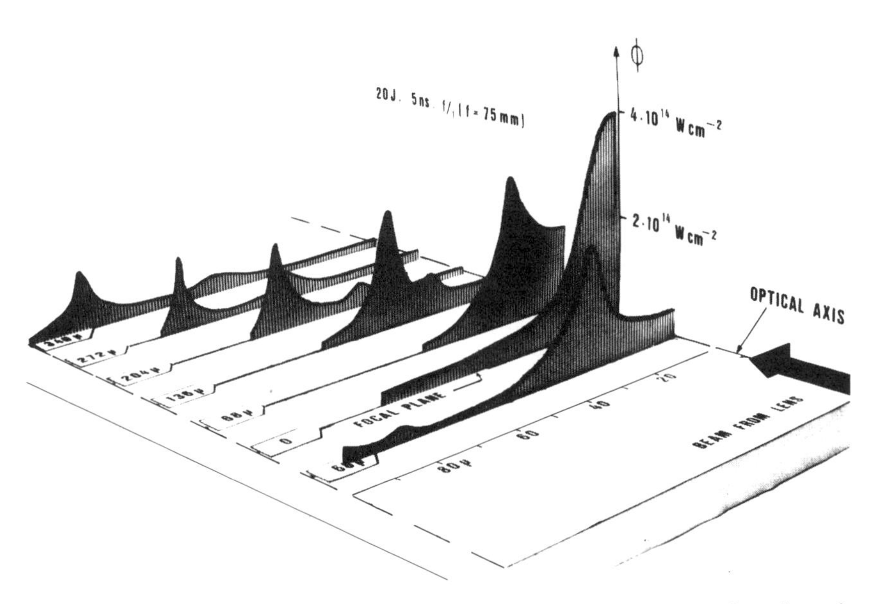

Figure 1.6 Measurement of (time integrated) spatial distribution of the laser intensity in a focus by Eidmann et al. [73].

while a complicated intensity field is observed in the off-focal regions. This consideration is even more important for laser breakdown in gases, where the whole focal region must be taken into account.

The mentioned experimental results are mostly several years old, and one may argue that they may lack accuracy when compared to present-day experiments. Even so it seems that some of the basic properties of the older measurements are often not fully taken into account when interpreting new measurements. The pluralism and curiosity of more recent experiments has still not been lessened. Comparing different measurements of the reflectivity of irradiated targets at various laser intensities, a very confusing scattering of results can be seen, Fig. 1.7 [74]. This is an example of what happens when the results from different authors working with different targets and different laser parameters are compared. It should be mentioned that, although this type of strong scatter of results may have been reduced during the last few years, a large amount of variation still exists.

To illustrate the complexity of the experimental results in the field of laser plasma interaction from more recent papers, it should be mentioned that, for neodymium glass laser intensities of 10^{13} W/cm^2 and above, several special properties have been detected. There was the observation of backscattering of higher harmonics of the laser frequency, indicating the action of parametric instabilities, in analogy to microwave experiments. There was the observation of the half frequency radiation in the backscattered radia-

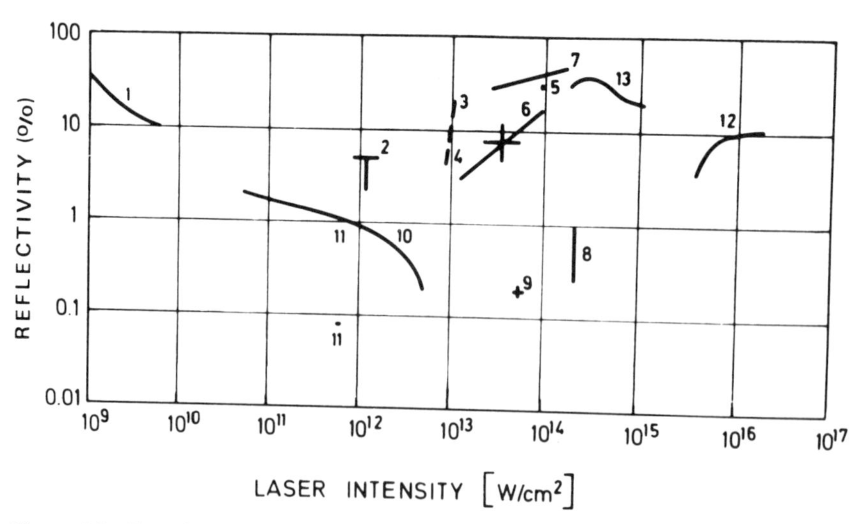

Figure 1.7 Experimental reflectivity (5) from laser produced plasma as a function of laser intensity. Data are from various authors with reference to [1, p. 4].

tion, and there was the result that no uniform temperature was present in in the focus of laser-plasma interaction. After measuring very contradictorily varying temperatures from x-ray signals, Eidmann [75] was able to analyze the emitted x-ray spectra in such a way that, apart from an expected plasma temperature of few hundred electron volts, another "temperature" of keV was shown to exist (Fig. 1.8). This was due to the intensively radiating focal region and was not an effect of different spatial properties. This elevated "temperature" turned out to be due to anomalous nonlinear processes and could reach values of 200 keV [76] and more, up to 8 MeV "temperature," appropriate for pair production [77].

Another interesting result is the fact that the initially observed different groups of expanding plasma, apart from fast ions with nonlinear behavior [72], were still seen in further observations. It was found later that the fast ions are of a few 10 keV energy [78], and that much more than 50% of the irradiated laser energy can go into the fast ions [79]. One example of the fast ions is shown in Fig. 1.9, where the probe signal indicates a clear separation of the ions by their charge number in a linear way with ion energies of several 100 keV. The detection of MeV ions was a further step [80] where, however, the inclusion of a relativistic self-focusing mechanism was necessary for an interpretation [81]. The mentioned linear dependence of the ion energy

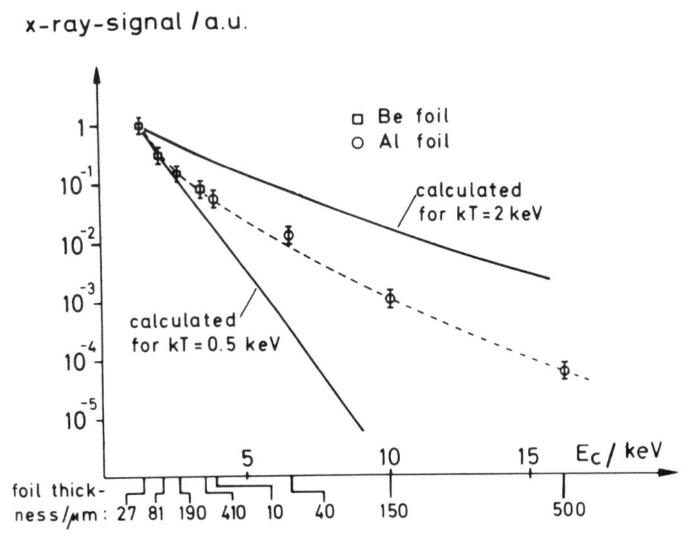

Figure 1.8 Emitted x-rays from the plasma as a function of cutoff energy of four Be and Al foils of different thicknesses. The different slopes correspond to "two different temperatures" discovered by Eidmann [75].

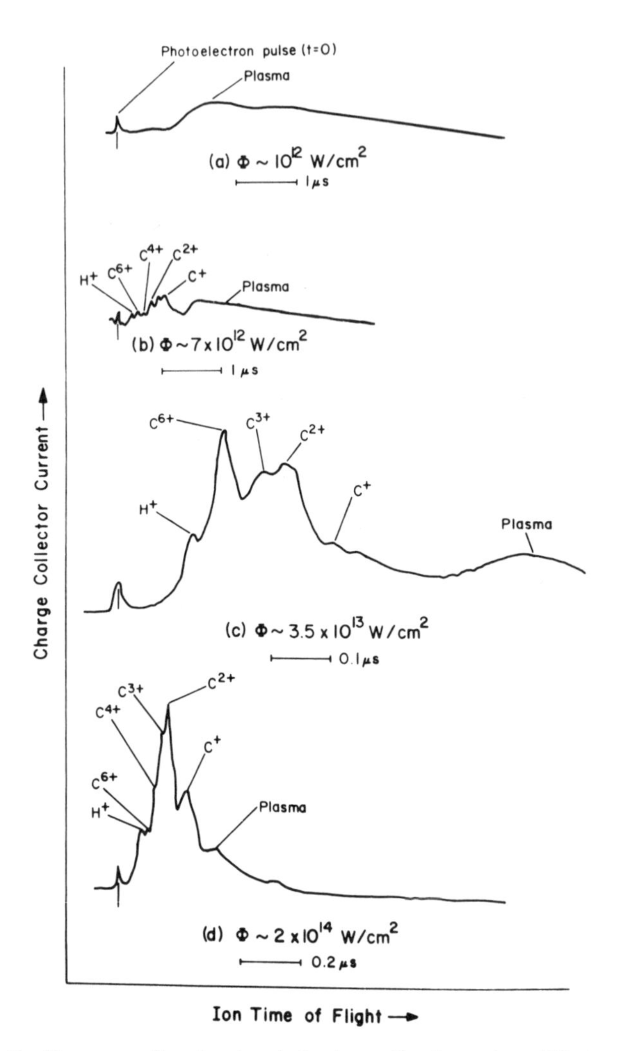

Figure 1.9 Oscillograms of probe signals for ion collection. when CO_2 laser radiation of the given intensity is incident Al and C targets. The peaks correspond to various ions of different ionization with energies of several 100 keV. measured by Ehler [76].

on the ion charge was a relation seen in several experiments, especially in the case of 100 keV ions and above. The analysis of the fast ions in the keV range, however, led to another modification if the light was incident obliquely on the plane targets [82]. It was discovered that one group of the ions behaved fully independent of the laser polarization (fast ions), while the other group showed a strong dependence on the polarization (ultrafast ions).

The most up to date experimental techniques for analyzing the properties of laser produced plasmas are able to show the most astonishing properties. It is possible to measure the spatial variation of densities of plasmas with a resolution of several micrometers and in temporal resolution down to picoseconds. One example can be seen in Fig. 1.10 [83]. Diagnostics of this kind were the basis for detecting the compression of plasma in the center of spherically irradiated gas filled glass balloons for producing genuine thermonuclear reaction in the center of the pellets [84] (see Fig. 1.11).

Another unique result is the fact that the irradiation of very thin gold foils, which are transparent to optical radiation, causes an anomalously high absorption when irradiated by very intense short laser pulses [86]. The foils are then not transparent. For oblique incidence a special behavior has been seen, the Yamanaka effect [86], and these anomalies have been confirmed by the fact [87], that the irradiation of an aluminium layer of

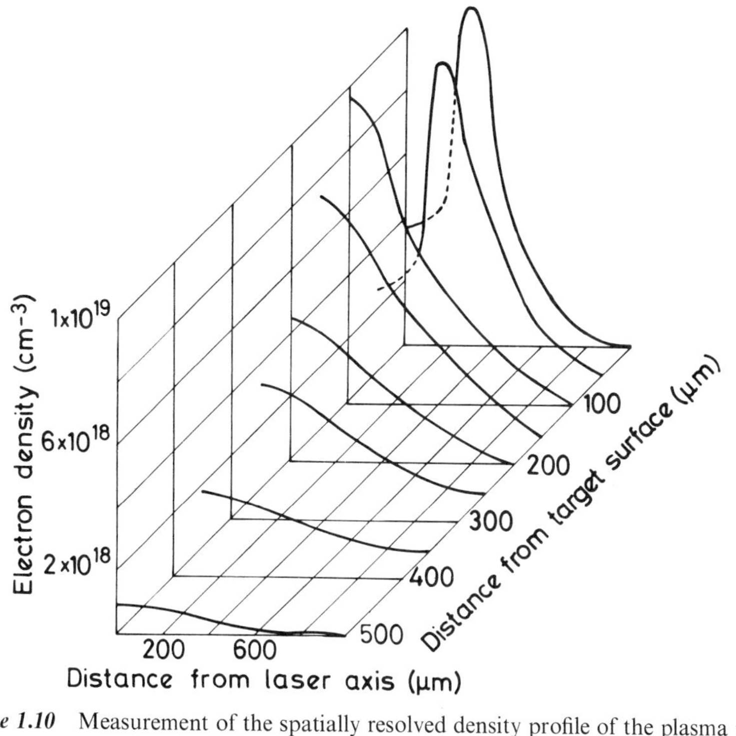

Figure 1.10 Measurement of the spatially resolved density profile of the plasma produced by a CO_2 laser pulse from a spherical target 25 psec after the start of the interaction. The generation of the density minimum was characteristic for the process. After Donaldson and Spalding [83].

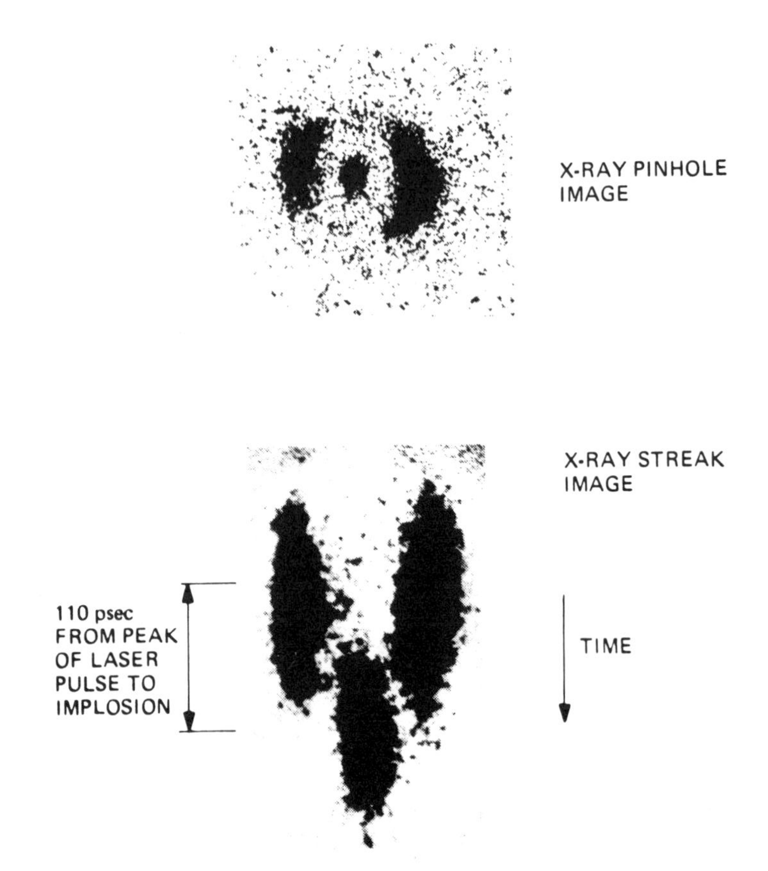

Figure 1.11 Example of high spatial and temporal resolution in the diagnostics of the diametral irradiation of two wide aperture Nd glass laser beams of 7.6 J and 100 psec duration each on a glass microballoon of 0.88 μm wall thickness and 88 μm diameter filled with 10 atm. D_2+T_2 gas. The upper picture is a time integrated x-ray pinhole photograph indicating radiation from the laser interaction at the glass surface and such from the center due to plasma compression resulting in 2×10^4 fusion neutrons. The lower picture is the x-ray streak image of a diameter showing the motion of the pellet periphery toward the center on time and the about 50 psec delayed radiation from the core after the compression has been achieved, after Evans, Key, et al. [85].

1000 Å thickness on a quartz substrate did not show silicon lines in the spectrum.

These very unexpected and very complex properties of the laser produced plasmas underscore the very complicated nonlinear processes involved. We shall nevertheless first describe the theory of the linear gasdynamics, and then plasma laser interaction processes, on the basis of which the theory for several nonlinear processes will be derived. There is definitely no complete theory possible for all the known or as yet undiscovered, unexpected, and anomalous processes. But we hope that a certain guideline can be given for a better understanding of the present developments. The complexity of the physics has to be taken into account to see how probable it will be to proceed with the already quite successful attempts to compress and heat plasma for thermonuclear reactions, as well as to understand the borderlines of low-intensity interaction for material treatment. If material treatment is taken into account the irradiation with laser intensities from CO_2 or neodymium glass lasers above 10^{12} W/cm^2 is definitely anomalous, as seen from the fact that most of the irradiated energy is transferred into a certain amount of very high energetic ions. These ions are interesting for use in accelerators or nuclear fusion. For the application of evaporation of large quanties of materials, however, these mechanisms may be disadvantageous.

TWO

Elements of the Microscopic Plasma Theory

A plasma can be defined in various ways. It has been called "the fourth state of matter" as distinct from the solid, liquid, and gaseous states. More than 99% of the cosmos consists of plasma: in the stars and to a large extent in the interstellar matter. Since there are no such sharp distinguishing marks as a melting point or boiling point, but only the fact that all matter is ionized at high temperatures (above 10,000°K all matter is ionized to some degree), the following definition can be given. *Plasma is a physical state of high electrical conductivity and mostly gaseous mechanical properties.* The term "mostly" allows for the fact that, for example, a metal or a semiconductor can have plasma properties, although its mechanical properties of compressibility and rheology are that of a solid. In the case of the high-temperature matter in the dense interior of a star, the plasma properties are evident, but the compressibility may be the same as for solids or liquids due to quantum effects (Fermi–Dirac degeneracy). The plasma may be defined alternatively as a *medium whose dielectric properties are determined only by free charges* (and not by dipoles).

In this case we neglect all bound states of electrons in atoms or molecules. We are discussing, then, the "fully ionized plasma." These are of interest at high-intensity laser interaction with plasmas. The range between the first irreversible damage in materials by lasers and the full ionization is then neglected. As self-focusing and other nonlinear effects are acting very quickly after the first damage threshold, the generated plasmas are then nearly fully ionized. This was justified also posteriorily, when Mulser included the ionization equilibrium (Saha equation) in very extensive hydrodynamic calculations [88] even at moderate neodymium glass or ruby laser intensities of 10^9 to 10^{10} W/cm^2, the dynamics of the generated plasmas was the nearly same whether the Saha equilibrium was included or not.

A fully ionized plasma is a gas consisting only of electrons and positive ions of a certain charge Z. The most basic description of such a plasma uses the single-particle equation, which are then the $3N$ equations of motion of each of the N plasma particles

$$m_n \frac{d^2 x_n}{dt^2} = f_{xn}\left(x_i, \ldots, z_N, \frac{\partial x_1}{\partial t}, \ldots \frac{\partial z_n}{\partial t}\right); \qquad n = 1, \ldots, N \tag{2.1a}$$

$$m_n \frac{d^2 y_n}{dt^2} = f_{yn}\left(x_i, \ldots, z_N, \frac{\partial x_1}{\partial t}, \ldots \frac{\partial z_n}{\partial t}\right); \qquad n = 1, \ldots, N \tag{2.1b}$$

$$m_n \frac{d^2 z_n}{dt^2} = f_{zn}\left(x_i, \ldots, z_N, \frac{\partial x_1}{\partial t}, \ldots \frac{\partial z_n}{\partial t}\right); \qquad n = 1, \ldots, N \tag{2.1c}$$

The particle coordinates x_n, y_n, and z_n and the derivatives with respect to time t are given by the masses m_n and the forces f_{xn}, f_{yn}, and f_{zn} depending on all coordinates and masses of all the N particles. In general, this task is not directly soluble though numerical solutions have been made for 50,000 or more single plasma particles by so called simulation codes [89]. However, even in such cases, the full description of the Coulomb collision forces has to be reduced to certain approximations, so that the results are not fully general and the interpretation has to be made with restrictions. The forces f will then be determined by the coordinates x_j of particles within a certain distance only for producing Coulomb forces. Particles at further distances will be ignored. Further, the first derivation of the coordinates on time had to be ignored in the forces f.

For the simulation of a plasma with 50,000 single particles, the impressive number of 150,000 differential equations had to be solved by the computer. Without going into detail, it should be mentioned that the treatment of laser plasma interaction resulted in the same properties as the following described hydrodynamic calculations resulted. Starting from a linear ramp of a plasma density, the acting laser radiation caused net forces to the plasma [90] completely equivalent to the earlier derived macroscopic nonlinear forces that will be described in the following in detail.

For the following treatment of the macroscopic hydrodynamic theory, several properties of the microscopic description will be used. This is the reason for the following subsections, which are basically related to the history of the development of plasma physics.

2.1 Plasma Frequency and Debye Length

The plasma state was discovered by Langmuir in 1920 when he attempted to explain the fact that radio waves of about 10^7 Hz frequency were totally reflected by the upper atmosphere or the ionosphere, and, in this way, were guided around the globe by the plasma shell of the ionosphere. Without having direct measurements (made later by balloons and satellites) Langmuir concluded that the upper region of atmospheric gas was ionized. He derived a characteristic frequency ω_p, the plasma frequency, for the electrostatic oscillations of the electrons in a plasma corresponding to the reflection of the waves.

Fig. 2.1 describes, in the upper part, an electron density in equilibrium given by a cell distance dx. The lower part shows a disturbance by distances $d\xi$, which causes a change of the electron density n_e of

$$\frac{dn_e}{n_e} = -\frac{d\xi}{dx} \tag{2.2}$$

Following Poisson's electrostatic potential equation ϕ, or the electric field $\mathbf{E}$, generated by a charge density ρ in Gaussian cgs units, the equation for the displaced electrons arrives at

$$\frac{d\mathbf{E}}{dx} = -4\pi e dn_e = +4\pi e n_e \frac{d\xi}{dx} \tag{2.3}$$

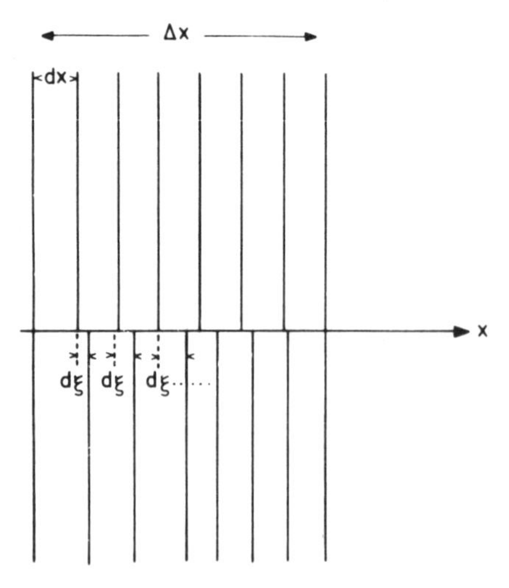

Figure 2.1 Displacements of electrons of homogeneous density (upper part) by $d\xi$ to generate electrostatic oscillations with the plasma frequency ω_p.

The electron charge is $e=4.803\times10^{-10}$ cm$^{3/2}$ sec^{-1}g$^{1/2}$ [91]. The equation of motion for the electrons is

$$m\frac{d^2\xi}{dt^2}=-e\mathbf{E}=-4\pi n_e e^2\xi(t) \tag{2.4}$$

using the electron mass $m=0.9109\times10^{-27}$ g. The solution for the differential equation (2.4), the undamped oscillation equation, is

$$\xi(t)=\text{const}\exp(i\omega_p t) \tag{2.5}$$

where n_e is given in cm^{-3}:

$$\omega_p^2=\frac{4\pi e^2 n_e}{m};\qquad \omega_p=5.65\times10^4\sqrt{n_e} \tag{2.6}$$

In the case where 30 m radio waves (10^7 Hz) were totally reflected by the ionosphere, Langmuir was able to calculate an electron density $n_e=1.23\times10^6$ cm^{-3}. The process of wave reflection will be seen later, when the dielectric constant based on ω_p will be discussed. A density $n_e=10^{21}$ cm^{-3} is obtained from (2.6) for the 1.78×10^{15} Hz radiation frequency of the neodymium glass laser. The 1.78×10^{14} Hz frequency of the CO_2 laser corresponds to a "cutoff" density of $n_{ec}=10^{19}$ cm^{-3}, while an excimer laser of 1200 Å corresponds to a cutoff density $n_{ec}=7.8\times10^{22}$ cm^{-3}, which is close to the atomic density in solids.

There exists a characteristic length λ_D corresponding to the radian plasma frequency ω_p for a wave velocity equivalent to the electron thermal velocity v_e. This is distinct by the factor 2π from the usual procedure of multiplying an oscillation frequency by a wavelength to obtain the velocity of a wave or transport process. Instead of the thermal average velocity of the electrons, a value given by the average energy $KT/2$ (T=temperature; K=Boltzmann constant$=1.38\times10^{-16}$ cm^2 sec^{-2}) per degree of freedom must be used. With these minor modifications one arrives at

$$\lambda_D=\frac{v_e}{\omega_p}=\left[\frac{KT}{4\pi n_e e^2}\right]^{1/2} \tag{2.7}$$

$$\lambda_D(\text{cm})=6.9\left[\frac{T(K)}{n_e(\text{cm}^{-3})}\right]^{1/2}=743\left[\frac{T(\text{eV})}{n_e(\text{cm}^{-3})}\right]^{1/2} \tag{2.8}$$

This length is identical with that derived by Debye [92] based on Millner's work [93] for the theory of electrolytes, and is called Debye length. For the plasma state it is the limit λ_D for space charge neutrality. Only over dimensions greater than the Debye length can space charge neutrality be assumed in the macroscopic theory.

A very instructive insight into the meaning of the Debye length can be

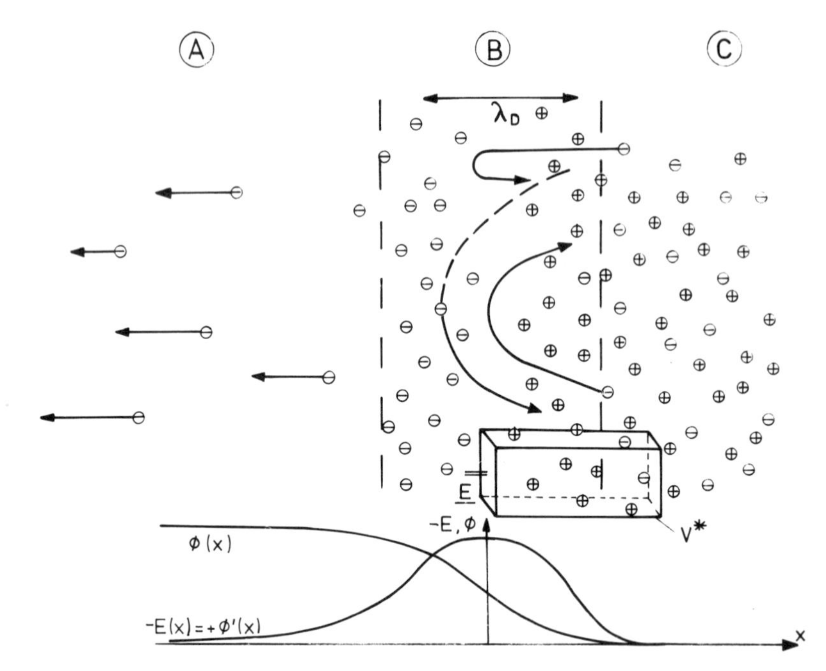

Figure 2.2 Between the vacuum range A and the space charge neutral plasma interior C, the plasma surface sheath is depleted by the escape of fast electrons until such a strong space charge is built up that the following fast electrons from the plasma C are electrostatically returned into C. The electric field $\mathbf{E}(x)$, due to the space charge separation in B, and its potential are given.

gained, if it is derived in another way from the surface properties of a plasma (Fig. 2.2). Between the interior of electrically neutral plasma and vacuum, a Debye sheath B is created due to the fact that the thermal electrons have a much higher velocity than the ions because of their much smaller masses. The number of electrons leaving the sheath is eventually limited by the equal and opposite ion chare built up, which electrostatically returns fast electrons to the plasma interior. The thermal energy KT corresponds then to an electrical potential given by an electrical field $\mathbf{E}$ times a length λ_D.

$$KT = e\phi = e\lambda_D\mathbf{E} \tag{2.9}$$

$\mathbf{E}$ can be derived from the charges in the interface, where the integration of the volume integral (always for 1 cm^2 cross section) results in

$$4\pi\lambda_d n_e e = \oint n^2\mathbf{E}d^2a = n^2\mathbf{E}(\times 1\ \mathrm{cm}^2) \tag{2.10}$$

Here use was made of the refractive index n, which for the case considered, can be put equal to unity. The field in C is zero and grows monotonically from the interfaces BC to AB. In the surface integral (2.10) only the value at AB is effective. The reflection of electrons from C occurs due to the negative charge in A. Expressing the right-hand side of Eq. (2.9) with the **E** from Eq. (2.10), is found

$$\lambda_D = [KT/4\pi e n_e]^{1/2} \tag{2.11}$$

This is a length for the thickness of the sheath, which is equal to the Debye length of Eq. (2.7).

Considering a plasma of a certain temperature expanding in vacuum, it must be taken into account that its surface will build up an electrostatic sheath with a net space charge where the space charge neutrality equations of plasma dynamics will not be valid. The space charge will then cause an electrostatic explosion of the ions, which is called ambipolar expansion, where the ions gain energies up to a factor times thermal energy equal to the *square root of the ratio of the ion mass to the electron mass*. This generation of fast ions in laser irradiated plasmas has been detected [94] where the number of the ions accelerated by the mechanism is smaller than the number of ions in the Debye sheath of the plasma surface. If the number of fast ions (fast in the sense of much faster than thermal) is much larger than the number of ions in the Debye sheath, another acceleration process has to be assumed, and this will be discussed in the following sections.

2.2 Plasmons

The quanta of the plasma oscillations are called plasmons. Their energy, E, is given by

$$E = \hbar\omega_p = 3.73 \times 10^{-11} (n_e)^{1/2} [\text{eV}] \tag{2.12}$$

using Planck's constant $\hbar = h/2\pi$, where $h = 6.67 \times 10^{-27}$ erg·sec, and $[n_e] = \text{cm}^{-3}$. The action of these plasmons is seen in solids. Electron beams of about 50 keV energy, transmitted through a thin film of metal whose density of conduction electrons near 10^{23} cm^{-3} is equivalent to plasmon energies near 10 eV, have energy losses of the order of the plasmon energies. Historically, the energy losses were measured first by Ruthemann [95] and repeated by Möllenstedt [96] and others, while the plasmon explanation was given later by Bohm and Pines [97]. It is remarkable that this plasmon interaction also works for the electrons of insulators. One can say that the transmitted fast 50 keV electrons "see" the insulator electrons as if they were free oscillating plasma electrons.

It is worth indicating at this point that the plasmon action is present in a plasma, where the plasma densities and the well-known parameters are assumed to be fully classical quantities. The fact is that, in such a classical plasma as a stationary arc discharge, the classical Maxwellian energy distributions of the electrons can be modified quantum mechanically. Fig. 2.3 shows the spectrum of a ruby laser beam scattered in a stationary hydrogen arc plasma of 7 eV temperature and 10^{16} cm^{-3} electron density [98, 99].

Instead of the expected smooth profile of the 90° scattered light, there was a superposition of maxima and minima which correspond to distances of half the plasma frequency. Taking into account that Thomson scattering is based on that *half part* of the oscillation energy due to the quivering

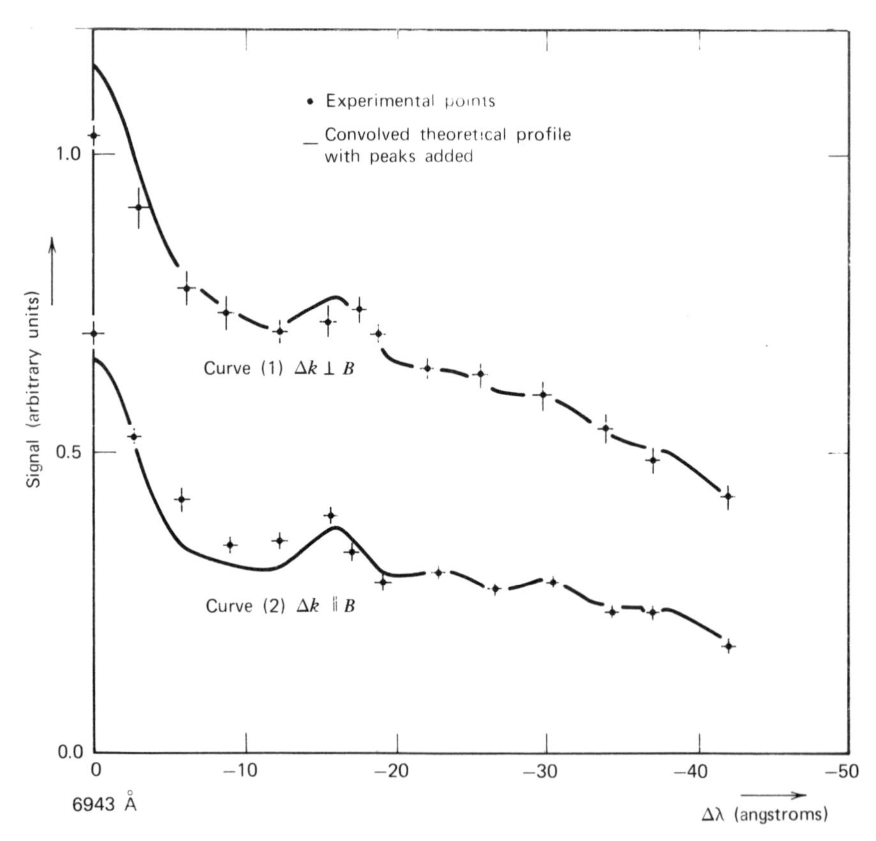

Figure 2.3 Scattering spectrum of laser light from an arc [98] where a modification by maxima and minima is due to the plasma frequency. After Ludwig and Mahn [98].

electrons in the laser field, the scattering corresponds to a leak of intensity, where the electron energy in the plasma is equal to the plasmon energy. This is very easily understood. Since the experiments on the photoemission from metals [100] confirming the volume photoeffect [101] in contrast to the earlier assumed surface photoeffect [102], it has been evident that the usual mean free path of electrons in a metal is reduced from more than 150 lattice distances to only a few if the electron has the energy of the plasmons. The same happens in the case of the arc plasma. Normally the electrons keep their Maxwellian energy distribution, unless an electron is scattered into an energy of a plasmon. Then it will lose its energy very quickly so that the Maxwellian energy distribution (Fig. 2.4) will have minima [1, p. 28].

The half-width of the minima in the energy distribution of Fig. 2.4 will be discussed. It must be related to a characteristic time Δt by a quantization relation

$$\Delta t \cdot \Delta \varepsilon = \hbar. \tag{2.13}$$

The question is what time Δt should be taken. Taking the time the electron needs, at its thermal velocity, to cross the distance of a Debye length will result in the meaningless equality of $\Delta\varepsilon$ and $h\omega_p$, the plasmon energy. A more reasonable result is obtained by starting from the remarkable fact, derived experimentally by Thomas from photoemission [100] that the mean free path of an electron of energy equal to the plasmon energy is a few ion distances. Assuming 10 distances and taking into account the spread of

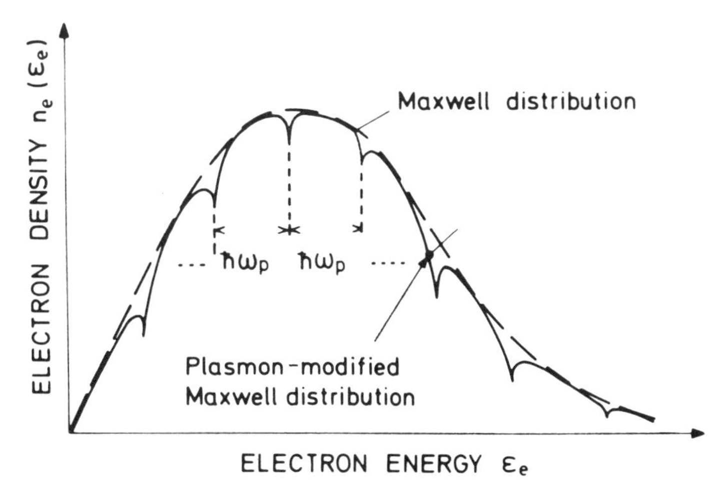

Figure 2.4 Maxwellian energy distribution for electrons and the quantum modification due to the plasmons [74].

energy in the quantum minima of the Maxwell distribution, the thermal time of flight of electrons for this condition leads to a width of the quantum minima in the Maxwellian distribution of

$$\Delta\varepsilon = h\sqrt{2KT/m}\, n_e^{1/3} \tag{2.14}$$

which, in the case of the arc result of Fig. 2.3, is 2.2 meV.

The plasmon energy is in the same case 3.74 meV, which is quite reasonable in comparison. The width of the minima in Fig. 2.3 corresponds to this relation. A further increase of accuracy leads to the assumption of a mean free path for thermal electrons with the plasmon energy of 3 to 10 ion distances, which is in quite good agreement with the experiments of Thomas [100].

This example is used to demonstrate that the quiet and stationary plasma state of an arc of low density is more complicated than was assumed before by using a classical Maxwellian energy distribution for the electrons. The experiments mentioned [98] viewed with the interpretation given here would be a clear indication of a very complicated quantum modification of the plasma state.

2.3 Polarization Shift of H-like Lines in Plasmas

We saw in subsection 2.1 how the Debye length determines the distance within which no space charge equilibrium can be assumed in a plasma, and in subsection 2.2 how the quantization of the plasma frequency can cause a modification of the Maxwellian electron distribution. This subsection considers the modification of the energy of levels of electrons bound in atoms or ions of charge Z, if these are located in plasmas. The electrostatic energy

$$\varepsilon_D \doteq Ze^2/\lambda_D \tag{2.15}$$

within the sphere of the diameter of a Debye length will modify the energy states of the bound electrons, resulting in a shift of the spectral lines which was called plasma polarization shift by Yaakobi and Goldsmith [103].

The energy ε_D, Eq. (2.15) also appeared as the "decrease of the ionization energy" measured when a bound electron is being ionized in a plasma of electron density n_e and temperature T [104]. This decrease is not of the value e^2/λ_d, but its part to each electron in the Debye sphere. Otherwise energy could be produced by adiabatic expanding and compressing of plasma at ionization and recombination of electrons in bound states. The use of the line shift and broadening is an important tool for a direct measurement of the density of laser compressed plasmas [105].

A similar electrostatic method is used to describe the quantum state of an electron in a Coulomb potential as it was successful to describe the decrease of the ionization energy of atoms in a plasma. The aim is to compare the energy ε_B with the energy of the electron when being bound to an ion or atom. If an electron within a Coulomb field $\mathbf{E}$ of Z protons $|\mathbf{E}|=Ze/r^2$, is confined within a radius r_0, the electrostatic energy released by the electron when coming from $r=\infty$ is from the Maxwellian stress tensor

$$\varepsilon_s=\frac{Z^{-1}}{8\pi}\int_{r_0}^{\infty}\mathbf{E}^2\,d^3\tau, \qquad d^3\tau=4\pi r^2\,dr$$
$$=\frac{Ze^2}{2r_0} \tag{2.16}$$

The quantization requires that confinement of an electron to a radius r corresponds to an increase of its momentum $\mathbf{p}$ (or energy $E=\mathbf{p}^2/2m$ causing a quantum pressure) given by

$$r=\frac{n\hbar}{\sqrt{2mE}}$$

where n are integers $n=1, 2, 3, \ldots, \infty$. Apart from a factor of order one, E can be considered as a Fermi–Dirac energy:

$$E_n=\frac{n^2\hbar^2}{2mr^2} \tag{2.17}$$

The fact is that the increase of ε_s on decreasing r_0 is slower than the increase of E on r. A stationary solution is obtained when both energies are equal, $E(r_0=r_s)=\varepsilon_s(r_0=r_s)$ if

$$r_s=\frac{n^2\hbar^2}{me^2Z}=\frac{n^2r_{\text{Bohr}}}{Z} \tag{2.17a}$$

where the Bohr radius $r_s=r_{\text{Bohr}}$ results for $n=1$; $Z=1$. It is well known from the solutions of the Schrödinger equation that the radius of an electron in the state $n=2$ is $4r_s$ as given in (2.17a). Using r_s, the energy $E=E_n$ in Eq. (2.17)

$$E_n=\frac{Z^2me^4}{2\hbar n^2}=\frac{Z^2 13.6\text{ eV}}{n^2} \tag{2.18}$$

arrives in the ionization energy of hydrogen for $n=1$, $Z=1$ of 13.59 eV. In the case of an electron in the potential of a He^{2+} ion, Eq. (2.18) arrives in the ionization energy of He^+ at $n=1$ of 54.38 eV and similar for higher Z in agreement with measurements. The terms

$$E_m-E_n=Z^2R\left(\frac{1}{m^2}-\frac{1}{n^2}\right) \tag{2.19}$$

where R is the (energetic) Rydberg constant $me^4/(2\hbar^2)$, define the energy level of transitions where, for example, $E_1 - E_2 = 10.24$ eV follows from the Lyman series for hydrogen ($Z = 1$) with $m = 1$ and $n = 2$.

This model easily answers the question why an electron does not "fall into the proton" at electrostatic contraction of the electron radius r: the rise of the "quantum energy" E or of the quantum pressure is faster than the gained electrostatic energy and is balanced at the Bohr radius. The electron has no orbital motion as in the usual quantum mechanical result (in difference to Bohr's model). This quantum pressure model has the advantage that the polarization energy ε_D per electron, Eq. (2.15) can be introduced immediately as correction to ε_s in a plasma, resulting then in the electrostatic energy

$$\varepsilon_s^{\text{plasma}} = \varepsilon_s - \varepsilon_D = \frac{Ze^2}{2r} - \frac{Ze^2}{\lambda_D} = \frac{Ze^2}{2r}\left(1 - \frac{2r}{\lambda_D}\right) \tag{2.20}$$

The model has the advantage of distributing the energy for $n = 2$ either into a nonorbiting localized state or one with one h into a localized and with the second h into an orbiting state with three possible (orthogonal) axes resulting in the 4×2 possible occupations, which equal 8 states of the L-shell. The 18 states for the M-shell are given by one localized state, three states with orbiting states and three states with orbiting states $2h$, and two states with an orbiting "ellipse" in the sense of the Sommerfeld-Bohr model where the number 2 comes from the 2 degrees of freedom for the rotation of the direction of the ellipse. The ground states are then without orbiting and cannot emit radiation, which was a difficulty in the Bohr's model and was overcome in the Schrödinger model, while the orbiting states in the electrostatic model for $n \geqslant 2$ can simply account for the dipole emission and the spontaneous transition. The Schrödinger model needed the second quantization of the field for achieving spontaneous transition.

Equating electrostatic energy in a plasma, Eq. (2.20) with the quantum energy E, Eq. (2.17), results in

$$\frac{n^2\hbar^2}{2mr^2} = \frac{Ze^2}{2r} - \frac{Ze^2}{\lambda_D}$$

and hence in the algebraic solution for

$$r = \frac{\lambda_D Z}{4}\left[1 - \left(1 - \frac{8n^2\hbar^2}{\lambda_D e^2 Z^2 m}\right)^{1/2}\right] \tag{2.21}$$

The energy state E_n for the quantum number n is then from Eq. (2.17) using Eq. (2.21):

$$E_n = \frac{Z^2}{\hbar^2} R \frac{A^2}{[1 - (1 - 2A)^{1/2}]^2}; \qquad A = \frac{2n^2\lambda_c}{\pi Z \lambda_D \alpha} \tag{2.22}$$

where the Compton wavelength $\lambda_c = h/mc$, the fine structure constant $\alpha = 1/137$, and the mean electron velocity v_e was used. The polarization shift $E_n - E_m$ is of the same order as the low-density approximation [103]. The advantage of the result (2.22) is an immediate inclusion of the plasma temperature.

The strongest shift of the levels occurs where the limiting case $A = A^* = \frac{1}{2}$ in Eq. (2.22) is given. This means if $2A$ is close to one,

$$2A = \frac{4n^2 \lambda_c}{\pi \lambda_D \alpha Z} \leqslant 1$$

The limiting case corresponds to the highest possible $\bar{n} = n^*$ in a plasma. This is the "drowning of spectral lines" (Margenau et al. [103]) which we have derived now on the basis of the Debye energy and which is temperature dependent. It limits to the Inglis–Teller continuum which is determined by the Stark broadening due to the (temperature independent) microfield (Holtzmark potential). The line shift based on the result of Eq. (2.22) is in excellent agreement with Lyman lines in difference to other models, as shown by Henry [106].

2.4 Cyclotron Frequency

An important behavior of free charges in a plasma (having an energy described by a general distribution function, or special thermal distribution (see next section) with a temperature T, or by a quantum modified distribution, is their motion in a magnetic field $\mathbf{H}$, which is assumed spatially homogeneous and temporally constant within this subsection. Without losing generality the velocity $\mathbf{v}$ of the particle of charge e and mass m can be split into one component $\mathbf{v}_p$ parallel and one component $\mathbf{v}_s$ perpendicular to the magnetic field $\mathbf{H}$. The Lorentz force leads to the following equations of motion for the particle:

$$m \frac{d\mathbf{v}}{dt} = \frac{e\, \mathbf{v} \times \mathbf{H}}{c} \tag{2.23}$$

$$m \frac{d\mathbf{v}}{dt} = 0 \tag{2.24}$$

A force-free motion of the particle parallel to the magnetic field follows. For the motion perpendicular to the magnetic field is obtained

$$m \frac{d\mathbf{v}_s}{dt} = \frac{e\, \mathbf{v}_s \times \mathbf{H}}{c} \tag{2.25}$$

Without losing generality the s component of the velocity can be expressed by an angular velocity vector **u**, which has a direction parallel to **H** and a modulus (absolute value) of ω_c:

$$\mathbf{v}_s = \mathbf{u} \times \mathbf{r}; \qquad \mathbf{H} \| \mathbf{u} \| \mathbf{r} \tag{2.26}$$

Because of the constancy of **H**, the angular velocity **u** is constant too, and so Eq. (2.24) results in:

$$\frac{m\mathbf{u} \times d\mathbf{r}}{dt} = \frac{e\mathbf{v}_s \times \mathbf{H}}{c} \tag{2.27}$$

$$m\mathbf{u} \times \mathbf{v}_s = \frac{e(\mathbf{u} \times \mathbf{r}) \times \mathbf{H}}{c}$$

$$m\mathbf{u} \times (\mathbf{u} \times \mathbf{r}) = \frac{e(\mathbf{u} \times \mathbf{r}) \times \mathbf{H}}{c}$$

$$m\mathbf{u}(\mathbf{u} \cdot \mathbf{r}) - m\mathbf{r}\mathbf{u}^2 = \frac{e\mathbf{r}(\mathbf{u} \cdot \mathbf{H})}{c} - \frac{e\mathbf{u}(\mathbf{r} \cdot \mathbf{H})}{c}$$

The first term on the left-hand side is zero, because **r** is perpendicular to **u**, and the last term on the right-hand side is zero because **H** is perpendicular to **r**. As **u** is parallel to **H**, the angular frequency ω_c, the modulus of **u**, is calculated to be

$$\omega_c = e|\mathbf{H}|/mc \ [\text{cgs}] = eH/m \ [\text{MKQS}] \tag{2.28}$$

This is called the cyclotron frequency, the gyro frequency, or the Larmor frequency.

The particle moves free of force along the magnetic field and rotates with a frequency ω_c around the magnetic field lines. The radius of this rotation is

$$r_L = v_s/\omega_c = v_s mc/e|\mathbf{H}| \ [\text{cgs}] = vm/e|\mathbf{H}| \ [\text{MKQS}] \tag{2.28a}$$

This is the gyration radius or the Larmor radius. The trapping of electrons and ions of a plasma on the lines of a magnetic field is exploited in the confinement of plasma for thermonuclear fusion. The problem is then to use closed magnetic fields as, for example, in a toroidal solenoid. As the initial condition of field homogeneity is then not possible, a drift can be prevented by shearing of the magnetic field. The particles would remain trapped in the field lines, if they did not undergo collisions, which makes them diffuse across the magnetic field lines. It has been discovered, however, that instead of following the classical description of diffusion, the plasma diffuses much faster [107], either according to Bohm diffusion, or the slower Pfirsch–Schlüter diffusion. The highly complex problems inherent in plasma confinement by magnetic fields for thermonuclear fusion will not be discussed here.

2.5 Collisions

An important quantity, which is used in macroscopic hydrodynamic plasma theory and which is taken from the microscopic theory, both classically and quantum mechanically, is the collison frequency of the plasma particles. It is shown now how a very primitive model for collisions is valid and how it reproduces the main properties of the collision process in quite good agreement with the most sophisticated theoretical models. Following Fig. 2.5, the Coulomb interaction of an electron with a positive ion can be described by the hyperbolic trace of the electron, where the deflection angle ϕ corresponds to an electron of initial velocity v_e which, if undeflected, passes by the positive charge at a distance r_0, which is called the "impact parameter".

The Coulomb force **f**, acting between the electron and ion, is given by the distance **r** between the electron and the ion, the latter is assumed to have a charge Z

$$\mathbf{f}=\frac{-Ze^2\mathbf{r}}{r^3} \tag{2.29}$$

The main interaction between the particles is during time t

$$t=\frac{r_0}{v} \tag{2.29a}$$

when a change of electron momentum

$$\Delta(mv)=|\mathbf{f}t|=\frac{Ze^2}{r_0 v} \tag{2.30}$$

occurs. It is the aim to calculate the interaction for a 90° scattering event. The change of the momentum is then equal to the initial momentum of the electron.

$$\Delta(mv)\simeq mv\simeq\frac{Ze^2}{r_0 v}; \qquad r_0=\frac{Ze^2}{mv^2} \tag{2.31}$$

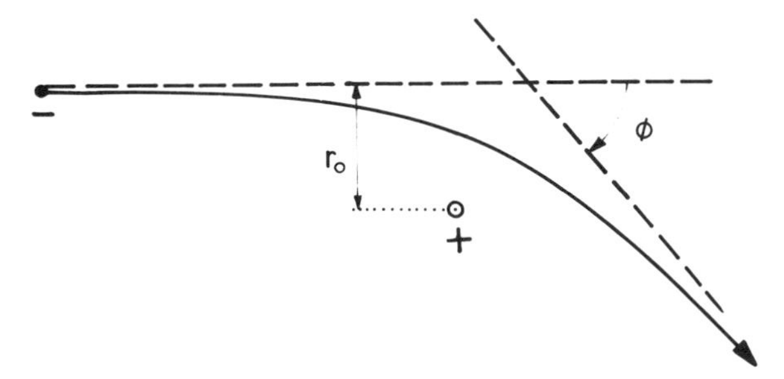

Figure 2.5 Coulomb collision of an electron and an ion.

The resulting cross section is

$$\bar{p} = \pi r_0^2 = \frac{Z^2 \pi e^4}{m^2 v^4} \tag{2.32}$$

The electron-ion collision frequency ν_{ei} is then for a plasma with an ion density $n_i = n_e/Z$:

$$\nu_{ei} = n_i \bar{p} v = \frac{Z n_e \pi e^4}{m^2 v^3} \tag{2.33}$$

The average velocity v of the electrons can be expressed by the electron temperature T_e leading to a collision frequency

$$\nu_{ei} = \frac{Z n_e \pi e^4 3^{-3/2}}{m^{1/2} (KT)^{3/2}} \tag{2.34}$$

For the collision process the most probable velocity v_m, $(m v_m^2/2 = 3KT/2)$, and not the velocity corresponding to the average energy $(E = KT)$, has to be used. Though a very crude description of the 90° scattering has been used, the calculated collision frequency of the electrons with the ions is in very good agreement with the exact classical calculation of Spitzer, and Härm [108], which takes into account small angle scattering. These authors included also the collisions for electrons with electrons, which resulted in a correction factor $\gamma_e(Z)$ varying between 0.5 for hydrogen and 1 for large Z (so that this correction will be neglected mostly in the following). Their calculation arrived at an electron collision frequency

$$\nu_e = \frac{n_e}{(KT)^{3/2}} \frac{Z \pi^{3/2} e^4 \ln \Lambda}{m^{1/2} 2^{5/2} \gamma_e(Z)} \tag{2.35}$$

where the Coulomb logarithm $\ln \Lambda$ is used, given by

$$\Lambda = \frac{\lambda_D}{r_0(90^\circ)} = \frac{3}{2Ze^3} \left(\frac{K^3 T^3}{\pi n_e} \right)^{1/2} \tag{2.36}$$

The Coulomb logarithm varies between 5 and 20, so that the difference between (2.35) and the primitive calculation (2.34) is really surprisingly small. The numerical value of Spitzer's electron collision frequency is

$$\nu_e = 2.72 \times 10^{-5} \frac{Z n_e}{\gamma_e(Z) T_e^{3/2}} \ln \left(1.55 \times 10^{10} \frac{T_e^{1/2}}{Z n_e^{1/2}} \right) \tag{2.37}$$

where the electron temperature T_e is given in electron volts and the electron density n_e is given in cm^{-3}.

The electrical resistivity $1/\sigma$ of the plasma can now be calculated from the collison frequency ν_e. Starting from the mean free path l of the electrons, their thermal velocity v is

$$l = \frac{v}{\nu_e} \tag{2.38}$$

the velocity $\mathbf{v}_D$ gained by drift in an electric field $\mathbf{E}$

$$\frac{d\mathbf{v}_D}{dt} = \frac{e\mathbf{E}}{m} \tag{2.39}$$

The average drift velocity between two collisions is then

$$\langle \mathbf{v}_D \rangle = \frac{e\mathbf{E}}{2m\nu_e} \tag{2.40}$$

which can be used to calculate the current density $\mathbf{j}$

$$\mathbf{j} = en_e\mathbf{v}_D = \sigma\mathbf{E} \tag{2.41}$$

$$\mathbf{j} = \frac{n_e e^2 \mathbf{E}}{2m\nu_e} = \sigma\mathbf{E}$$

The definition of the electrical conductivity σ from Ohm's law and (3.40) are used. The Ohmic conductivity in a plasma, using the simplified collision frequency (2.34), is then

$$\sigma = \frac{3(KT)^{3/2}}{2Z\sqrt{m/3\pi}e^2} \tag{2.42}$$

which is numerically

$$\sigma = \frac{T^{3/2}}{Z} \times 1.93 \times 10^8 \text{ cgs}$$

$$\sigma = \frac{T^{3/2}}{Z} \times 2.14 \times 10^{-4} \Omega^{-1} \text{ cm}^{-1} \tag{2.43}$$

In both cases, the temperature is in degrees Kelvin. If the temperature is given in electron volts, the numerical value of the conductivity is

$$\sigma = \frac{T^{3/2}}{Z} \times 268 \ \Omega^{-1} \text{ cm}^{-1}$$

The electrical resistivity is given by

$$1/\sigma = \frac{Z}{T^{3/2}} \times 4.66 \times 10^3 \ \Omega \text{ cm}$$

or (2.44)

$$1/\sigma = \frac{Z}{T^{3/2}} \frac{\ln \Lambda}{\gamma_e(Z)} 3.08 \times 10^3 \ \Omega \text{ cm} \qquad ([T] = K)$$

where in the last equation the Spitzer collision frequency is used, Eq. (2.35).

It is remarkable that the electrical conductivity in fully ionized plasmas agrees with these values, confirming the correctness of the Coulomb collision frequency derived above. A comparison with the quantum mechanically derived collision frequencies is possible by considering the high collision frequency given by the optical constants from inverse bremsstrahlung theory. This will be discussed later in connection with the theory of the refractive index of plasmas.

The electrical resistivity of plasma, Eq. (2.44), does not depend on the electron density. Comparing with the electrical resistivity of metals, for example, of aluminium as a good conductor with $\sigma = 36 \times 10^4\ \Omega^{-1}\ \text{cm}^{-1}$, the plasma conductivity reaches that of metals when the electron temperature is in the region of 20 to 100 eV (corresponding to about 200,000 to 1,000,000 °K).

At this point the case of metals is briefly mentioned. In metals, the nearly free conduction electrons will undergo Coulomb collisions. When at the beginning of this century Drude and Lorentz were developing their electron theory of metals, they calculated the Coulomb collision frequency in a similar way as shown here for plasmas. Using the room temperature of the metals in Eq. (3.42), the conductivities were about 10^6 times less than measured. This disappointing result was due to the fact that the quantum mechanical properties of the electrons had not been taken into account. Planck's discovery of the atomistic nature of all quantities which have the dimension of a unit of action (quantization) leads to the result that, if an electron is concentrated within a volume of length x, its momentum p along each direction has to be so large to result in an energy E, as given in Eq. (2.17).

The exact Fermi energy E_F, which is the corrected energy E of Eq. (2.17) with respect to spherical geometry

$$E_F = \frac{h^2}{2m} n_e^{2/3} \frac{(3/\pi)^{2/3}}{4} \tag{2.45}$$

This energy is valid also for plasmas of high density. If any plasma is compressed to densities such that the Fermi energy of the electrons is larger than the thermal energy, the electrons are called Fermi–Dirac degenerate (see Appendix A). The compression of a degenerate plasma is adiabatic with respect to an increase of the Fermi energy. The compression is against their Fermi pressure. It results in a compressibility that is equal to the compressibility of a solid within an order of magnitude. The fact that an electron in an atom or in a solid cannot be compressed as easily as in a dilute gas is due to the Fermi pressure.

The Fermi energy of electrons in a metal is about 6 to 20 eV. Thus it is understandable that, when conduction electrons interact with the electric field, the electrons act as if they have a "temperature" of about 100,000° K.

If in this way the electrical conductance based on the Coulomb collisions, Eq. (2.42), is still about 10 times less than measured values for the best metals, this can be ascribed to the so-called effective mass (see Appendix A) of the conduction electrons, that differs from the mass of the electrons in a dilute plasma that is equal to the vacuum electron mass m. This effective mass indeed varies within the order of magnitude necessary to give high conductance. Another point is that the Coulomb logarithm in the more accurate equation (2.42) must be determined. The theory restricts the validity of the Coulomb logarithm to values between 2 and 3 as a lower limit. For very high electron densities in metals and temperatures $T \sim 10$ eV, the Coulomb logarithm has to be corrected within one order of magnitude.

The usual derivation of ν_e, Eq. (2.37) was based on the collision integral of the Boltzmann equation (see following Section). While Spitzer and Härm [108] included a quantum correction in the Coulomb logarithm (which is rather small and was neglected here), the need of a similar correction for the total value of ν_e was lost in the complicated procedures with the Boltzmann integral. The simplified 90° collision model as used in this section, led immediately to a quantum generalization of the impact parameter

$$r_0 = \frac{r_{\text{Bohr}}}{2Z} \frac{1}{(1+4T/T^*)^{1/2}-1} = \begin{cases} r_{\text{Bohr}} T^*/(4ZT) = Ze^2/(3KT);\ \text{if } T \ll T^* \\ \dfrac{r_{\text{Bohr}}}{4Z}\sqrt{\dfrac{T^*}{T}} = \dfrac{h}{2mr};\ \text{if } T \gg T^* \end{cases} \tag{2.46}$$

where the Bohr radius $r_{\text{Bohr}} = \hbar^2/me^2$, the fine structure constant $\alpha = e^2/\hbar c$, and a critical temperature

$$T^* = 4Z^2 mc^2\alpha^2/(3K) = Z^2 4.176 \times 10^5 (^\circ K) \tag{2.47}$$

was used. For $T \ll T^*$ we have classical properties (see Equation (2.31)), otherwise quantum collisions which will lead to a collision frequency

$$\bar{\nu}_{ei} = \sqrt{\frac{3kT}{m}} \frac{\pi^2_{\text{Bohr}} n_e}{4Z^2(\sqrt{1+4T/T^*}-1)^2} = \begin{cases} \dfrac{Z\pi e^4}{3^{3/2} m^{1/2}} \dfrac{n_e}{(KT)^{3/2}} = \nu_{ei};\ \text{if } T \ll T^* \\ \nu_{ei} T/T^*;\ \text{if } T \gg T^* \end{cases} \tag{2.48}$$

where ν_{ei} is the classical value (7.34). This quantum correction had not been included in the numerous calculations published before. The corrections can be neglected in most of the following cases where the nonlinear electrodynamic forces will dominate over the collision induced processes.

THREE

Kinetic Theory

Despite the fact that the following discussion of the laserplasma interaction follows the macroscopic (hydrodynamic) description, this section will sketch the kinetic theory of plasmas for some background understanding. The reader looking more for the experimental or engineering aspects may skip this section.

The kinetic theory or the theory of kinetic equations is a link between the single-particle description of the microscopic theory and of the macroscopic continuum theory of hydrodynamic quantities such as particle densities, net velocities, and temperatures. Just the definition of temperature as a macroscopic quantity involves complications if no thermal equilibrium has been achieved and the microscopic state shows several "temperatures" of plasma components.

3.1 Distribution Functions

For a straightforward understanding, an ad hoc description of the elements of the kinetic theory is given in this subsection. We first have to underscore that distribution functions do not necessarily relate—in a philosophical sense—to probabilities, uncertainties, or inaccuracies. If one has to find the average value of an ensemble of N integers a_i

$$M = \frac{\sum_{i=1}^{N} a_i}{N} \tag{3.1}$$

If there are several of the integers of the same numerical value, one can order the numbers in microensembles $j=1,\ldots,n$, where f_j determines the number of elements in each microensemble. The task of finding the average value (3.1) is then solved by

$$M=\frac{\sum_{i=1}^{n} f_i a_i}{\sum_{i=1}^{n} f_i} \tag{3.2}$$

Therefore a set of f_i or, continuously, a distribution function $f(x)$ can then determine the average value of a quantity $a(x)$

$$\langle a\rangle=\frac{\int f(x)a(x)\,dx}{\int f(x)\,dx} \tag{3.3}$$

The same procedure with distribution functions is used to reduce the microscopic N-particle problem with its $3N$ differential equations (2.1) by the following way: The configuration space is subdivided into cells of a volume $d^3r=dx_1\,dx_2\,dx_3$ about $\mathbf{x}$ and the velocity space is subdivided into cells $d^3w=dw_1\,dw_2\,dw_3$ about $\mathbf{w}$. Instead of describing each of the N particles by their differential equation (2.1) in time, one can ask, how many particles are present in the mentioned cells, depending on time t, given by the distribution function

$$f(x_1,x_2,x_3,v_1,v_2,v_3,t)d^3x\,d^3v \tag{3.4}$$

The density of the particles (number of particles per cubic centimeter) is then from this definition with the index e or i (electrons or ions)

$$n_{e,i}(\mathbf{r},t)=\iiint_{-\infty}^{+\infty} f(\mathbf{r},\mathbf{w},t)\,d^3w \tag{3.5}$$

Similar to Eq. (3.3) we find the value of any physical quantity $Q(\mathbf{r},\mathbf{w},t)$ depending on the velocity $\mathbf{w}$ at a point $\mathbf{r}$ at a time t as

$$\overline{Q(\mathbf{r},t)}=\frac{\iiint_{-\infty}^{+\infty} Q(\mathbf{r},\mathbf{w},t)f(\mathbf{r},\mathbf{w},t)\,d^3w}{\iiint_{-\infty}^{+\infty} f(\mathbf{r},\mathbf{w},\;)\,d^3w} \tag{3.6}$$

The average velocity of the N particles at $\mathbf{r}$ and t is then

$$\mathbf{v}(\mathbf{r}, t) = \frac{\displaystyle\iiint_{-\infty}^{+\infty} \mathbf{w} f(\mathbf{r}, \mathbf{w}, t)\, d^3w}{\displaystyle\iiint_{-\infty}^{+\infty} f(\mathbf{r}, \mathbf{w}, t)\, d^3w} \tag{3.7}$$

If $Q(\mathbf{w})$ does not depend on $\mathbf{r}$ and t, we find the following relations using Eq. (3.5):

$$\int Q(\mathbf{w}) \frac{\partial f}{dt}\, d^3w = \frac{\partial}{\partial t} \int Q(\mathbf{w}) f\, d^3w \tag{3.8}$$

$$= \frac{\partial}{\partial t}(n\bar{Q})$$

$$\int Q(\mathbf{w})\mathbf{w} \cdot \nabla f\, d^3w = \nabla \cdot \int Q(\mathbf{w})\mathbf{w}\mathbf{f}\, d^3w = \nabla(n\overline{\mathbf{w}Q}) \tag{3.9}$$

where the vector operator $\nabla = (\partial/\partial x_1;\ \partial/\partial x_2;\ \partial/\partial x_3)$ is called "del" or nabla. Using a general vectorial function $\mathbf{F}$ and the operator $\nabla_w = (\partial/\partial w_1;\ \partial/\partial w_2;\ \partial/\partial w_3)$

$$\int Q(\mathbf{w})F(\mathbf{r}, \mathbf{w}) \cdot \nabla_w f\, d^3w = -\int f \nabla_w \cdot \{\mathbf{F}(\mathbf{r}, \mathbf{w})Q(\mathbf{w})\}\, d^3w$$

and using the fact at partial integration of $f(\pm\infty)=0$ [for reasons of convergence of (3.5)],

$$\int Q(\mathbf{w})F(\mathbf{r}, \mathbf{w}, t) \cdot \nabla_w f\, d^3w = -n\nabla_w \cdot (\overline{FQ}) \tag{3.10}$$

We arrive at a kinetic equation for the distribution function f by the very trivial statement that, if there are no changes, there is no explicit dependence on time, or—in other words—the total derivative

$$\frac{d}{dt} f = 0 \tag{3.11}$$

has to be zero. Taking into account all seven variables of f, Eq. (3.4), we find by partial differentiation from (3.11)

$$\frac{d}{dt} f = \frac{\partial}{\partial t} f + \frac{\partial}{\partial x_1} f \frac{\partial x_1}{\partial t} + \frac{\partial}{\partial x_2} f \frac{\partial x_2}{\partial t} + \frac{\partial}{\partial x_3} f \frac{\partial x_3}{\partial t}$$
$$+ \frac{\partial}{\partial w_1} f \frac{\partial w_1}{\partial t} + \frac{\partial}{\partial w_2} f \frac{\partial w_2}{\partial t} + \frac{\partial}{\partial w_3} f \frac{\partial w_3}{\partial t} \tag{3.12}$$

or using the velocity vector $\mathbf{w}=(\partial x/\partial t; \partial y/\partial t; \partial z/\partial t)$, and the force density $\mathbf{F}$, where

$$\mathbf{F}=m\frac{\partial}{\partial t}\mathbf{w} \tag{3.13}$$

if all particles have the same (or averaged) mass m, we arrive at

$$\frac{\partial f}{\partial t}+\mathbf{w}\cdot\nabla f+\frac{\mathbf{F}}{m}\cdot\nabla_w f=0 \tag{3.14}$$

This equation is a *kinetic equation* and is called the *Vlasov equation*, if the force is due to electrical quantities ($\mathbf{E}$=electric and $\mathbf{H}$=magnetic field strength, c=velocity of light)

$$\mathbf{F}=e\left(\mathbf{E}+\frac{1}{c}\mathbf{w}\times\mathbf{H}\right) \tag{3.14a}$$

resulting in

$$\frac{\partial f}{\partial t}+\mathbf{w}\cdot\nabla f+\frac{e}{m}\left(\mathbf{E}+\frac{1}{c}\mathbf{w}\times\mathbf{H}\right)\cdot\nabla_w f=0 \tag{3.15}$$

If f does depend explicitly on the time, the total differentiation in Eq. (3.11) is not zero but equal to the change due to collisions (index c):

$$\frac{\partial f}{\partial t}+\mathbf{w}\cdot\nabla f+\frac{\mathbf{F}}{m}\nabla_w f=\left(\frac{\partial f}{\partial t}\right)_c \tag{3.16}$$

which is the Boltzmann equation. The whole problem is then concentrated in the collision term on the right-hand side. An approximation for collisions with neutral atoms is the Krook collision term

$$\left(\frac{\partial f}{\partial t}\right)_c=\frac{f_n-f}{\tau} \tag{3.17}$$

where f_n is the distribution function of the neutral atoms and τ is the averaged collision time.

For Coulomb collisions, Eq. (3.16) can be approximated using binary collisions by the Fokker–Planck equation

$$\frac{df}{dt}=\sum_{n=1}^{\infty}\frac{(-1)^n}{n!}\frac{\partial^n}{\partial_{vi_1}\partial v_{i_2}\cdots\partial_{vi_n}}(\bar{\alpha}_{(m)}f) \qquad (i_n=1, 2, 3, \ldots, n) \tag{3.18}$$

where the $\bar{\alpha}$'s are the Fokker–Planck coefficients [109].

Without discussion we mention here Liouville's theorem: for a conservative system, f is constant along a dynamic trajectory. This means that the volume $d^3r\,d^3w$ which is taken by a number of particles, does not change in time if the particles are not interacting.

While the kinetic equations provide the possibility of treating a plasma without having reached thermal equilibrium, a central meaning has the distribution function for equilibrium. If one normalizes f at each point of certain particle density $n(\mathbf{r}, t)$, the factor f_n in

$$f(\mathbf{r}, \mathbf{v}, t) = n(\mathbf{r}, t)\hat{f}_M(\mathbf{r}, \mathbf{v}, t) \tag{3.19}$$

is called the Maxwellian distribution (or Maxwell–Boltzmann distribution), see Appendix B,

$$\hat{f}_M = \left(\frac{m}{2\pi KT}\right)^{3/2} \exp(-\mathbf{w}^2/v_{\text{th}}^2) \tag{3.20}$$

using Boltzmann's constant K and the average thermal velocity $v_{\text{th}} = (2KT/m)^{1/2}$. Equation (3.20) takes into account positive and negative velocity components. Based on the scalar magnitude of the velocity w one defines another distribution $g(w)$ by

$$\int_0^{\infty} g(w)\, dw = \int_{-\infty}^{+\infty} f(\mathbf{w})\, d^3w \tag{3.21}$$

where

$$g(w) = 4\pi n(\mathbf{r}, t)\left(\frac{m}{2\pi KT}\right)^{3/2} w^3 \exp(-w^2/v_{\text{th}}^2) \tag{3.22}$$

to be distinguished from f.

3.2 Loss of Information

The description of the kinetic equations in the preceding subsection is sufficient to understand the derivation of the macroscopic equations in the next following subsection. In this subsection, some basic problems of the kinetic theory should be mentioned. Boltzmann's basic question was the description of the collisions even in the most simplified way where the collision time can be neglected. The derivation of the Boltzmann equation can be based on the more general Liouville distribution function for the N particles

$$F(\mathbf{r}_1, \mathbf{r}_2 \cdots \mathbf{r}_N, \mathbf{w}_1, \mathbf{w}_2 \cdots \mathbf{w}_N, t) \tag{3.23}$$

which gives the joint probability of finding particle 1 at $\mathbf{r}_1$ with velocity $\mathbf{w}_1$, and particle 2 at $\mathbf{r}_2$ with velocity $\mathbf{w}_2$, and particle N at $\mathbf{r}_N$ with $\mathbf{w}_N$. A single-particle distribution function f defines the probability of finding a particle at $\mathbf{r}$, with a velocity $\mathbf{w}$, at a time t from integrating (3.23) over all but the first particle coordinate:

$$f(\mathbf{r}, \mathbf{w}, t)=\int d^3r_2 \cdots d^3r_N d^3w_1 \cdots d^3w_N F(\mathbf{r}_1 \cdots \mathbf{r}_N, \mathbf{w}_1 \cdots \mathbf{w}_N) \qquad (3.24)$$

If the particles move independently, a factorization is possible:

$$F(\mathbf{r}_1 \cdots \mathbf{r}_N, \mathbf{w}_1 \cdots \mathbf{w}_N, t)=f(\mathbf{r}_1, \mathbf{w}_1, t) \cdots f(\mathbf{r}_N, \mathbf{w}_N, t)=f^N(\mathbf{r}, \mathbf{w}, t) \qquad (3.25)$$

F satisfies the Liouville equation

$$\frac{\partial F}{\partial t}+[F, H]=0 \qquad (3.26)$$

where the Poisson bracket of F with the Hamiltonian H of the complete system

$$[F, H]=\sum_{i=1}^{N} \frac{\partial H}{\partial p_i}\frac{\partial F}{\partial q_i} - \frac{\partial H}{\partial q_i}\frac{\partial F}{\partial p_i} \qquad (3.27)$$

remembering Hamilton's equations

$$\dot{q}_i=\frac{\partial H}{\partial p_i}; \qquad \dot{p}_i=-\frac{\partial H}{\partial q_i}$$

and stepping back from the generalized coordinates q_i to the Cartesians $\mathbf{x}_i$ and from the generalized momenta p_i to velocities $\mathbf{w}$; or accelerations or forces (as in Eq. (3.13)), Eq. (3.27) can be written

$$[F, H]=\sum_{i=1}^{N} \mathbf{w}_i \cdot \nabla F+\frac{\mathbf{F}}{m} \cdot \nabla_{w_i} F \qquad (3.28)$$

rewriting (3.26) in

$$\frac{\partial}{\partial t} F+\mathbf{v}_i \cdot \nabla F+\frac{\mathbf{F}}{m} \cdot \nabla_{w_i} F=0 \qquad (3.29)$$

Integration over $N-1$ coordinates as in Eq. (3.24), results in the collisionless Boltzmann equation

$$\frac{\partial f}{\partial t}+\mathbf{v} \cdot \nabla f+\frac{\mathbf{F}}{m} \cdot \nabla_v f=0 \qquad (3.30)$$

Following the result of Blatt and Opie [110], there are fundamental insufficiencies about the loss of information about the initial state when the system is described by the kinetic equations. The approximation (3.25) is an especially strong restriction. But even there are considerable doubts as to whether the Liouville equation (3.26) with a general F represents any real thermal system adequately [111]. The careful derivation of the kinetic model [110] is further made on the basis of minor approximations appropriate to a very dilute gas, ignoring effects of ternary and higher order collisions.

In laser produced plasmas, the ternary process at the usual high densities are of importance (see, e.g., J. M. Dawson [112]) and the predominance of three-body recombinations [113] was observed even in the early experiments [114] of laser produced plasmas. The basic problems involved are described within a consequent derivation of the Langrangean and Hamilton theory on the basis of the d'Alembert's principle [115].

These arguments should be taken into account if doubts are considered against the macroscopic hydrodynamic models: the kinetic models still are not free of doubts, especially for the high densities of laser produced plasmas. A comeback to the single-particle description of the microscopic models based on Eq. (2.1) is difficult because of computer capacity, the necessary neglection of low-range interactions, an approximative description of the Coulomb collisions with respect to small-angle scattering, and the approximation of collective effects.

3.3 Derivation of Macroscopic Equations

In this subsection macroscopic equations will be derived from the integration of the Boltzmann equation (3.16)

$$\frac{\partial f}{\partial t}+\mathbf{w}\cdot\nabla f+\frac{\mathbf{F}}{m}\nabla_w f=\left(\frac{\partial f}{\partial t}\right)_c \tag{3.31}$$

Each term will be integrated over the whole velocity space d^3w, if we assume that there are no velocity dependent forces. The forces are then holonomic following the definition of H. Hertz [115].

Integration from $-\infty$ to $+\infty$ of the first term of (3.31) results in

$$\int\frac{\partial f}{\partial t}\,d^3w=\frac{\partial}{\partial t}\int f\,d^3w=\frac{\partial}{\partial t}n \tag{3.32}$$

following Eq. (3.5). Integration of the second term of (3.31) based on Eq. (3.9) with a quantity $Q=1$ results in

$$\int\mathbf{w}\cdot\nabla f dw=\nabla\cdot(\overline{n\mathbf{w}})=\nabla\cdot\overline{n\mathbf{v}} \tag{3.33}$$

where the velocity of the particles was split in the drift velocity $\mathbf{v}$ and the random (thermal) velocity $\mathbf{u}$

$$\mathbf{w}=\mathbf{v}+\mathbf{u} \tag{3.34}$$

with

$$\bar{\mathbf{u}}=0 \tag{3.35}$$

Integration of the third term of (3.16) using Eq. (3.10) with $Q=1$ arrives at

$$\int \frac{\mathbf{F}}{m}\cdot\nabla_w f\, d^3w = -\int f\nabla_w\cdot\mathbf{F}\, d^3w = -n\overline{\nabla_w\cdot\mathbf{F}}=0$$

which is zero because we assume velocity independent (holonomic) forces $\mathbf{F}$. The right-hand side of Eq. (3.31) results in the integration

$$\int_{-\infty}^{+\infty}\left(\frac{\partial f}{\partial t}\right)_c d^3w=0 \tag{3.36}$$

because collisions cannot change the total number n of particles per cubic centimeter in average.

Summarizing Eqs. (3.32) to (3.36) results in the velocity integral of the Boltzmann equations

$$\frac{\partial}{\partial t}n+\nabla\cdot n\mathbf{v}=0 \tag{3.37}$$

which is the hydrodynamic equation of continuity (equation of conservation of mass).

Another integration of the Boltzmann equation over the velocity space after multiplying with the momentum ($m\mathbf{w}$) will lead to the hydrodynamic equation of motion (equation of conservation of momentum). From Eq. (3.31) we arrive then at

$$\int m\mathbf{w}\frac{\partial}{\partial t}f\, d^3w+\int m\mathbf{w}\mathbf{w}\cdot\nabla f\, d^3w+\int m\mathbf{w}\frac{\mathbf{F}}{m}\cdot\nabla_w f\, d^3w=\int m\mathbf{w}\left(\frac{\partial f}{\partial t}\right)_c d^3w \tag{3.38}$$

We use, as before, Eqs. (3.7) to (3.10), where

$$Q(\mathbf{w})=m\mathbf{w} \tag{3.39}$$

The first term A is from Eq. (3.7)

$$A=\int m\mathbf{w}\frac{\partial}{\partial t}f\, d^3w=\frac{\partial}{\partial t}(\overline{nm\mathbf{w}}) \tag{3.40}$$

where the separation into drift velocity $\mathbf{v}$ and random velocity $\mathbf{u}$, Eq. (3.34) leads to

$$A=mn\frac{\partial}{\partial t}(\overline{\mathbf{v}+\mathbf{u}})+(\overline{\mathbf{v}+\mathbf{u}})\frac{\partial}{\partial t}mn \tag{3.41}$$

and using $\bar{\mathbf{u}}=0$ to

$$A=mn\frac{\partial}{\partial t}\mathbf{v}+\mathbf{v}\frac{\partial}{\partial t}mn \tag{3.42}$$

The second term B from Eqs. (3.8), (3.34), and (3.39) arrives at

$$\begin{aligned} B &= \int m\mathbf{w}\mathbf{w}\cdot\nabla f\, d^3w \\ &= \boldsymbol{\nabla}\cdot n\overline{\mathbf{w}m\mathbf{w}} = \boldsymbol{\nabla}\cdot nm\overline{(\mathbf{v}+\mathbf{u})(\mathbf{v}+\mathbf{u})} \\ &= \boldsymbol{\nabla}\cdot nm[\overline{\mathbf{v}\mathbf{v}}+\overline{\mathbf{u}\mathbf{u}}+\overline{\mathbf{v}\mathbf{u}}+\overline{\mathbf{u}\mathbf{v}}] \end{aligned} \tag{3.43}$$

where the last two dyadic terms are zero ($\bar{\mathbf{u}}=0$).

Furthermore,

$$B=\bar{\mathbf{v}}\bar{\mathbf{v}}\cdot\boldsymbol{\nabla}mn+mn\overline{\mathbf{v}\cdot\boldsymbol{\nabla}\mathbf{v}}+nm\overline{\mathbf{v}\boldsymbol{\nabla}\cdot\mathbf{v}}+\nabla\cdot mn\overline{\mathbf{u}\mathbf{u}} \tag{3.44}$$

For the following addition of A and B it is important that the second terms of (3.42) and the first term of (3.44) are zero

$$\mathbf{v}\frac{\partial}{\partial t}mn+\mathbf{v}\mathbf{v}\cdot\boldsymbol{\nabla}mn=\mathbf{v}\frac{d}{dt}mn=0 \tag{3.45}$$

because of the conservation of mass within one volume element at substantial motion (total differentiation for explicit time).

The third term on the right-hand side of Eq. (3.44) of B is

$$mn\mathbf{v}\boldsymbol{\nabla}\cdot\mathbf{v}=-\mathbf{v}\frac{\partial}{\partial t}mn-m\mathbf{v}\mathbf{v}\cdot\boldsymbol{\nabla}n \tag{3.46}$$

using (3.37) which describes adiabatic heating after compensating the first term of B (3.44) against the last term of (3.46). The fourth term in Eq. (3.44) is

$$\boldsymbol{\nabla}\cdot mn\mathbf{u}\mathbf{u}=\boldsymbol{\nabla}\cdot 2mn\tfrac{1}{2}\mathbf{u}^2\mathbf{1}=\boldsymbol{\nabla}nKT \tag{3.47}$$

using the unity tensor

$$\mathbf{1}=\mathbf{i}_1\mathbf{i}_1+\mathbf{i}_2\mathbf{i}_2+\mathbf{i}_3\mathbf{i}_3 \tag{3.48}$$

and interpreting the energy of random motion $nm\mathbf{u}^2/2$ as inner energy.

The third term C in Eq. (3.38) is derived using Eqs. (3.9), (3.10), and (3.39):

$$C=\int m\mathbf{w}\frac{\mathbf{F}}{m}\cdot\boldsymbol{\nabla}_w f\, d^3w=-n\boldsymbol{\nabla}_w\cdot\mathbf{F}\mathbf{w}=-n\mathbf{F}\cdot\boldsymbol{\nabla}_w\mathbf{w}-n\mathbf{w}\boldsymbol{\nabla}_w\cdot\mathbf{F} \tag{3.49}$$

The second term in the last expression vanishes because the forces $\mathbf{F}$ should depend on the velocity $\mathbf{w}$ as given by Eq. (3.14a), where $\boldsymbol{\nabla}_w\times\mathbf{w}=0$,

$$\boldsymbol{\nabla}_w\cdot\mathbf{F}=0 \tag{3.50}$$

The tensor of the first term in the last expression of Eq. (3.49) is

$$\begin{aligned} \boldsymbol{\nabla}_w\mathbf{w} &= \mathbf{i}_1\mathbf{i}_1\frac{\partial}{\partial w_1}w_1+\mathbf{i}_1\mathbf{i}_2\frac{\partial}{\partial w_1}w_2+\mathbf{i}_1\mathbf{i}_3\frac{\partial}{\partial w_1}w_3 \\ &+\mathbf{i}_2\mathbf{i}_1\frac{\partial}{\partial w_2}w_1+\mathbf{i}_2\mathbf{i}_2\frac{\partial}{\partial w_2}w_2+\mathbf{i}_2\mathbf{i}_3\frac{\partial}{\partial w_2}w_3 \\ &+\mathbf{i}_3\mathbf{i}_1\frac{\partial}{\partial w_3}w_1+\mathbf{i}_3\mathbf{i}_2\frac{\partial}{\partial w_3}w_2+\mathbf{i}_3\mathbf{i}_3\frac{\partial}{\partial w_3}w_3 \end{aligned}$$

where all nondiagonal terms vanish for Cartesian coordinates; therefore, using Eq. (3.48)

$$\nabla_w \mathbf{w} = \mathbf{1}$$

and from (3.49) we arrive at

$$C = -n\mathbf{F}\cdot\mathbf{1} = -n\mathbf{F} \tag{3.51}$$

the forces in plasmas should be

$$\mathbf{F} = Ze\mathbf{E} + \frac{Ze}{c}\,\mathbf{v}\times\mathbf{H} + \mathbf{F}_g \tag{3.52}$$

where Z is the number of charges for the equation of ions, while $Z=1$ for electrons, and $\mathbf{F}_g$ are forces of gravitation, Coriolis forces, and others.

The last term D in Eq. (3.38)

$$D = \int m\mathbf{w}\left(\frac{\partial f}{\partial t}\right)_c d^3w \tag{3.53}$$

expresses the net momentum per volume transferred to the ions by collisions with the electrons

$$D = \mathbf{P}_{ie} \tag{3.54}$$

If there are no asymmetric velocity distributions of the electrons,

$$\mathbf{P}_{ie} = 0 \tag{3.55}$$

This is not the case, for example, if an electron beam is fired into the plasma.

Putting together the results of A, B, C, and D, we arrive at

$$mn\frac{\partial}{\partial t}\mathbf{v} + mn\mathbf{v}\cdot\nabla\mathbf{v} = \nabla(nKT) + nmZe\left[\mathbf{E} + \frac{1}{c}\mathbf{v}\times\mathbf{H}\right] + mn\mathbf{F}_g \tag{3.56}$$

where the second term of A, (3.42), and the first term of B, (3.44), canceled because of (3.45) and the first terms and second term of B, (3.44). The macroscopic (hydrodynamic) equation of motion is then

$$mn\frac{\partial}{\partial t}\mathbf{v} + mn\mathbf{v}\cdot\nabla\mathbf{v} = -\nabla nKT + mnZe\left[\mathbf{E} + \frac{1}{c}\mathbf{v}\times\mathbf{H}\right] + mn\mathbf{E}_g \tag{3.57}$$

or

$$mn\frac{d}{dt}\mathbf{v} = \mathbf{f} \tag{3.58}$$

where $\mathbf{f}$ is the force density.

From Eq. (3.56) we derive immediately the Bernoulli equation for a vortex-free stationary ($\partial/\partial t = 0$) motion without external forces $\mathbf{F}_g$ and fields $\mathbf{E}$ and $\mathbf{H}$:

$$\nabla\left(nKT + \frac{mn}{2}\mathbf{v}^2\right) = 0 \tag{3.59}$$

or using the (static) pressure $p=nKT$ and the density $\rho=nm$, integration of Eq. (3.59) results in Bernoulli's equation

$$p+\frac{\rho}{2}\mathbf{v}^2=\text{const} \tag{3.60}$$

3.4 Landau Damping

Landau [116] studied the behavior of a distribution function of electrons in a plasma when being changed slightly from the thermal equilibrium. Without any dissipation or energy transfer, a damped oscillation will occur for returning into the undisturbed state. This can be seen in the following way. The plasma is assumed to be field free ($\mathbf{E}_0=\mathbf{H}_0=0$) apart from small fields due to the perturbation. The equilibrium thermal electron distribution should be a Maxwellian $f_0(\mathbf{v})$, Eq. (3.20), and the perturbation should be described by the linear approximation f,

$$f(\mathbf{r},\mathbf{v},t)=f_0(\mathbf{w})+f_1(\mathbf{r},\mathbf{w},t) \tag{3.61}$$

No collisions should be present; therefore, Eq. (3.61) will result in a first-order Vlasov equation from (3.14)

$$\frac{\partial f_1}{\partial t}+\mathbf{v}\cdot\nabla f_1-\frac{e}{m}\mathbf{E}_1\cdot\nabla_v f_0=0 \tag{3.62}$$

where the electric field $\mathbf{E}_1$ is due to the perturbation, and f_1 has been neglected compared with f_0 in the last term of (3.62). If f_1 and $\mathbf{E}_1$ are described as plane waves in the x direction

$$f_1=f_{10}\exp i(kx-\omega t); \qquad E_x=E_{x0}\exp i(kx-\omega t) \tag{3.63}$$

using a wave number $k=\omega/c$ and a radian frequency ω where c is the velocity of the wave. Eq. (3.62) with Eq. (3.63) becomes

$$-i\omega f_1+ikv_x f_1=\frac{e}{m}E_x\frac{\partial f_0}{\partial v_x} \tag{3.64}$$

or

$$f_1=\frac{ieE_x}{m}\frac{\partial f_0/\partial x}{\omega-kv_x} \tag{3.64a}$$

Using Poisson's equation

$$\nabla\cdot\mathbf{E}=ikE_x=-4\pi e n_1=-4\pi e\int f_1\,d^3v \tag{3.65}$$

with the disturbation n_1 of the density. Substituting (3.64) and E_x from

(3.65) in the integral in (3.65) results in

$$1 = -\frac{4\pi e^2}{km} \int_{-\infty}^{+\infty} \frac{\partial f_0/\partial v_x}{\omega - kv_x} \, d^3v \tag{3.66}$$

The integral has to be finite, but as any real ω would cause a divergence, the only possibility is to assume a complex ω whose imaginary part will contribute to a damping of an oscillation.

The evaluation of the integral (3.66) with Cauchy's residue theorem is not very easy, as the integral along the half-circle for infinite $|v|$ of complex values v_x is not zero. Only for a small imaginary part of ω, can the following approximation be achieved from (3.66):

$$1 = \frac{\omega_p^2}{\omega^2} + i\pi \frac{\omega_p^2}{k^2} \frac{\partial f_0}{\partial v}\bigg|_{v=v\phi} \tag{3.67}$$

where v_ϕ is given from $\mathrm{Re}(\omega) - k \cdot \mathrm{Re}(v_x) = 0$ and ω_p is the plasma frequency. The resulting frequency is then

$$\omega = \omega_p \left[1 + i\frac{\pi}{2} \frac{\omega_p^2}{k^2} \left(\frac{\partial f_0}{\partial v} \right)_{v=v\phi} \right] \tag{3.68}$$

The result shows that a disturbance of the distribution of the electrons results in an oscillation with a frequency ω_p (as we have seen in Section 2.1 in a very direct way), and, if the perturbation has been generated once, its oscillation will be damped by the plasma without an exchange of energy. A plasma therefore has a self-stabilizing property

If collisions are included, the Landau damping may modify the usual absorption process. These effects are strong only if the wavelength of the electromagnetic wave $\lambda = \omega/2\pi c$ is close to the Debye length λ_D of the plasma. Thinking of laser wavelengths near the visible range and plasmas of solid-state density and above, even for extremely high temperatures of 10^5 eV, the Debye length is 100 times less than λ.

Considering this section about the kinetic theory on which the macroscopic theory can be based, we can summarize as follows. The most general (microscopic) description of a plasma as an N-particle system is limited by computing capacity and/or restricting simplifications on the range of forces and so on. The kinetic theory using distribution functions results in an irreversible loss of information. Even with the general Liouville distribution, the Liouville equation (3.28) contains basic insufficiencies as has been shown by Blatt in 1959 [111]. Taking these reservations into account, the derivation of macroscopic (hydrodynamic) equations (Section 3.3) is convincing but limited. The Landau damping was an example of how deviations from an average value are automatically damped down to the equilibrium state, showing how the equilibrium is stable. There might be difficulties at strong deviations. Meixner [117] finds from the Chapman–Enskog solution [118] of Boltzmann's equation (3.31) with a restriction of the local equilibrium of a system that temperature variations ΔT must be less than the average temperature in a cell of the mean free path l, of the particles.

There was a discussion on whether the kinetic theory limits a derivation of the macroscopic hydrodynamics if conditions of irreversible thermodynamics appear. This could lead to criticism of hydrodynamic computer results for very short times of laser interaction with plasma. It must be remembered that highly irreversible processes can be described by hydrodynamic codes in full agreement with experiments. A splendid example is the compression of thermonuclear plasma in a theta-pinch. The irrelevance of the mentioned criticism is perhaps due to the well-known fact that the limit for reversible processes can be very much extended into the irreversible range. The role of fluctuations on entropy production and the formulation of the optical absorption by Onsager coefficients [119] should be mentioned in this connection.

FOUR

Hydrodynamics

Macroscopic plasma theory is a combination of electromagnetic theory with the hydrodynamic properties of plasma. There are various hydrodynamic descriptions of plasma in literature, yet starting in the 1940s with the one-fluid model of Alvén, very general macroscopic equations for fully ionized plasma are the two-fluid equations derived by Schlüter, which will be discussed in Section 6. To make the reader more familiar with the background of hydrodynamics, this section will consider some significant properties.

4.1 Euler's Equation of Motion

The hydrodynamic equation of motion is the field-theoretical generalization of Newton's single-particle equation of motion:

$$\bar{m}\mathbf{a}=\mathbf{F}=-\nabla\phi \tag{4.1}$$

The product of mass $\bar{m}$ times the acceleration $\mathbf{a}$ of a body equals the force $\mathbf{F}$, which can be expressed by the gradient of a potential ϕ. In the case of a fluid, there is a velocity field $\mathbf{v}(x, y, z, t)$, of which the temporal derivative corresponds to the acceleration; a mass density field $\rho(x, y, z, t)$, which corresponds to the mass m; and a force density, which is given by the gradient of the pressure field $p(x, y, z, t)$. Consider the fluid as being composed of electrons of mass m, density n_e, temperature T_e, and ions of mass m_i, density n_i, and charge Z. Assuming space charge neutrality $n_i=n_e/Z$, the mass density field is given by

$$\rho(x, y, z, t)=m_i n_i(x, y, z, t)+mn_e(x, y, z, t) \tag{4.2}$$

The pressure field (also as function of x, y, z, t) is

$$p=n_e KT_e+n_i KT_i \approx (1+Z)n_i KT_e, \qquad \text{if } T_e \approx T_i \tag{4.3}$$

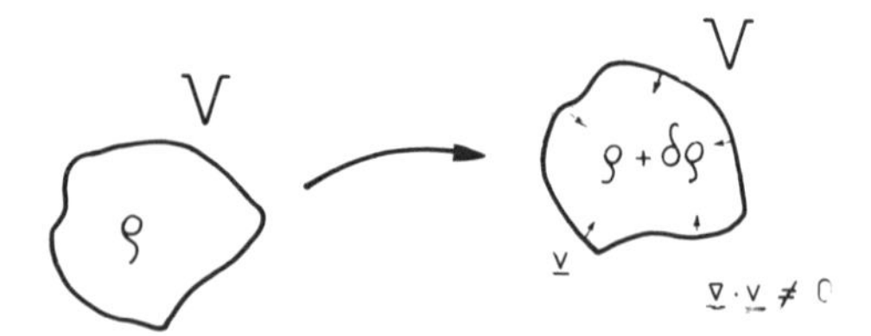

Figure 4.1. Geometrical variation with conservation of mass for the derivation of the equation of continuity.

The equation of motion corresponding to Newton's Eq. (4.1), the Euler equation, is

$$\rho\frac{d\mathbf{v}}{dt} = -\nabla p + \eta\nabla^2\mathbf{v} \tag{4.4}$$

The last term on the right-hand side is added to the original Euler equation and is called the Navier–Stokes term. This includes the hydrodynamic viscosity η, which determines the internal friction of the fluid. The operation on the left-hand side of Euler's equation (4.4) is

$$\rho\frac{d\mathbf{v}}{dt} = \rho\frac{\partial\mathbf{v}}{\partial t} + \rho\frac{\partial\mathbf{v}}{\partial x}\frac{dx}{dt} + \rho\frac{\partial\mathbf{v}}{\partial y}\frac{dy}{dt} + \rho\frac{\partial\mathbf{v}}{\partial z}\frac{dz}{dt}. \tag{4.5}$$

We combine the last three terms to $\rho\mathbf{v}\cdot\nabla\mathbf{v}$, which is a nonlinear term. Remember that the spatial "del" operator is $\nabla = \mathbf{i}_x\partial/\partial x + \mathbf{i}_y\partial/\partial y + \mathbf{i}_z\partial/\partial z$, where the unit vectors $\mathbf{i}_x$, $\mathbf{i}_y$, and $\mathbf{i}_z$ are of the modulus 1 and are in the direction of the Cartesian coordinates x, y, and z, respectively. Using this way of writing, the Euler equation is then

$$\rho\frac{\partial\mathbf{v}}{\partial t} + \rho\mathbf{v}\cdot\nabla\mathbf{v} = -\nabla p \tag{4.6}$$

where the viscosity has been dropped.

4.2 Bernoulli's Stationary Solution

From Euler's equation (4.6), Bernoulli's equation can be derived as a stationary solution. Because of the then necessary time independence:

$$\frac{\partial}{\partial t} = 0 \tag{4.7}$$

the Euler equation (4.6) reduces to

$$\nabla\mathbf{v} + \nabla p = 0 \tag{4.8}$$

By using the vector identity:

$$\mathbf{v}\cdot\nabla\mathbf{v}=\tfrac{1}{2}\nabla\mathbf{v}^2-\mathbf{v}\times(\nabla\times\mathbf{v}) \tag{4.9}$$

The last term can be dropped for vortex-free motion, giving:

$$\frac{\rho}{2}\nabla\mathbf{v}^2+\nabla p=0 \qquad 4.10)$$

and under the further special assumption that the spatial variation of the density ρ is zero,

$$\rho=\text{const} \tag{4.11}$$

Eq. (4.10) can be written as:

$$\nabla\left(\frac{\rho}{2}\mathbf{v}^2+p\right)=0 \tag{4.12}$$

which, when integrated, yields the Bernoulli equation

$$\frac{\rho}{2}\mathbf{v}^2+p=\text{const} \tag{4.13}$$

Note that the Bernoulli equation is a very particular case of the general Euler hydrodynamic equation, not only because of the *stationary* condition, Eq. (4.7), but also because of the neglect of the last term in Eq. (4.9) (*vortex-free motion*) and by condition (4.11), which corresponds to a chemical homogeneity and *incompressibility* of the fluid. Instead of this general derivation, which is appropriate for a theoretical method of deduction, Bernoulli's equation (4.13) can be derived primitively by compensating the kinetic pressure or the first term in Eq. (4.13), by the hydrostatic pressure, or second term on the left-hand side of Eq. (4.13).

4.3 Equation of Continuity

The Euler equation is one of the three basic equations of hydrodynamics and corresponds to the conservation of momentum. The next basic equation is the conservation of mass, which is sometimes called the equation of continuity. It can be derived from the following geometrical consideration. Take a constant volume V with a density ρ, which is moved during a time interval dt along the large arrow of Fig. 4.1. After the time dt the volume V has changed its density to the value $\rho+\delta\rho$. The increase of mass in the volume V can only be due to material streaming into the volume V; therefore it has a negative divergence of the velocity field (expressed by $-\nabla\cdot\mathbf{v}$)

$$\frac{d\rho}{dt}=-\rho\nabla\cdot\mathbf{v} \tag{4.14}$$

Gauss' law, expressing the converging or diverging velocity field at a closed surface by the divergence ($\nabla \cdot \mathbf{v}$) of a volume integral within a closed area, gives

$$\oiint \mathbf{v}\, d^2\mathbf{a} = \iiint \nabla \cdot \mathbf{v}\, d^3\tau$$

The left-hand side of equation (4.14) can be expressed by the partial differentiation:

$$\frac{\partial \rho}{\partial t} + \mathbf{v} \cdot \nabla \rho = -\rho \nabla \cdot \mathbf{v} \tag{4.15}$$

and using the differentiation relation:

$$\nabla \cdot \rho \mathbf{v} = \rho \nabla \cdot \mathbf{v} + \mathbf{v} \cdot \nabla \rho \tag{4.16}$$

Eq. (4.15) can be rewritten as:

$$\frac{\partial \rho}{\partial t} + \nabla \cdot (\rho \mathbf{v}) = 0 \tag{4.17}$$

which is the equation of continuity [see Eq. (3.37)]. For the special case of incompressible fluids, Eq. (4.11), the special formulation of the equation of continuity is obtained:

$$\frac{\partial \rho}{\partial t} + \rho \nabla \cdot \mathbf{v} = 0 \tag{4.18}$$

which is related to a fast change of the density within a very large volume. It will be seen in the following that hydrodynamic waves are specially related to this process. It should be noted that a large number of problems in hydrodynamics and aerodynamics can be studied on the assumption of an incompressible flow.

4.4 Compressibility

For the following derivation of acoustic waves (in the next subsection) compressibility will be now discussed. By definition, compressibility is the relation of a change δV in the volume V of a medium which is generated by the change ∂p of the pressure p. With an initial volume V_0,

$$V(p) = V_0 + \delta V \tag{4.19}$$

the change expressed by a variation δp of the pressure p is

$$V(p) = V_0 - \frac{\partial V}{\partial p} \delta p$$

or the relative change of the volume V is

$$\frac{\delta V}{V} = -\frac{1}{V}\frac{\partial V}{\partial p}\delta p \tag{4.20}$$

The definition of compressibility is the proportionality factor between the relative change of the volume against a variation δp of the pressure:

$$\kappa = -\frac{1}{V}\frac{\partial V}{\partial p} \tag{4.21}$$

This is nothing more than a definition for which a reasonable explanation was given by the preceeding equations. The connection must now be found with the thermodynamic quantity of the adiabatic compression. Referring to thermodynamics, the connection between pressure p and volume V in an adiabatic compression (i.e., for a change without exchange of heat or energy with any other medium outside) is

$$pV^{\gamma} = \text{const} \tag{4.22}$$

where the exponent $\gamma = c_p/c_v$ is the ratio of the specific heat c_p for constant pressure to that for constant volume c_v. Thermodynamics derives this ratio from the number F of the degrees of freedom of particles in the medium:

$$\gamma = \frac{c_p}{c_v} = \frac{F+2}{F} \tag{4.23}$$

For the case of a fully ionized plasma there are $F = 3$ degrees of freedom, as in the case of rare gases, where the particles are single atoms.

From Eq. (4.22) is found:

$$\frac{\partial V}{\partial p} = -\frac{1}{\gamma}\frac{\text{const}^{1/\gamma}}{p^{(1/\gamma)+1}} \tag{4.24}$$

and from (4.21) by using Eqs. (4.22) to (4.24), finally:

$$\kappa = -\frac{1}{V}\frac{\partial V}{\partial p} = \frac{1}{\gamma p} \tag{4.25}$$

the compressibility for an adiabatic change of state, expressed in term of the pressure.

4.5 Acoustic Waves

The description of acoustic waves uses the Euler equation (4.4) in its linearized form:

$$\frac{\partial \mathbf{v}}{\partial t} = -\frac{1}{\rho}\nabla p \tag{4.26}$$

and the equation of continuity (4.17) with the assumption, that $|\mathbf{v}\cdot\nabla\rho| \ll |\nabla\cdot\mathbf{v}|$ (quasi incompressible)

$$\frac{\partial\rho}{\partial t} = -\rho\nabla\cdot\mathbf{v} \tag{4.27}$$

where the relation of the density ρ to particle densities is given by Eq. (4.2). The variation of the density ρ with the pressure p

$$\rho = \rho_o + \frac{\partial\rho}{\partial p}\,\delta p \tag{4.28}$$

can be expressed by definition through the compressibility κ from Eq. (4.21)

$$\rho = \rho_0[1 + \kappa(p - p_0)] \tag{4.29}$$

The variations $\delta p = p - p_o$ and $\delta\rho = \rho - \rho_o$ can be expressed by their initial values (index o) and their instantaneous values (without index), so that:

$$\delta p = \frac{1}{\rho_0\kappa}\,\delta\rho \tag{4.30}$$

or by differential description:

$$\nabla p = \frac{1}{\rho_0\kappa}\,\nabla\rho \tag{4.31}$$

Substituting this into the Euler equation (4.26), it is found:

$$\frac{\partial\mathbf{v}}{\partial t} = -\frac{1}{\rho}\frac{1}{\rho_0\kappa}\nabla\rho = -\frac{1}{\rho_0\kappa}\nabla\ln\rho \tag{4.32}$$

From the equation of continuity, Eq. (4.27), is found the approximation:

$$\frac{1}{\rho}\frac{\partial\rho}{\partial t} = \frac{\partial}{\partial t}\ln\rho = -\nabla\cdot\mathbf{v} \tag{4.33}$$

Differentiation of Eq. (4.32) by t and of Eq. (4.33) by ∇ leads to:

$$\frac{\partial^2\mathbf{v}}{\partial t^2} = -\frac{1}{\rho_0\kappa}\,\nabla\frac{\partial}{\partial t}\ln\rho = +\frac{1}{\rho_0\kappa}\nabla^2\mathbf{v} \tag{4.34}$$

or

$$\nabla^2\mathbf{v} - \rho_o\kappa\frac{\partial^2\mathbf{v}}{\partial t^2} = 0 \tag{4.35}$$

The solutions of the wave equation (4.35) are, for example, plane waves of a radiation frequency ω

$$\mathbf{v} = \mathbf{v}_o\exp(\pm i\mathbf{k}\cdot\mathbf{r} - i\omega t)$$

where the wave vector k determines the direction of the wave propagation and $|\mathbf{k}| = \omega/c_s$ gives a result for the phase velocity c_s:

$$c_s^2 = \frac{1}{\kappa \rho_0} \tag{4.36}$$

This velocity is the velocity of sound or the ion acoustic velocity. Substituting the compressibility from Eq. (4.25) leads to

$$c_s = \sqrt{\frac{\gamma p}{\rho_o}} \tag{4.37}$$

The same sound equation could have been obtained if the expression of $\ln \rho$ had been eliminated from the equation of motion and the equation of continuity to give:

$$\nabla^2 \ln \rho - \kappa \rho_o \frac{\partial^2 \ln \rho}{\partial t^2} = 0 \tag{4.38}$$

where again the same wave equation is reproduced as for the velocity with the same wave velocity c_s.

4.6 Equation of Energy

In addition to the conservation equations of momentum (Euler's equation of motion) and of mass (equation of continuity), the equation of energy conservation is needed to arrive at the complete set of differential equations for uniquely solving hydrodynamic problems. This equation is of the type:

$$\frac{\partial}{\partial t} \frac{\rho}{2} \mathbf{v}^2 = -\frac{\partial}{\partial t} n_i K T (1+Z) - \nabla \cdot (\kappa_T \nabla T) + W \tag{4.39}$$

where the left-hand side describes the temporal change of the kinetic energy of the fluid to be compensated by the change of internal energy (first term on the right-hand side), by thermal conduction, characterized by the thermal conductivity κ_T, and by any power density W of energy exchange by radiation and so on.

The expression for the power density W for energy transfer to the plasma will include the linear or nonlinear absorption constant. This is derived from the optical linear, nonlinear, or relativistic refractive index or from an effective collision frequency due to parametric instabilities or by an effective dynamic nonlinear absorption process. These steps will be discussed later in the contents of specific applications.

Any additional potentials are not to be included in Eq. (4.39), if these are independent of the time. This related to gravitational potentials or such of other static fields. In the case of laser produced plasmas, electrodynamic net potentials can change in time. These components will then have to be included in Eq. (4.39).

FIVE

Self-Similarity Model

The hydrodynamic equations of the preceding section can be used to analyze the gasdynamic expansion of a laser produced plasma of spherical shape into a vacuum. The transfer of the laser energy to the plasma is assumed to be fully symmetric, given by the spatially constant power density $W(t)$ in the energy equation (4.39). All specific conditions as to how this fast power transfer and equilibration can be realized are neglected and will be the topic of subsequent sections. Agreement with experiments at not too high laser intensities and not too short pulses for a wide range of parameters will justify the assumptions.

The expansion of a spherical plasma into vacuum can be described by a relatively simple model where the radius R is found as a function of time. The plasma temperature T has, at a time t_0 an initial value T_0, the radius R of the plasma an initial value R_0 and the velocity of expansion $\partial R/\partial t$ an initial value $\dot{R}_0$. During the expansion, an adiabatic transfer of thermal energy into kinetic energy of expansion will occur. The complete radial symmetric hydrodynamic calculation, for example, by Fader [120], with any initial radial velocity profile $v(r, t=0)$ and an ion density profile $n_i(r, t=0)$, resulted after some time t into a solution where n_i became a Gaussian density profile, while the velocity became a linear profile

$$v(r, t) = \frac{v_0(t) r}{R} \tag{5.1}$$

and the temperature dropped adiabatically. The fact that the density and velocity profiles remained similar is the reason for using 'self-similar' expansion. This has nothing to do with the similarity laws of hydrodynamics, for example, the Reynold's number, which are relations of dimensions and quantities of a fluid characteristic for the hydrodynamic motion.

Historically, the self-similarity model was used in the expansion of the universe [121] where (5.1) was derived [122]. The relation to the case of laser produced plasmas was underlined by Lengyel and Salvat [123]. This model was used for thermonuclear plasmas, for example, by Zeldovich and Raizer [124], especially for the calculation of laser plasmas by Basov and Krokhin [125], Dawson [112], and for the optimization of nuclear fusion gains [126].

Instead of the very global derivation of the self-similarity model [112, 125], here a general derivation from the hydrodynamic equations [127] is followed to show the limitations and restrictions of the model and to discuss a classical error because the global consideration was not based on hydrodynamics. Finally, some applications to laser produced plasmas will be considered.

5.1 Hydrodynamic Derivation

We use the definitions (4.2) and (4.3) and hydrodynamic equations of conservation (4.6), (4.17), and (4.39) in the following way. The equation of continuity;

$$\frac{\partial n_i}{\partial t}+\nabla\cdot(n_i\mathbf{v})=0 \tag{5.2}$$

the equation of motion:

$$\frac{d}{dt}n_i\mathbf{v}\left(1+Z\frac{m_e}{m_i}\right)m_i=-\nabla p \tag{5.3}$$

and the equation of energy conservation:

$$\frac{\partial}{\partial t}\frac{n_i m_i}{2}\left(1+Z\frac{m_e}{m_i}\right)\mathbf{v}^2=-\frac{\partial}{\partial t}(1+Z)n_i KT+W \tag{5.4}$$

where the terms Zm/m_i are very small compared with unity. Equations (5.2), (5.3), and (5.4) are the general hydrodynamic equations to solve n_i, v, and T in space and time for given initial conditions.

In the case of a spherical plasma with radial symmetry resulting in only a spatial dependence on the radial coordinate r with the radial velocity component v_r, we get the equation of continuity from Eq. (5.2):

$$\frac{\partial}{\partial t}n_i+\frac{\partial}{\partial t}n_i v_r+\frac{2n_i}{r}v_r=0 \tag{5.5}$$

Using the equation of state

$$p=n_i(1+Z)KT \tag{5.6}$$

in Eq. (5.3) one gets the equation of motion for the spherical case:

$$\frac{d}{dt} n_i m_i \left(1+Z\frac{m_e}{m_i}\right)\mathbf{v}_r = -\frac{\partial}{\partial r} n_i(1+Z) \geq T \tag{5.7}$$

and finally the equation of energy conservation from Eq. (5.4):

$$\frac{\partial}{\partial t}\frac{n_i m_i}{2}\left(1+Z\frac{m_e}{m_i}\right)v_r^2 = -\frac{\partial}{\partial r}(1+Z)n_i KT - \kappa\frac{1}{r^2}\frac{\partial}{\partial r}\left(r^2\frac{\partial}{\partial t}T\right)+W \tag{5.8}$$

Then, the three basic equations (5.5), (5.7), and (5.8) have to be solved together with the initi al conditon at a time $t=t_0$

$$T(r, t_0); \qquad n_i(r, t_0); \qquad v_r(r, t_0) \tag{5.9}$$

to find solutions for the three equations T, n_i, and v_r as a function of r and t.

Now the formulas of the self-similarity model, which were used by Dawson [112], are derived starting from the general radially symmetric hydrodynamic equations. The equation of motion (5.7) is multiplied with v_r

$$\frac{1}{2}\frac{d}{dt} n_i m_i \left(1+Z\frac{m_e}{m_i}\right)\mathbf{v}_r^2 = -v_r\frac{\partial}{\partial r}p \tag{5.10}$$

and integrated over the whole volume of the spherical plasma with the radius R of the plasma surface. This leads to

$$-\int_0^R v_r\left(\frac{\partial p}{\partial r}\right)4\pi r^2 dr = \frac{1}{2}\frac{d}{dt}\int_0^R n_i m_i\left(1+Z\frac{m_e}{m_i}\right)v_r^2 4\pi r^2 dr \tag{5.11}$$

It is essential to point out that the procedure of integration in Eq. (5.11) and the following equations give a loss of information. Instead of details of $n_i(r, t)$, only averaged values of the functions under the integrals can be expected.

If the pressure within the spherical plasma is assumed to be a constant value p from $r=0$ until $r=R-\varepsilon$ and decreases to $p=0$ at $r=R$, then the left-hand side of Eq. (5.11) can be written

$$\begin{aligned} -\lim_{\varepsilon\to 0} v_r 4\pi R^2 \int_{R-\varepsilon}^R \frac{\partial p}{\partial r} dr &= -v_r 4\pi R^2(p(R)-p(R-\varepsilon)) \\ &= 4\pi R^2 p v_r(R) \end{aligned} \tag{5.12}$$

Constant p at constant T gives an averaged constant n_i in the inner of the sphere. Assuming additionally a linear velocity profile,

$$v_r(r) = v_{r0}\frac{r}{R} \qquad (0<r<R) \tag{5.13}$$

which expresses the properties of a similarity expansion, then the right-hand side of Eq. (5.11) leads to

$$\frac{1}{2}\frac{d}{dt}\int_0^R n_i m_i \left(1+Z\frac{m_e}{m_i}\right) v_r^2 4\pi r^2 dr = \frac{1}{2}\bar{M}\frac{d}{dt}v_r(R)^2 \tag{5.14}$$

with the abbreviation of an averaged mass $\bar{M}=\frac{3}{5}n_i m_i[1+Z(m/m_i)]$ with respect to the spherical expansions. Combining Eqs. (5.14) and (5.12) leads to

$$4\pi R^2 p\frac{dR}{dt}=\frac{1}{2}\bar{M}\frac{d}{dt}v_r(R)^2=\frac{1}{2}\bar{M}\frac{\partial}{dt}\left(\frac{dR}{dt}\right)^2 \tag{5.15}$$

A further integration over the volume is of interest when similar assumptions to those for n_i are used as in Eq. (5.15), making n_i constant in the averaged sense within the plasma sphere and using a linear velocity profile Eq. (5.13). In this case, one gets from the left-hand side of Eq. (5.8)

$$\frac{\partial}{\partial t}\int_0^R \frac{n_i m_i}{2}\left(1+Z\frac{m_e}{m_i}\right) v_r^2 4\pi r^2\, dr = \frac{\partial}{\partial t}\frac{n_i m_i}{2}\left(1+Z\frac{m_e}{m_i}\right)\frac{v_0}{R^2}4\pi\int_0^R r^4\, dr \tag{5.16}$$

Using

$$\tfrac{3}{5}n_i m_i\left(1+Z\frac{m_e}{m_i}\right)v_0=\bar{M}v_0=\tfrac{2}{3}p \tag{5.17}$$

which by definition led to the pressure p.

One finds for the integral of the left-hand side of Eq. (5.8)

$$\frac{\partial}{\partial t}\frac{2}{3}4\pi p R^3 \tag{5.18}$$

Integrating the right-hand side of Eq. (5.8) over the plasma volume and assuming a constant temperature $[(\partial/\partial t)T=0]$, one obtains an equation of energy conservation with the total power W absorbed in the plasma after subtraction of the radiation loss

$$4\pi p R^2\frac{\partial}{\partial t}R=-\tfrac{3}{2}K\frac{dT}{dt}\left[\frac{4\pi}{3}R^3(1+Z)n_i\right]+W \tag{5.19}$$

Equations (5.6), (5.15), and (5.19) are the same that Dawson derived from the basis of a phenomenological combination of gas kinetic laws and are the basic equations of the self-similarity model. The equations determine the time dependence of the plasma radius $R(t)$ and the power transfer $W(t)$ to the plasma. The two essential points in the derivation from the hydrodynamic equations are first that the energy transfer from the radiation has to be assumed in such a way that the plasma temperature can be assumed spatially constant at each instant, and second that a linear velocity profile is valid, Eq. (5.13). The properties of an averaged density n_i, an averaged mass M, and a steep pressure decrease at the surface are consequences of the integration procedures; these are mathematically correct but they involve a loss of details of mostly unnecessary information. Therefore, the self-similarity model contains the constant density profile as an averaged value only.

The problem is different from that of the preceeding result when *details* of the density profile $n_i(r)$ at varying times must be evaluated. This leads to a mathematical contradiction. To reach a statement of the actual density profile, Haught and Polk [128] started from a linear (not Gaussian) decrease of the density $n_i(r)$ from the center of the plasma to the surface. Simultaneously, a linear velocity profile, as given by Eq. (5.13) was erroneously assumed to be conserved. It is easy to show that both linear profiles are not conservable in time. Substituting $n_i = n_{i0}(1 - r/R)$ with the normalization $n_{i0} \sim R(t)^{-3}$ with a spatially invariant T in the equation of motion one finds:

$$\frac{dv_r}{dt} = \frac{(1+Z)KT}{[1+Z(m_e/m_i)]m_i R(1-r/R)}; \tag{5.20}$$

this gives an acceleration profile varying nonlinearly with r even with a pole at $r/R = 1$. Should, at any time t_0, the velocity profile be linear and the density profile is linear for $t > t_0$, then the velocity profile must change its linearity at $t > t_0$ due to Eq. (5.20).

The motivation for the controversial view of Haught [128] about the self-similarity model is not trivial if one looks at the question of the pressure profile from the ad hoc assumptions of the not easily understandable self-similarity model [112]. The analysis given here [127] of a derivation of Eqs. (5.15) and (5.19) from the general hydrodynamic equations has the advantage—as always in general theory—of demonstrating how the mentioned pressure problem has been solved by the justified assumptions of the averaging procedure at Eq. (5.12).

One way in which a linear velocity profile (self-similarity motion) can be observed under the condition of spatially constant (but temporally varying) temperatures, can be verified if the density profile is Gaussian [129] and [130]. Such a profile

$$n_i(r, t) = \frac{n_0}{\pi^{3/2} l^3} \exp\left(-\frac{r^2}{l^2}\right) \tag{5.21}$$

where the length $l(t)$ is only a function of t, produces an acceleration in the equation of motion (5.7):

$$\frac{\partial^2 r}{\partial t^2} = -\frac{KT(1+Z)}{m_i[1+Z(m_e/m_i)]n_i}\frac{\partial}{\partial r} n_i = \frac{KT(1+Z)}{m_i[1+Z(m_e/m_i)]}\frac{2r}{l^2} \tag{5.22}$$

If the starting velocity $v_r(r, t=0)$ is zero or linear with r, the acceleration conserves this property. The Gaussian density profile expands similarly. The limitation of the model with the Gaussian profile is given by the fact that the real plasma has a definite surface and a finite expansion velocity, while the Gaussian profile distributes the plasma to any distance. It is interesting that the Gaussian density profile was observed by interferometry of plasma, produced by lasers from thin foils [131], and that a direct numerical calculation, based on the hydrodynamic equations of an initial linear density profile, assimilated a Gaussian profile after a certain time [120]. The function $l(t)$ behaves like $R(t)$ from Eqs. (5.15) and (5.19).

5.2 Laser Irradiation with Varying Pellet Radius

As an example of the application of the self-similarity model of Eqs. (5.6), (5.15), and (5.19) some experimental results [132] are interpreted. For the model some generalizations with respect to the energy transfer from the radiation to the plasma were used. A further step is to include the special condition of the experiment of a varying diameter for a spherical plasma, due to the heating and expansion during irradiation within a constant laser focus. One can go to analytical expressions, demonstrating immediately the physical properties. A numerical application is based on a numerical stable iteration procedure.

As evaluated by Dawson [112], the treatment of Eqs. (5.6), (5.15), and (5.19) for an energy input $W_1 = W$ from a laser of frequency ω starting at a time t_0 and remaining constant gives the solutions, derived also by Basov and Krokhin [125], for the time dependence of the laser heated sphere of the radius R and temperature T

$$R^2 = \left[R_0^2 + \frac{10}{9} G\right] \tag{5.23}$$

$$KT = \frac{W_1 t}{3N_i(1+Z)} \frac{2R_0^2 + \frac{5}{9}G}{R_0^2 + \frac{10}{9}G} \tag{5.24}$$

where R_0 is the initial radius of the target before t_0 and

$$G = \frac{W_1 t^3}{N_i m_i} \tag{5.25}$$

The total number of ions N_i was used

$$N_i = n_i(t_0) \frac{4\pi}{3} R_0^3 \tag{5.26}$$

The model used, [112, 125], implies a Z constant in time, which is quite reasonable.

A modification in the time variation of the input power W is made, taking into account that the plasma sphere expands and changes its cross section for energy transfer. A power density in the focus, which is spatially constant, has a time dependence as a step function and is constant for $t > t_0$ as assumed. The further assumption that all energy incident within the cross section of the plasma is transferred to the plasma as long as the plasma is overdense, $\omega_p > \omega$ [see Eq. (3.13)], with the ion density n_i, leads to

$$\omega_p^2 = \frac{4\pi e^2}{m_e} Z n_i \tag{5.27}$$

The details of the energy transfer from the irradiated plasma corona to the whole plasma are assumed to be fast enough, and all laser energy within the cross section of the target is taken as contributing to W. The power input then gives the formula:

$$W(t) = \frac{R(t)^2}{R_0^2} W_1 \tag{5.28}$$

Here W_1 is the constant power input of the initial cross section, if the laser power density is constant in time and within the focal area. Using this time dependent, $W(t)$, one cannot solve Eqs. (5.6), (5.15), and (5.19) immediately. An iteration is used as done in Eqs. (5.23) and (5.24), a first iteration $R_1(t)$, $T_1(t)$ using $W = W$ and a second iteration $R_2(t)$, $T_2(t)$ using $W = R(t)W/R$, and so on. The second iteration is then (avoiding the index 2):

$$R^2 = R_0^2 + \frac{10}{9} G\left(1 + \frac{1}{18R_0^2} G\right) \tag{5.29}$$

$$KT = \frac{W_1}{3N_i(1+Z)} \frac{2R_0^2 + 5G/3 + (80G^2/81R_0^2) - (200G^3/1458R_0^4)}{R_0^2 + (10G/9)(1 + G/18R_0^2)} \tag{5.30}$$

The difference of this solution compared with the solutions (5.23) and (5.24) is obvious when we evaluate the time t_{TP} at which the plasma becomes transparent ($\omega_p = \omega$). In the case of ruby laser radiation ($\omega = 2.7 \times 10^{15}$ Hz) and an initial density $n_0 = 6 \times 10^{22}$ cm^{-3}, as is the case for solid hydrogen or solid aluminium, from Eq. (5.23) with Eq. (5.25) such an R is found, for which $(R/R_0)^3 = n_0/n_{c0}$. With the cutoff density, given by Eq. (5.27) at $n_i = n_{c0}$ for $\omega_p = \omega$,

$$t_{TP}^{(1)} = \left(\frac{7.78 r_0^2 N_i m_i}{W_1}\right)^{1/3} \tag{5.31}$$

For the solution with varying cross section, from Eq. (5.29) there follows in the same way

$$t_{TP} = \left(\frac{7.15 r_0^2 N_i m_i}{W_1}\right)^{1/3} \tag{5.32}$$

The higher energy input in this case makes the plasma transparent a little earlier.

A greater difference can be seen in plasma temperature. The maximum temperature of the plasma is reached when the total energy input $W_1(t)$ is so fast that the expansion of the plasma is negligible. In this case, the temperature is [112]:

$$T = T_{max} = \frac{2}{3KN_i(1+Z)} \tag{5.33}$$

Considering the temperature T at the time $t = t_{TP}$, one finds in the case of Eq. (5.24) the first-order temperature at the time of transparency

$$T^{(1)} = 0.32 T_{max} \tag{5.34}$$

and in the case of varying diameter

$$T = 0.583 T_{max} \tag{5.35}$$

The temperature is differing nearly by a factor of two and is larger, although $t_{TP} < t_{TP}^{(1)}$.

In order to evaluate the time dependence of the plasma parameters of self-similarity expansion, a numerical program to solve $R(t)$ and $T(t)$ for a more general input power W in Eqs. (5.6), (5.15), and (5.19) is based on the following assumptions.

Evaporation, ionization, and recombination of the plasma were neglected. The focal region is approximated by a boxlike intensity profile constant in space. The intensity has a time dependence of variable forms (rectangular, triangular, symmetric, or with steepened rise time, tailored with triangular initial prepulse and triangular main pulse), expressed by $W_1(t)$ given by the laser power within the area of the cross section of the plasma $t = t_0$. The geometry is given by a factor

$$f = \begin{cases} R^2/R^2 & \text{for } R < R_F \\ R_F^2/R^2 & \text{for } R \geqslant R_F \end{cases} \tag{5.36}$$

with the focus diameter R_F taking into account, that the plasma can reach a larger R than the focus radius R_F before $\omega_p < \omega$ is reached. The energy transfer due to absorption, given by the optical absorption constant K, is approximated with respect to the spherical geometry by

$$g = \begin{cases} 1 & \text{for } n_i \geqslant n_{c0} \\ 1 - \dfrac{1-(1+2KR)\exp(-2KR)}{2K^2R^2} & \text{for } n_i < n_{c0} \end{cases} \tag{5.37}$$

The power input into the plasma with negligible radiation losses is then

$$W(t) = W_1(t) g(t) f(t) \tag{5.38}$$

The time dependent functions $R(t)$ and $T(t)$ are solved from the following system of equations, where the total mass of the plasma is $M=4\pi R_0^3 n_0 m_i/3$

$$R^2(t)=R_0^2+\frac{20}{3M}\int_0^t d\tau\int_0^\tau d\tau'\int_0^{\tau'} d\tau'' W\left[R(\tau''),\, T(\tau'')\right] \qquad (5.39)$$

and

$$KT(t)=\frac{M}{5N_i(1+Z)}\left\{-\left(\frac{dR}{dt}\right)^2+\frac{10}{3M}\int_0^t d\tau W[R(\tau),\, T(\tau)]\right\} \qquad (5.40)$$

The solution is verified by iteration with the first step $R_1(t)T_1(t)$ as described before, put into $W(R_1(t),\ T_1(t))$ and Eqs. (5.39) and (5.40) solved to find the second iteration $R_2(t)$, $T_2(t)$ and with these values $W(R_2(t),\ T_2(t))$ is used to solve the third iteration $R_3(t)$, $T_3(t)$, and so on. The iteration was performed until R- and T-values differ by less than 10^{-4} from the values of the previous iteration.

In the following examples, the necessary number of iterations of this procedure is given by the result that the temperature $T(t)$ decreased to zero at large t compared with the laser pulse length t_L, defining $W_1=0$ at $t>t_L$ in Eq. (5.38)

5.3 Numerical Example

The results of the self-similarity model in the formulation described here will be compared with experiments. A first comparison is performed with experiments [132] where single aluminum balls of 50 to 150 μm radius were irradiated by focused laser pulses of 2 to 5 J energy and 15 to 35 nsec pulse length. A numerical solution of the time dependence of the ball radius $R(t)$ and the temperature $T(t)$ for a case measured [132] is shown in Fig. 5.1. The velocity of the plasma surface $v^{th}_{\max}(t)=dR/dt$ is evaluated and also the maximum ion energy $\varepsilon^{th}_{\max}(t)=m_i v^2/2$. In addition, the amount of the totally absorbed energy was evaluated, taking into account the geometry of the growing plasma within the laser focus and the absorption constant. The results show that the calculated absorption of energy is equal to the measured values within 15%, and the dependence of the amount of the absorbed energy on the ball radius and the pulse length shows the same systematic variation as seen from the measurements.

The final maximum ion energy of the expanding plasma has been measured from side-on framing camera pictures [132]. These values fit very well the theoretical slope with a triangular pulse shape, the parameters of which were determined by the measured pulse shape (Fig. 5.2). The index "th" in $\varepsilon^{\mathrm{th}}_{\max}$ expresses the inner thermal part of the created plasma studied,

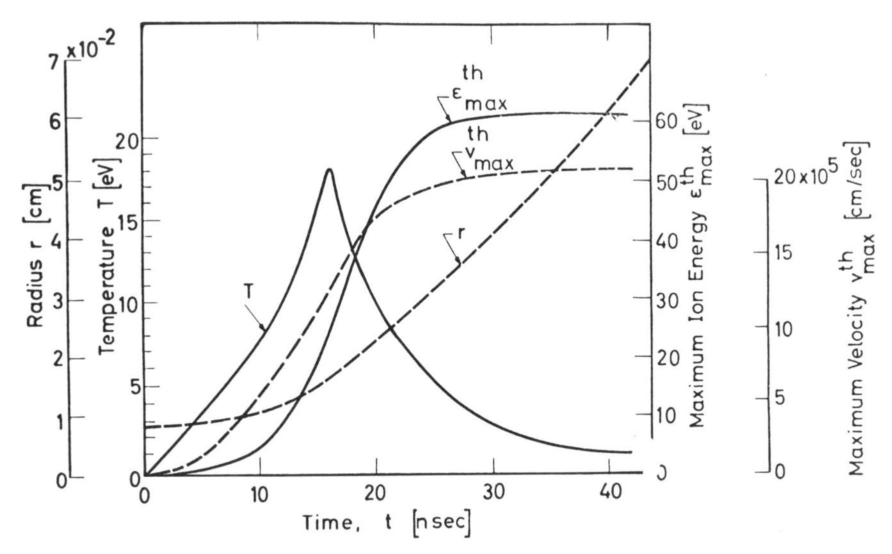

Figure 5.1 Numerical calculation by iteration of Eqs. (5.39) and (5.40) to evaluate the time dependence of the plasma radius R, temperature T, maximum velocity v_{max} and ion energy of the plasma surface for an aluminium ball of 80 μm radius, irradiated by a laser pulse of 3.4 J energy and a rectangular pulse length of 16 nsec.

while an outer part of the plasma [132] has properties of a nonlinear surface mechanism (see Fig. 1.4), which is due to nonlinear processes and will be discussed later.

The result of Eqs. (5.34) and (5.35) indicates a higher temperature of the plasma if the self-similarity model is applied in the way described. This increase of the temperature was observed by Thomson scattering experiments [132]. Another possibility is that recombination mechanisms increase the electron temperature.

5.4 Application to Foils

It should not surprise us that the self-similarity model is so successful in explaining the experiments and the gasdynamics for interaction of medium intensity laser radiation (10^{10} to 10^{12} W/cm^2, ruby or Nd glass lasers) with spherical targets. The intensities are so low that nonlinear effects, if present, are at least not dominant, and the conditions of a fully gasdynamic behavior are realized to a large extent. Nevertheless, it is not clear from the beginning, whether the energy transfer to the whole overdense pellet can occur fast enough (for about 10 nsec total irradiation time) to fulfill the conditions of

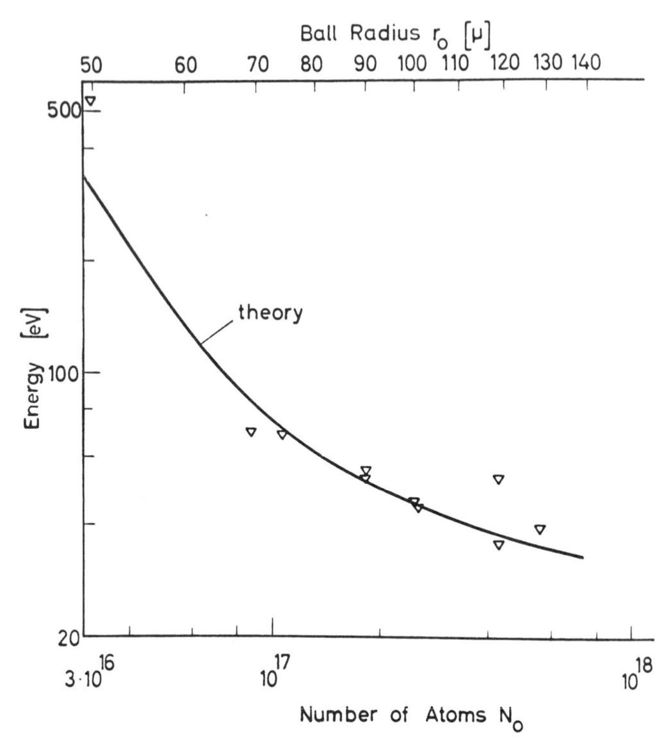

Figure 5.2 Measured maximum ion energies of plasmas produced from aluminium balls of varying ball radius with irradiation by laser pulses of about 70 MW and 30 nsec pulse length (∇) compared with theoretical values, based on the self-similarity model (curve).

the model. The excellent agreement subsequently achieved between theory and experiment confirms the validity of the assumptions for the energy deposition.

A greater surprise is the fact that the self-similarity model reproduces the thermal expansion properties of plasmas produced by laser irradiation of thin foils. Here, a complication of the energy transfer should be expected, due to the interactions of the plasma with the nonirradiated cold foil material. Nevertheless, a reasonably good agreement with the self-similarity model is possible.

The experiments consist of the production of thin foils of solid hydrogen for measuring the transmitted ruby laser pulse [133]. Initially, light passes through the solid target, but is then blacked out sharply by the generated plasma. When the plasma is assumed to expand according to the self-similarity model to smaller electron densities, at a certain time t_{TP}, the plasma becomes transparent [when $\omega_p(n_e(t_{TP}))\leq\omega$], which can be measured from the onset of the transmission of light.

For solid hydrogen and ruby laser radiation, from Eq. (5.31) the transparency time t_{TP} is evaluated, using r_0 for the thickness of the foil we have

$$W_1=\frac{W_0 r_0^2}{r_F^2}=I\pi r_0^2 \tag{5.41}$$

W_0 is the input laser power at the front of the layer, where the laser beam is focused to a radius r_F, and I is the laser intensity. Using the density $\rho=0.1$ g/cm^3 of solid hydrogen and Eq. (5.31) leads to

$$t_{TP}=\left(\frac{7.78(4/3)r_0^3\rho}{I}\right)^{1/3} \tag{5.42}$$

In the experiment [133] with $I=2.4\times10^{12}$ W/cm^2, Eq. (5.42) results in

$$t_{TP}=3.50\times10^{-7}r_0 \qquad \text{for } W_0=200 \text{ MW} \tag{5.43}$$

$$t_{TP}=4.75\times10^{-7}r_0 \qquad \text{for } W_0=500 \text{ MW} \tag{5.44}$$

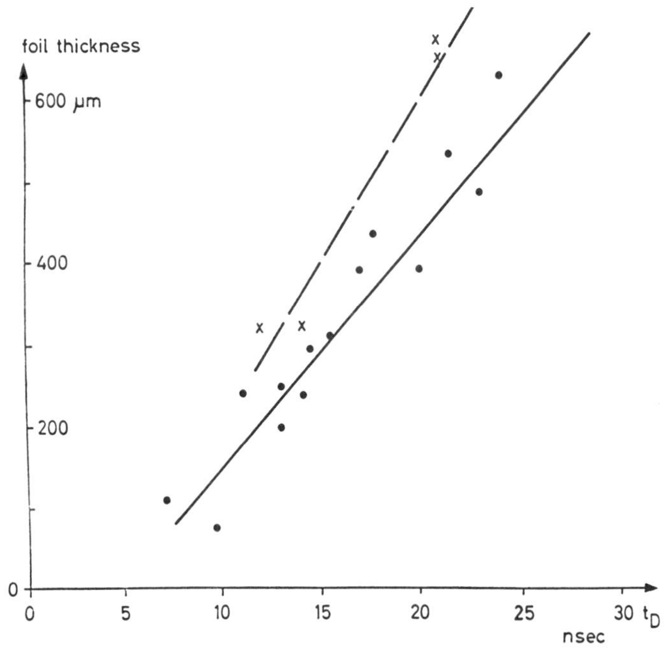

Figure 5.3 Measured values (•, ×) [133] of the delay time of transparency t_{TP} for solid hydrogen foils of given thickness compared with the calculated (lines) transparency time t_{TP} by the self-similarity model [127], when t_D-t_{TP} is 5 nsec. The laser intensities were 200 MW (———) and 500 MW (- - - - - -).

measuring foil thicknesses r_0 in centimeters and transparency time t_{TP} in seconds. To compare these values with the measured transparency time, we have to add to t_{TP} the time between the beginning of the laser pulse and the creation of the plasma, which is about 5 nsec (Fig. 5.3). In agreement with analogous experiments, Eqs. (5.43) and (5.44) fit the measurements as shown in Fig. 5.3 very well. At this point, nothing has been said about the details of the process achieving the very fast deposition of the laser energy to the plasma for justifying the self-similarity model. It has to be taken into account that self-focusing and related complicated dynamics may be responsible for the fast energy transfer.

It is indeed a surprise that the self-similarity model with homogeneous heating fits the experimental values of t_{TP} so well. The numerical calculation using a shockfront type heating [134] of plasma arrived at 30 times longer transparency times. This result induces scepticism about several shock wave models developed for explaining the gasdynamic laser plasma interaction for laser intensities of 10^{10} to 10^{12} W/cm^2 of about 1 μm wavelength, pulses of 1 to 30 nsec duration and targets up to 0.4 mm diameter or characteristic size. While the plane shock wave calculation will be correct for such experimental conditions, the experiments might provide different conditions of a fast energy transfer due to self-focusing and other mechanisms [135].

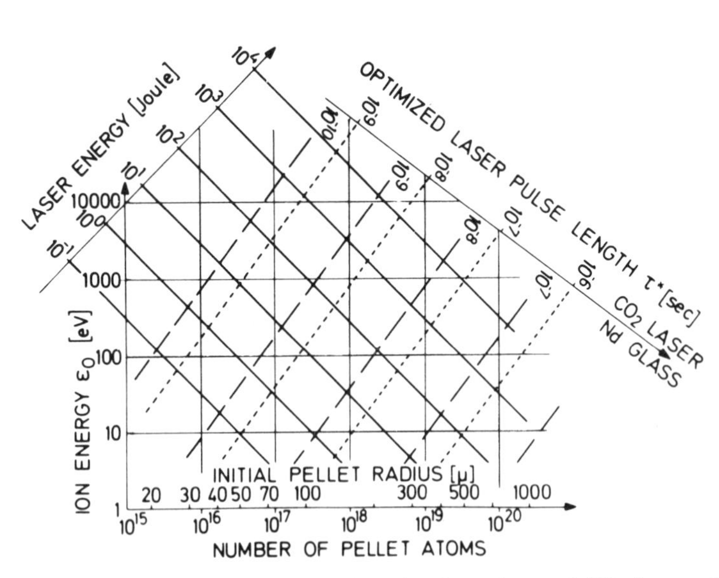

Figure 5.4 Energy and pulse length τ^* of neodymium-glass and CO_2 laser radiation for heating a solid deuterium pellet of given initial radius to averaged ion energies ε_0, derived from the self-similarity model of homogeneous heating [74, p. 37].

Finally, a diagram has been derived from the self-similarity model for constant pellet radius [74, Fig. 5.3] for calculating maximum temperature, Eq. (5.24) corresponding to the ion energy ε_0 after subsequent free expansion if $t=t^*$ is the time of achieving transparency $t_{TP}=t^*$, Eq. (5.31) at irradiation by a constant power W, onto a solid D_2 pellet of initial radius R_0 (corresponding to a number N_i of atoms). $Wt^*\pi R_0^2$ is then equal to the laser energy E_L for the irradiation up to the time t^*. The transparency depends on the wavelength (cutoff density) and results in different optimized times τ^* for CO_2 laser or Nd glass irradiation. The nomogram in Fig. 5.4 is read in the following way as an example. One may start from the basic diagram by using a pellet radius of, for example, 350 μm (10^{19} atoms) for a maximum temperature (averaged ion energy) of 100 eV. From this print one can read a necessary laser energy of 3 kJ for 10 nsec pulse length for Nd glass or for 45 nsec pulse length for CO_2 laser irradiation. The laser intensity is then 1.02×10^{15} W/cm^2 for Nd glass and 2.2×10^{14} W/cm^2 for CO_2. These intensities might be a little above the threshold where nonlinear processes start and for which the self-similarity model might be questionable.

SIX

Plasma Dynamics and Lorentz Theory

In the two preceding sections the three basic plasma hydrodynamic equations have been described. These have been used in simplified form to determine the expansion of a spherical plasma with the self-similarity model. In this section, the mechanical response of the plasma to electric and magnetic fields **E** and **H** will be discussed on the basis of the two-fluid equations [136]. These basic equations show that mechanical equations can lead to Ohm's law, which is an electrodynamic equation.

6.1 The Two-Fluid Equation of Motion

Schlüter [136] started from the Euler equations for the electron and ion fluids in a plasma.* The indices e and i denote the electron and ion parameters, respectively.

$$m_i n_i \frac{d\mathbf{v}_i}{dt} = Z n_i e\mathbf{E} + n_i \frac{Ze}{c}\mathbf{v}_i \times \mathbf{H} - \nabla n_i K T_i - m n_e \nu_{ei}(\mathbf{v}_i - \mathbf{v}_e) + \mathbf{K}_i \tag{6.1}$$

$$m n_e \frac{d\mathbf{v}_e}{dt} = -n_e e\mathbf{E} - n_e \frac{e}{c}\mathbf{v}_e \times \mathbf{H} - \nabla n_e K T_e + m n_e \nu_{ei}(\mathbf{v}_i - \mathbf{v}_e) + \mathbf{K}_e \tag{6.2}$$

The force densities on the right-hand side of Eqs. (6.1) and (6.2) arise from the electric field **E**, the Lorentz force $\mathbf{v} \times \mathbf{H}$, and the pressure $p = n_{i,e}KT$.

*A generalization to a three-fluid model with remaining neutral atoms was a subsequent step [137]. These plasmas of partial ionization are of marginal importance for laser produced plasmas.

The ultimate term corresponds to the viscosity, where ν_{ei} is the electron ion collision frequency given by Eqs. (3.34) and (3.35). For any additional forces, such as, for example, the gravitational force $\mathbf{K}_i$ and $\mathbf{K}_e$ are used. The net velocity $\mathbf{v}$, as defined by Schlüter [136], is (see Appendix C)

$$\mathbf{v}=\frac{m_i\mathbf{v}_i+Zmv_e}{m_i+Zm} \tag{6.3}$$

and the current density $\mathbf{j}$ is

$$\mathbf{j}=e(n_i\mathbf{v}_i-n\mathbf{v}_e) \tag{6.4}$$

Addition of the Eqs. (6.1) and (6.2) and substitution of (6.3) and (6.4) and rearranging terms leads to an *equation of motion*, given by a force density $\mathbf{f}$

$$\mathbf{f}=m_in_i\frac{d\mathbf{v}}{dt}=-\nabla p+\frac{1}{c}\mathbf{j}\times\mathbf{H}+\frac{1}{4\pi}\left(\frac{\omega_p}{\omega}\right)^2\mathbf{E}\cdot\nabla\mathbf{E} \tag{6.5}$$

where p represents the total gasdynamic pressure in the plasma. The additional force densities due to gravitation and so on, $\mathbf{K}_i$ and $\mathbf{K}_e$, are neglected in Eq. (6.5). The last term was written originally [136] as:

$$\frac{1}{4\pi}\left(\frac{\omega_p}{\omega}\right)^2\mathbf{E}\cdot\nabla\mathbf{E}=\mathbf{j}\cdot\nabla\frac{1}{\omega^2}\frac{\partial\mathbf{E}}{\partial t} \tag{6.6}$$

For this substitution see Ref. [138]. The importance of generalizing the equation of motion (6.5) by more nonlinear terms to describe the laser-plasma interaction, will be shown in Section 8. The derivation of an equation of motion for a plasma without the nonlinear term in (6.5) has been shown by Spitzer [107], starting from kinetic theory (Boltzmann equation), see Section 3.

6.2 The Diffusion Equation (Ohm's Law)

In order to obtain an equation for the motion of the electrons relative to the ions, Schlüter [136] subtracted Eq. (6.1) from Eq. (6.2) to obtain (see Appendix C)

$$\frac{m}{e^2n_e}\left(\frac{d\mathbf{j}}{dt}+\nu\mathbf{j}\right)=\mathbf{E}+\frac{1}{c}\mathbf{v}\times\mathbf{H}+\frac{1}{en_ec}\mathbf{j}\times\mathbf{H}+\frac{c}{en_e}\frac{\nabla p}{1+1/Z} \tag{6.7}$$

Schlüter called this the "diffusion equation", which is a *generalized Ohm's law*, containing a relation between the current density $\mathbf{j}$ and the electrical field $\mathbf{E}$, which—in the sense of an acting electric field—must be extended by the Lorentz term $\mathbf{v}\times\mathbf{H}$, the Hall term $\mathbf{j}\times\mathbf{H}$, and an electron pressure term. Neglecting these last terms, a form of Ohm's law is obtained:

$$\frac{d\mathbf{j}}{dt}+\nu\mathbf{j}=\frac{\omega_p^2}{4\pi}\mathbf{E} \tag{6.8}$$

This is how it was originally formulated for a plasma by Langmuir, so the purely mechanical Euler equations (6.1) and (6.2) lead to the electrical relation (6.8), known as Ohm's law, where automatically the plasma frequency ω_p of the electrostatic plasma oscillations were determined [see Eq. (2.16)].

6.3 Electrodynamic Equations

The electric and magnetic fields **E** and **H** in the equations of motion (6.1) and (6.2) obey the electrodynamic equations derived by Maxwell. The integral formulation of Faraday's induction law, including the magnetic permeability μ, is

$$\oint \mathbf{E}\cdot ds = -\frac{1}{c}\frac{\partial}{\partial t}\iint \mu\mathbf{H}\cdot d^2\mathbf{a} \tag{6.9}$$

The induction of an electric field **E** along a closed loop is created by a temporal change of the magnetic flux **H** through this loop. With Stokes' law,

$$\oint \mathbf{E}\cdot ds = \iint \nabla\times\mathbf{E}\cdot d^2\mathbf{a}$$

Faraday's law (6.9) in differential form results in the first Maxwellian equation:

$$\nabla\times\mathbf{E} = -\frac{1}{c}\frac{\partial}{\partial t}\mu\mathbf{H} \tag{6.10}$$

The integral formulation of Ampere's law

$$\oint \mathbf{H}\cdot ds = \frac{4\pi}{c}\iint \mathbf{j}\cdot d^2\mathbf{a} \tag{6.11}$$

expresses the magnetic field **H**, generated along a closed loop, within which an electric current, given by the current density **j**, is produced. It can be formulated again by the use of Stokes' law in differential form, to which Maxwell added the dielectric displacement current, given by the dielectric constant ε, and arrived at his second equation:

$$\nabla\times\mathbf{H} = \frac{4\pi}{c}\mathbf{j} + \frac{1}{c}\frac{\partial}{\partial t}\varepsilon\mathbf{E} \tag{6.12}$$

The source equation of the electric field, given by a charge density ρ_e, is

$$\oiint \varepsilon\mathbf{E}\cdot d^2\mathbf{a} = 4\pi\iiint \rho_e\, d^3\tau \tag{6.13}$$

The integration of $\varepsilon\mathbf{E}$ is along the closed area determining the integration volume of ρ_e. Using Gauss' law

$$\oiint \varepsilon \mathbf{E} \cdot d^3\mathbf{a} = \iiint \nabla \cdot (\varepsilon \mathbf{E})\, d^3\tau$$

the differential form of (6.13) can be written as

$$\nabla \cdot (\varepsilon \mathbf{E}) = 4\pi\rho_e = 4\pi e(Zn_i - n_e) \tag{6.14}$$

With the electron charge $e = 4.803 \times 10^{-10}$ cgs, charge densities are the source of the **E** fields. Magnetic monopoles, as the source of magnetic fields, have not yet been observed, though Dirac's theory of 1933 [139] shows that their existence is possible. With the exclusion of this possibility, the source of the magnetic field is zero.

$$\nabla \cdot \mathbf{H} = 0 \tag{6.15}$$

It was of the ingenious discoveries of Maxwell to differentiate Eq. (6.10) by ∇ and Eq. (6.12) by $\partial/\partial t$ and eliminate **E** or **H** (assuming $\varepsilon = \mu = 1$ and $\mathbf{j} = 0$) to arrive at

$$\nabla^2 \mathbf{E} - \frac{1}{c^2}\frac{\partial^2}{\partial t^2}\mathbf{E} = 0 \tag{6.16}$$

This is a wave equation with the speed $c = 299{,}796$ m/sec, the speed of light. An equivalent can be written for **H**,

It was O. D. Chwolson, who in 1905 introduced Maxwellian theory into the textbooks for a *deductive* description of the electrodynamics: firstly to establish the Maxwellian equations, and secondly to treat electrostatics or magnetostatics separately for $\partial/\partial t = 0$, the quasistatic case for slowly varying **j** and the wave fields as the general case. It is surprising that this basic methodological scheme has not yet been introduced into all physics textbooks for university students even 100 years after Maxwell's death.

The treatment of media in the Maxwellian theory uses ε or μ as material constants. Lorentz described the phenomena with $\varepsilon = \mu = 1$, as in vacuum, and described all material phenomena by charge densities ρ_e and current densities **j**. The dielectric properties of insulators are then due to electric dipoles and currents. This way of description is preferable for plasmas. In the microscopic plasma theory, only currents **j** are present, while any space charge is balanced to zero for spatial dimensions exceeding the Debye length, Eq. (2.8), due to the good electrical conductivity of the plasma. Additionally, electric dipoles determine the dielectric properties. The only exception is higher frequency electromagnetic fields, where high-frequency oscillations of charge densities can be influenced.

The Lorentz theory of plasma uses the Maxwellian equations for vacuum and a current density, given by Ohm's law (6.8), derived from the equations of mechanics

$$\nabla \times \mathbf{E} = -\frac{1}{c}\frac{\partial}{\partial t}\mathbf{H} \tag{6.17}$$

$$\nabla \times \mathbf{H} = \frac{4\pi}{c}\mathbf{j} + \frac{1}{c}\frac{\partial}{\partial t}\mathbf{E} \tag{6.18}$$

If the quantities **E**, **H**, and **j** are of periodic time dependence with a frequency ω, we have

$$\begin{aligned} \mathbf{E} &= \mathbf{E}_r \exp(i\omega t) \\ \mathbf{H} &= \mathbf{H}_r \exp(i\omega t) \\ \mathbf{j} &= \mathbf{j}_r \exp(i\omega t) \end{aligned} \tag{6.19}$$

where $\mathbf{E}_r$, $\mathbf{H}_r$, and $\mathbf{j}_r$ only depend on spatial coordinates. Integration of (6.8) leads to

$$\mathbf{j} = \frac{\omega_p^2}{4\pi i\omega(1 - i\nu/\omega)}\mathbf{E} \tag{6.20}$$

and the time-independent Maxwellian equations

$$\nabla \times \mathbf{E}_r = -\frac{i\omega}{c}\mathbf{H}_r \tag{6.21}$$

$$\nabla \times \mathbf{H}_r = -\frac{i\omega_p^2}{c\omega(1 - i\nu/\omega)}\mathbf{E}_r + \frac{1}{c}\omega E_r \tag{6.22}$$

With the operation $\nabla \times$ on (6.22) and the substitution of $\nabla \times \mathbf{E}$ from (6.21), the following equation is obtained:

$$\nabla^2 \mathbf{H}_r + \frac{\omega^2 n^2}{c^2}\mathbf{H}_r - i\frac{\omega}{c}\mathbf{E}_r \times \nabla n^2 = 0 \tag{6.23}$$

with Eq. (6.15) and resubsititution according to Eq. (6.19) a wave equation is obtained.

$$\nabla^2 \mathbf{H} - \frac{n^2}{c^2}\frac{\partial^2}{\partial t^2}\mathbf{H} - \frac{1}{n^2}(\nabla \times \mathbf{H}) \times \nabla n^2 = 0 \tag{6.24}$$

or

$$\nabla^2 \mathbf{H} - \frac{n^2}{c^2}\frac{\partial^2}{\partial t^2}\mathbf{H} + 2(\nabla \mathbf{H}) \cdot \nabla \ln n - 2(\nabla \ln n) \cdot \nabla \mathbf{H} = 0 \tag{6.25}$$

with the phase velocity

$$c_\phi = c\sqrt{\mathrm{Re}(n)} \tag{6.26}$$

The complex constant n is the time independent complex refractive index and is related to the complex dielectric constant $\boldsymbol{\varepsilon}$ by

$$\boldsymbol{\varepsilon} = n^2 = 1 - \frac{\omega_p^2}{\omega^2(1 - i\nu/\omega)} \tag{6.27}$$

The first-order term of $\mathbf{H}$ in Eq. (6.24) is zero in a homogeneous plasma with $\nabla \cdot n^2 = 0$. For an inhomogeneous plasma, where n is a function of x, y, z due to the spatial dependence of n_e or T [Eqs. (2.33) or (2.34) with (2.35)], Eq. (6.24) can be written as:

$$\nabla^2 \mathbf{H} + \frac{\omega^2 n^2}{c^2} \mathbf{H} + 2(\nabla \mathbf{H}) \cdot \nabla \ln n - 2(\nabla \ln n) \cdot \nabla \mathbf{H} = 0 \tag{6.28}$$

This equation will be important in the later chapter on resonance absorption.

It should be noted that all the steps from Eqs. (6.2) to (6.28) presumed time independent n_e and T. When these quantities are time dependent, as is possible in plasma, the derivation is much more complex. In most cases of the study of transient behavior of laser plasma interactions, the time independence of n_e and T in the Maxwellian equations (N.B.—not in the mechanical equations!) is a reasonable approximation. However, there comes a point where a more general treatment is necessary, such as to treat the case of very short time interactions.

Again, with the same assumptions of a time independent n, $\nabla \times$ is operated on Eq. (6.21) and $\nabla \times \mathbf{H}$ is substituted from Eq. (6.22). The resulting equation is

$$\nabla^2 \mathbf{E} + \frac{\omega^2 n^2}{c^2} \mathbf{E} - \nabla \nabla \cdot \mathbf{E} = 0 \tag{6.29}$$

By using $\rho_e = 0$ in Eq. (6.14)

$$\boldsymbol{\varepsilon} \nabla \cdot \mathbf{E} + \mathbf{E} \cdot \nabla \boldsymbol{\varepsilon} = 0$$

Equation (6.29) can be written as:

$$\nabla^2 \mathbf{E} + \frac{\omega^2 n^2}{c^2} \mathbf{E} - \nabla \frac{2}{n} \mathbf{E} \cdot \nabla n = 0 \tag{6.30}$$

The wave equation results again by resubstitution from Eq. (6.19) to

$$\nabla^2 \mathbf{E} + 2(\nabla \mathbf{E}) \cdot \nabla \ln n - \frac{1}{c^2} \left[n^2 + 2 \left(\frac{c}{\omega} \right)^2 \nabla^2 \ln n \right] \frac{\partial^2}{\partial t^2} \mathbf{E} = 0 \tag{6.31}$$

This wave equation is still specialized for monochromatically oscillating fields (6.19) and for the time independent n only. On the other hand, it is more general than Eq. (6.16), first due to a first-order derivation, a spatial damping term which is determined by $\nabla \ln n$, and second due to a refractive index n, which is modified by a second-order term $\nabla^2 \ln n$. If for any further study, the time dependence of n has to be included, the time periodic dependence (6.19) has to be revised, as any Fourier superposition of single-frequency solutions would have to be further generalized.

6.4 Refractive Index of Plasma and Its Relation to Absorption

A discussion of the complex refractive index n, Eq. (6.24), will be given in this subsection representing the dispersion of electromagnetic waves in plasma by using the dependence of ω_p upon the electron density n_e, Eq. (2.6), and the collision frequency $\nu(n_e, T)$, Eq. (2.37) including the following nonlinear generalizations.

The complex optical refractive index n is given as the dispersion relation of electromagnetic waves in plasma, Eq. (6.27), where the real part n' and the imaginary part κ are evaluated algebraically:

$$n = n' + i\kappa = \left[1 - \frac{\omega_p^2}{\omega^2(1 + i\nu/\omega)}\right]^{1/2} \tag{6.32}$$

$$n' = \frac{1}{\sqrt{2}}\left[\sqrt{\left(1 - \frac{\omega_p^2}{\omega^2 + \nu^2}\right)^2 + \left(\frac{\nu}{\omega}\frac{\omega_p^2}{\omega^2 + \nu^2}\right)^2} + \left(1 - \frac{\omega_p^2}{\omega^2 + \nu^2}\right)\right] \tag{6.33}$$

$$\kappa = \frac{1}{\sqrt{2}}\left[\sqrt{\left(1 - \frac{\omega_p^2}{\omega^2 + \nu^2}\right)^2 + \left(\frac{\nu}{\omega}\frac{\omega_p^2}{\omega^2 + \nu^2}\right)^2} - \left(1 - \frac{\omega_p^2}{\omega^2 + \nu^2}\right)\right] \tag{6.34}$$

Sometimes n', the real part only, is called the refractive index. For a collisionless plasma ($\nu = 0$), both values are equivalent

$$n = n' = (1 - \omega_p^2/\omega^2)^{1/2} \qquad (\text{if } \nu = 0) \tag{6.35}$$

The cutoff density n_{ec} is that electron density n_e where the collisionless refractive index vanishes [$\omega_p = \omega$, and Eq. (2.6)]

$$n_{ec} = \frac{m\omega^2}{4\pi e^2} \tag{6.36}$$

Since $c_\varphi = 0$, Eq. (6.26), no propagation of transversal electromagnetic waves is possible. A plasma of this density causes total reflection.

The imaginary part of n, κ, is called the absorption coefficient. Its meaning is seen immediately from its relation to the absorption constant $\bar{K}$, which determines the attenuation of a laser intensity I at some depth x; if I_0 is the intensity at $x = 0$

$$I = I_0 \exp(-\bar{K}x)$$

The absorption constant is then

$$\bar{K} = \frac{2\omega}{c}\kappa \tag{6.37}$$

As demonstrated by the preceding equations, the optical properties of a plasma depend on the plasma frequency ω_p, Eq. (2.6), and, therefore, on the electron density n_e, the electron mass m (to the extent it can be changed relativistically), and the collision frequency ν. Here it is important to recognize in what sense the collision frequency is defined. In Eqs. (2.34) and (2.35) it was defined by changes in the motion of the electrons and ions due to Coulomb interaction, in which energy is exchanged. This kind of collision leads to equipartition and characterizes the thermal conductivity and the friction in plasmas, Eqs. (6.1) and (6.2). It also characterizes the exchange of energy, given by an equipartition within one component (e.g., in the electrons if for some reason, the velocity distribution is non-Maxwellian), or between electrons and ions, if there is no thermal equilibrium for some reason (e.g., if only the electrons are heated by incident laser radiation). The following steps show that this collision frequency can be identified with collisions for high-frequency processes.

In a numerical evaluation of n, $\bar{K}$, and n' [140] values for neodymium glass and CO_2 laser radiation were calculated. The agreement with plasma experiments can be considered as a simple proof of the equivalence of the dc and HF collision frequencies. Furthermore, the very simple derived dc collision frequency in Section 2, now used for the HF optical constant, is essentially the same—apart from a minor factor—because the quantum mechanical derivation of the absorption constant of a plasma is based on inverse bremsstrahlung. Indirectly, therefore, a quantum mechanical justification of the very primitive 90° collision frequency, Eq. (2.34), has been established.

The numerical evaluation of $\bar{K}$ and n' leads to the curves in Figs. 6.1 to 6.4. For lower plasma temperatures, the Coulomb logarithm reaches $\ln\Lambda = 1$, $\Lambda = 2.718$. Below this point, the collision theory is *not valid*. This restriction means, from Eq. (2.36), that

$$n_e \leqslant \frac{9K^3T^3}{4Z^2e^6\pi} = 2.42\times10^{20}\ T^3; \qquad [n_e] = \text{cm}^{-3}; \qquad [T] = \text{eV} \tag{6.38}$$

If the computation is extended to lower temperature, neglecting (6.38), all curves of $\bar{K}$ merge together. This expresses a density independent absorption. This case has to be excluded on the plots [140]. Another restriction is that the calculations assume a classical Boltzmann-type energy distribution for the electrons. The curves of Figs. 6.1 to 6.4 are drawn as lines for temper-

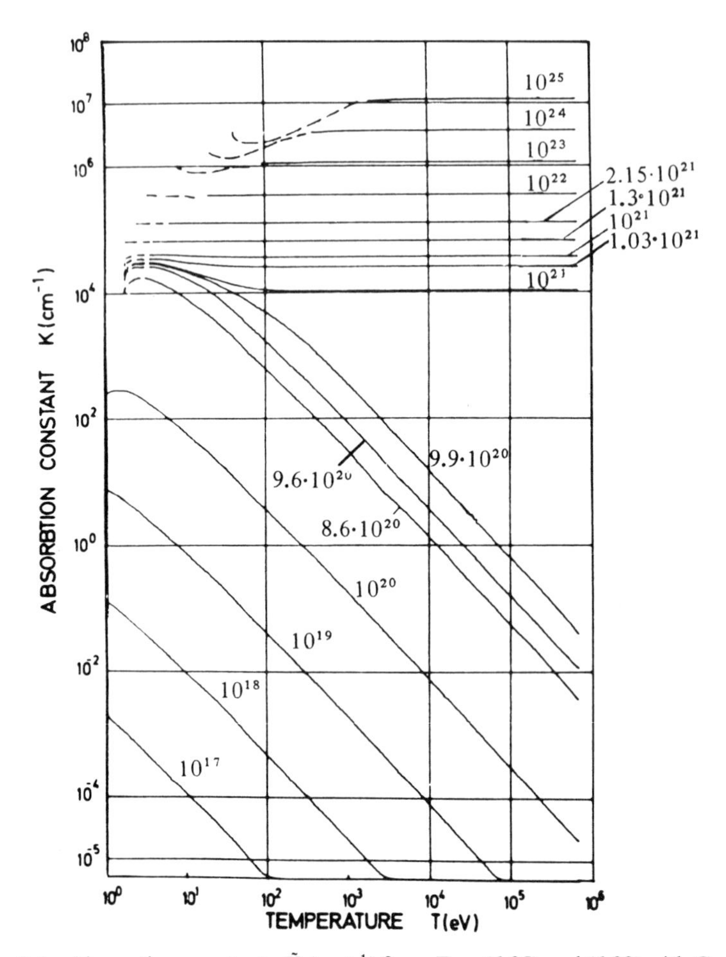

Figure 6.1 Absorption constant $\tilde{K}$ (cm^{-1}) from Eqs. (6.35) and (6.38) with Coulomb collision for neodymium-glass laser radiation in plasma with temperature T(eV) and density n_e (cm^{-3}) [140].

atures equal to or above 10 times the Fermi energy E_F (*Eq.* 2.47)

$$T > 10E_F = 10\frac{h^3}{2m}\left(\frac{3n_e}{8\pi}\right)^{2/3} = 3.65 \times 10^{-14} n_e^{2/3} \tag{6.39}$$

$[T] = \text{eV}$; $[n_e] = \text{cm}^{-3}$. For lower temperatures, the curves are dashed because it is not certain that this model is valid for Fermi–Dirac degenerate plasmas.

In quantum mechanical terms, the motion of an electron within a Coulomb field of an ion has continuous energy eigenvalues, between which

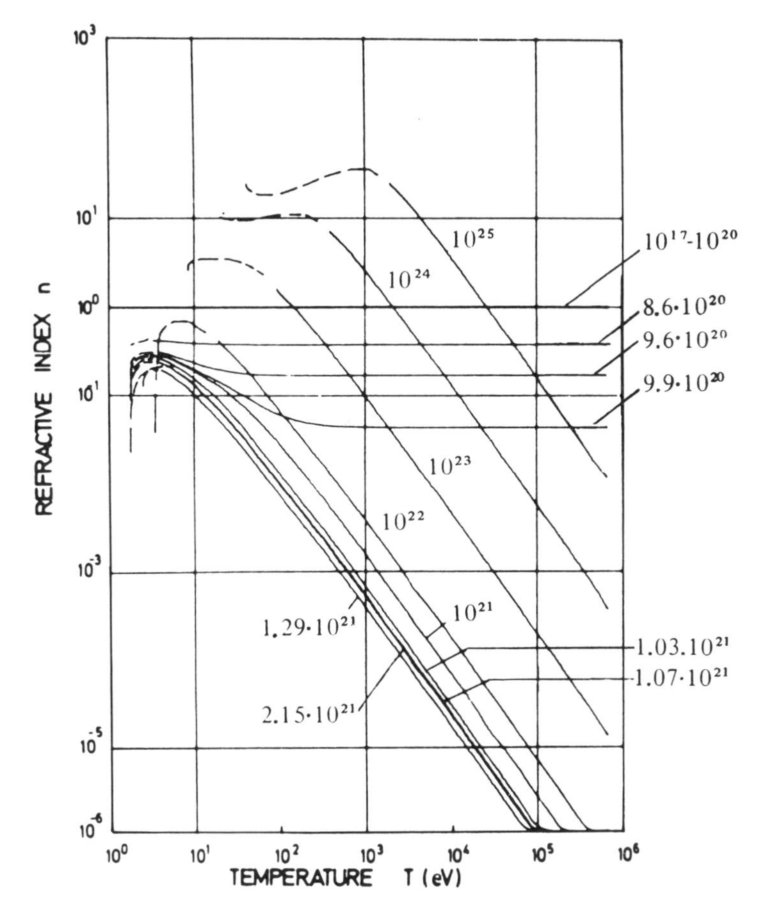

Figure 6.2 Variation of refractive index n' (real part), Eq. (6.34), for neodymium-glass laser radiation in plasma with temperature T(eV) and density n_e (cm^{-3}) [140].

the electron changes to a higher energy state by absorbing the energy of a photon. The exact quantum mechanical description [141] results in an absorption constant (index B for bremsstrahlung):

$$\bar{K}_B = \frac{8\pi^2}{3^{1/2}} \frac{Z^2 n_e n_i e^6 g(T}{c\omega^2 (2\pi mkT)^{3/2}} \tag{6.40}$$

T is the plasma temperature and $g(T)$ is the "Gaunt factor" [142], a value that corrects the point-mechanical description by a factor between 0.1 and 10. The comparison of the quantum mechanical process of inverse bremsstrahlung and that defined by the collisions is justified by the ratio of $\bar{K}_B$ to $\bar{K}$. This ratio is derived from Eq. (6.37) using the plasma collision frequency

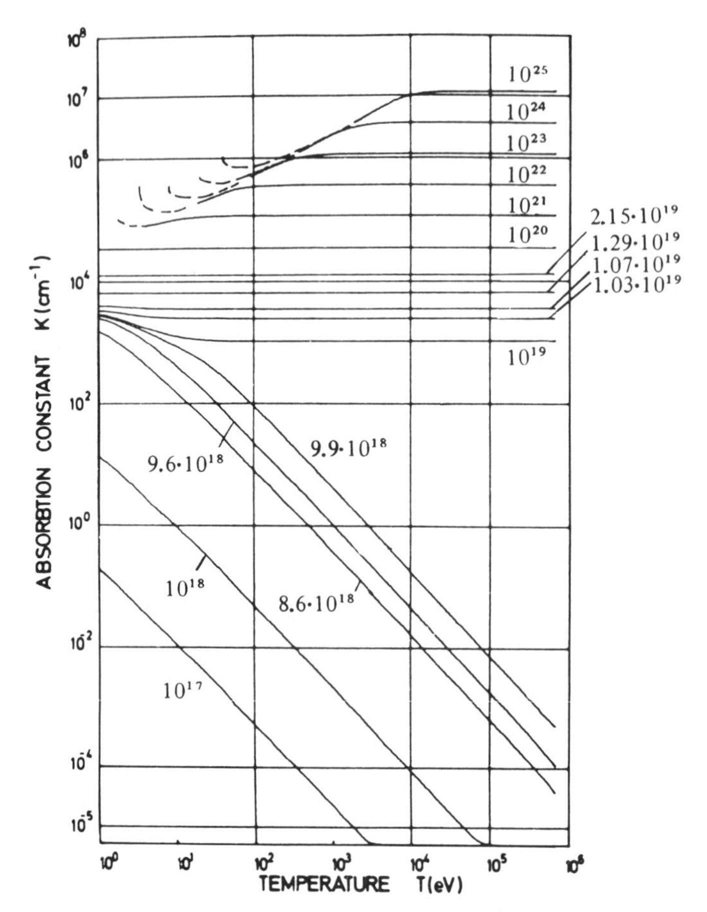

Figure 6.3 Dependence of absorption constant $\bar{K}$ (cm^{-1}) for CO_2 laser radiation in plasma on temperature T(eV) and density n_e (cm^{-3}) [140].

Eq. (2.35):

$$\frac{\bar{K}}{\bar{K}_B}=0.324\,\frac{\ln\Lambda}{\gamma_E(Z)g} \tag{6.41}$$

The validity of $\bar{K}$ in

$$\bar{K}=\frac{4\pi^2}{\gamma_E(Z)} \tag{6.42}$$

is restricted to plasma densities below the cutoff density, $\omega \gg \omega_p$, $n_e \ll n_{ec}$ (Eq. 2.6) and to relatively low collision frequencies $\nu \ll \omega$. For the following representation of the Gaunt factor, the validity is limited in a similar way

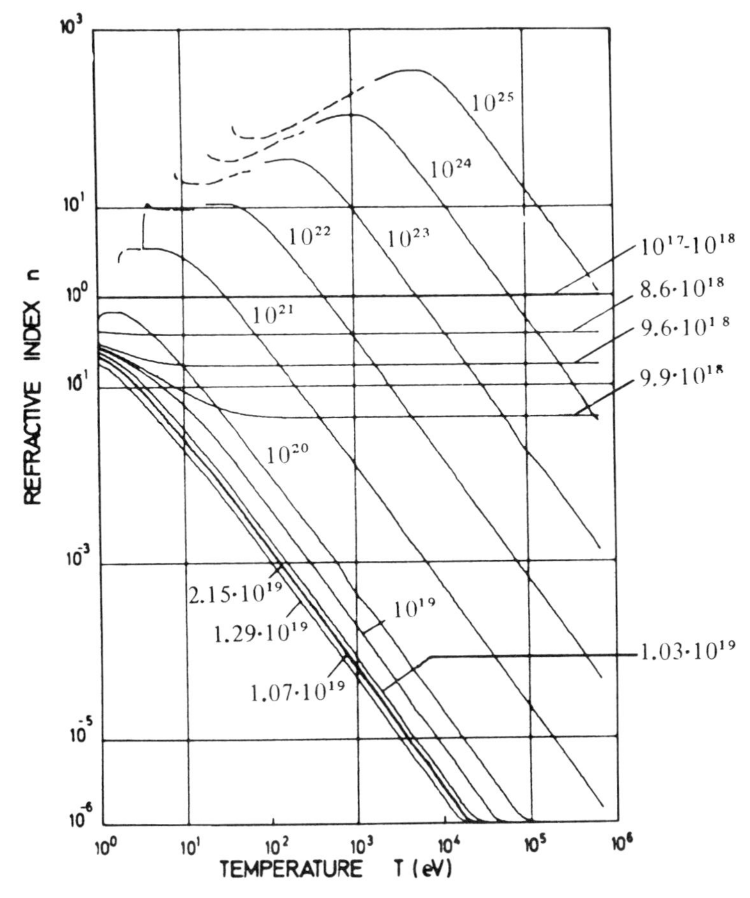

Figure 6.4 Refractive index (real part) n' for CO_2 laser radiation as a function of plasma temperature T (eV) and density n_e (cm^{-3}) [140].

(T in eV and n_e in cm^{-3}) [143]

$$g(T, n_e) = 1.2695(7.45 + \ln T - \tfrac{1}{3} \ln n_e) \tag{6.43}$$

If the value of the Coulomb logarithm is

$$\ln \Lambda = 3.45(6.69 + \ln T - \tfrac{1}{3} \ln n_e - \tfrac{2}{3} \ln Z) \tag{6.44}$$

the ratio becomes

$$\frac{\bar{K}}{\bar{K}_B} = \frac{0.88}{\gamma_E(Z)} \left(1 - \frac{0.66 + \frac{2}{3} \ln Z}{7.45 + \ln T - \frac{1}{3} \ln n_e}\right) \tag{6.45}$$

The agreement of $\bar{K}$ derived from Spitzer's value [140] with K_B, derived from quantum mechanics, can be seen for $Z=1$, $T=10^4$ eV, $n_e=10^{21}$ cm^{-3}, where $\bar{K}/\bar{K}_B=1.06$.

The absorption of laser radiation in a plasma was considered by Dawson and Oberman [144]. The plasma was described by the collisionless Vlasov equation (3.7) and included the interaction of the particles by phase mixing. This theory [140] was justified by a comparison with quantum mechanical theory in the same way as shown previously for the theory of absorption due to Coulomb collisions. For the absorption constant derived by Dawson and Oberman K_{DO}, the ratio is quite close to unity. Comparing this result with Eq. (6.41), the absorption theory based on Coulomb collisions is found to be closer to the inverse bremsstrahlung theory (e.g., for deuterium by a factor of 2.02).

The quantum mechanical absorption constants may be considered as the most probable values. The disadvantage of the theory is its limitation to low-density and high-temperature plasmas (low collision frequency). However, the laser-plasma interaction is important for plasmas near and above the cutoff densities. For these cases, the refractive index n' (real part) and the absorption constant $\bar{K}$ have been evaluated numerically for several interesting wavelengths (ruby, neodymium-glass, and CO_2 lasers and their second harmonics [140]).

The optical constants are very similar to those of metals and semiconductors. The low-density majority carriers follow very well the oblique lines of $\bar{K}$ in Figs 6.1 and 6.3. The high absorption constants for nearly visible radiation of 10^5 cm^{-3} and more are similar with that of metals for the superdense case ($\omega<\omega_p$). It is remarkable that, for high electron densities, the real part of n can grow to values of 10 and more [145].

For some purposes it is important to evaluate the absolute minimum value of the refractive index n, which can be much less than unity in a hot plasma near the cutoff density. From the exact value

$$|n|=\left[\left(1-\frac{\omega_p^2}{\omega^2+\nu^2}\right)^2+\left(\frac{\nu}{\omega}\frac{\omega_p^2}{\omega^2+\nu^2}\right)^2\right]^{1/4} \tag{6.46}$$

the minimum is found at

$$|n|_{\min}=\left[\left(\frac{\nu}{\omega}\right)^4+\left(\frac{\nu}{\omega}\right)^2\right]^{1/4} \qquad \left[\text{at } \omega_p^2=\frac{\omega^2+\nu^2}{1+(\nu/\omega)^2}\right] \tag{6.47}$$

If (as is usual) $\nu \ll \omega$, this can be written as

$$|n|_{\min}=\left(\frac{\nu}{\omega}\right)^{1/2}=\frac{a'}{T^{3/4}} \tag{6.48}$$

The constant a' in the preceding equation can be derived from Eqs. (2.6) and (2.35)

$$a' = \frac{\omega_p \pi^{3/4} m^{1/2} Z^2 \ln \Lambda}{8\pi\gamma_E(Z)(2K)^{3/2}} \tag{6.49}$$

for example, for deuterium ($Z=1$) and $\ln \Lambda = 10$

$a = 3.25\ (\mathrm{eV})^{3/4}$, for Nd-glass lasers

$a = 1.03\ (\mathrm{eV})^{3/4}$, for CO_2 lasers

6.5 Nonlinear and Relativistic Absorption

The assumption in the preceding discussion of the optical constants concerns n_e and T only. Now, derivations based on the fact that the oscillation energy $\varepsilon_{\mathrm{osc}}$ of the electrons (coherent quiver motion) in a laser field can exceed the thermal energy KT will be discussed. The absorption is dependent on the laser intensity and the whole Coulomb collision induced absorption process becomes nonlinear. A further generalization is necessary if the quiver motion arrives at such high energies, that relativistic effects (such as, e.g., changes of the electron mass) have to be taken into account.

The equation of motion of a single electron in an electromagnetic wave field is given in the nonrelativistic case ($|\mathbf{v}| \ll c$) by

$$m \frac{d}{dt} \mathbf{v} = e\mathbf{E} + \frac{e}{c} \mathbf{v} \times \mathbf{H} \tag{6.50}$$

Owing to $|\mathbf{v}| \ll c$ the Lorentz term can be neglected. If the elongation of the oscillation, induced by the real part of the electric field $\mathbf{E}$ of a laser varying with frequency ω (Eqs. 6.19), is much less than the wavelength $\lambda = 2\pi c/\omega$,

$$\mathbf{v} = \frac{e}{m} \frac{\mathbf{E}_r}{i\omega} \cos \omega t \tag{6.51}$$

The maximum value of the energy of the quiver motion is

$$\frac{m}{2} \mathbf{v}^2 = \frac{e^2}{2m} \frac{\mathbf{E}_r^2}{\omega^2} = \varepsilon_{\mathrm{osc}} \tag{6.52}$$

This is equal to the total oscillation energy of the electrons. The average value of the kinetic energy at quivering is

$$\varepsilon_{\mathrm{osc}}^{\mathrm{kin}} = \tfrac{1}{2}\varepsilon_{\mathrm{osc}} = \frac{e^2}{4m} \frac{\mathbf{E}_r^2}{\omega^2} \tag{6.53}$$

Inserting the plasma frequency ω_p, Eq. (2.6), and the cutoff density n_{ec},

Eq. (6.36), results in

$$\varepsilon_{\text{osc}} = \frac{\mathbf{E}_r^2}{8\pi n_{ec}}; \qquad \varepsilon_{\text{osc}}^{\text{kin}} = \frac{\mathbf{E}_r^2}{16\pi n_{ec}} \tag{6.54}$$

The definition of the energy density in electromagnetic fields leads to the relation of the field $\mathbf{E}_r$ to the laser intensity I:

$$|\mathbf{E}_r|\,(\text{cgs}) = \sqrt{8\pi I/c} = 2.91\times 10^{-5}\sqrt{I}\,(\text{erg/cm}^2\,\text{sec}) \tag{6.55}$$
$$|\mathbf{E}_r|(\text{V/cm}) = \sqrt{2I/\Omega_0} = 27.4\sqrt{I}\,(\text{W/cm}^2)$$

Ω_0 is the resistivity of vacuum (377 Ω). Using this, (6.54) in cgs units can be written as

$$\varepsilon_{\text{osc}} = \frac{I}{cn_{ec}}; \qquad \varepsilon_{\text{osc}}^{\text{kin}} = \frac{I}{2cn_{ec}} \tag{6.56}$$

To arrive at the optical constant n for $\varepsilon_{\text{osc}}^{\text{kin}} \gtrsim KT_{\text{th}}$ we have to take into account that the Coulomb collision frequency ν_{ei} must be modified to include the velocity due to the coherent quiver motion. Very simply, this means that we have to use an effective temperature T^*

$$T^* = T_{\text{th}} + \varepsilon_{\text{osc}}^{\text{kin}}/K \tag{6.57}$$

instead of the temperature T_{th} representing the mean energy of random thermal motion only. The absorption constant and collision frequency ν_{ei} (2.35) is then calculated to [146]

$$\bar{K}_{NL} = \frac{4\pi^4 e^6 n_e^2 n_i^2 Z^2 \ln\,\Lambda}{\sqrt{2}\,16(KT_{\text{th}} + \varepsilon_{\text{osc}}^{\text{kin}})^{3/2}} \tag{6.58}$$

where

$$\nu_{ei} = \frac{(m\pi)^{1/2}\omega_p^2 e^2 Z \ln\,\Lambda}{\sqrt{2}\,16(KT_{\text{th}} + \varepsilon_{\text{osc}}^{\text{kin}})^{3/2}} \tag{6.58a}$$

The index NL is used for "nonlinear". The Spitzer factor is then unity, as there is no electron–electron collision for coherent motion of the electrons in the laser field. This simple generalization is fully justified by the fact that the general quantum mechanical calculation, including the quiver motion, arrives at a very similar result (Rand [147])

$$\bar{K}_{\text{NL}}^{\text{Rand}} = \frac{\sqrt{\pi c}}{4}\, n_e^2 n_i^2 Z^2 e^3 \omega m^{1/2} \ln\left(\frac{32e^2 I}{mKT_{\text{th}} c\omega^2}\right)/I^{3/2} \tag{6.59}$$

In both cases, (6.58) and (6.59),

$$\bar{K} \sim (\varepsilon_{\text{osc}}^{\text{kin}})^{-3/2} \sim I^{-3/2}$$

The ratio of the two nonlinear absorption constants

$$\frac{\bar{K}_{\rm NL}^{\rm Rand}}{\bar{K}_{\rm NL}} = \frac{4 \ln(16\varepsilon_{\rm osc}^{\rm kin}/KT_{\rm th})}{\sqrt{2\pi}\,\pi \ln \Lambda} \tag{6.60}$$

is a constant, which is close to unity within one order of magnitude.

The relativistic generalization of the optical constants has to consider the single-particle motion of an electron in an electromagnetic field

$$\frac{d}{dt}\, m\mathbf{v} = e\mathbf{E} + \frac{e}{c}\, \mathbf{v} \times \mathbf{H} \tag{6.61}$$

Assuming a linear polarized laser field propagating along the x-direction ($\mathbf{E} = \mathbf{i}_y E_y$, $\mathbf{H} = \mathbf{i}_z E_z$), the following equations are obtained:

$$\frac{d}{dt} \frac{m_0 v_y}{[1-(v_x^2+v_y^2)/c^2]^{1/2}} = eE_y \cos \omega t - \frac{e}{c}\, v_x H_z \cos(\omega t + \phi) \tag{6.62}$$

$$\frac{d}{dt} \frac{m_0 v_x}{[1-(v_x^2+v_y^2)/c^2]^{1/2}} = \frac{e}{c}\, v_y H_z \cos(\omega t + \phi) \tag{6.63}$$

m_0 is the rest mass of the electrons and ϕ describes a possible phase between **E** and **H** in a medium. The spatial dependence of **E** and **H** is neglected. Furthermore, the second term on the right-hand side of Eq. (6.62) is neglected for the moment, but this point will be discussed later again. Using

$$\gamma = [1-(v_x^2+v_y^2)/c^2]^{1/2} = (1-\mathbf{v}^2/c^2)^{1/2} \tag{6.64}$$

Eq. (6.62) can be integrated using the wave number $k = \omega/c$

$$v_y = \frac{eE_y\gamma}{m_0\omega} \sin(-kx + \omega t) \tag{6.65}$$

With Eqs. (6.64) and (6.65) there follows from Eq. (6.63)

$$\frac{d}{dt}\{c^2 - \gamma^2[c^2 + e^2 E_y \sin^2(-kx+\omega t)/m_0^2\omega^2]^{1/2}\}\gamma^{-1}$$
$$= \gamma \frac{e^2\mathbf{E}_r^2}{m_0^2 c\omega}\{\tfrac{1}{2} \sin 2(-kx+\omega t) \cos \phi - \sin^2(-kx+\omega t) \sin \phi\} \tag{6.66}$$

Obviously the expression under the radical cannot become negative, therefore γ can vary only between

$$0 < \gamma^{*2} \leqslant \gamma^2 \leqslant 1 \tag{6.67}$$

where

$$\gamma^{*2}\left\{1 + \frac{e^2\mathbf{E}_r^2}{c^2 m_0^2 |n| \omega^2}\right\} = 1 \tag{6.68}$$

The minimum value of γ is

$$\gamma^* = (1 - \mathbf{v}_{\max}^2/c^2) = 1\Big/\left(1 + \frac{e^2\mathbf{E}_r^2}{m_0^2\omega^2c^2|n|}\right)^{1/2} \tag{6.69}$$

This can be used to evaluate the maximum value of the kinetic energy $\varepsilon_{\mathrm{kin}}$ directly from the oscillation (which can be identified by the oscillation energy $\varepsilon_{\mathrm{osc}}$ of the electron) without solving the equation of motion [Eqs. (6.62) and (6.63)]

$$\varepsilon_{\mathrm{osc}} = m_0c^2\left(\frac{1}{\gamma} - 1\right) = m_0c^2\left[\left(1 + \frac{e^2\mathbf{E}_r^2}{m_0^2\omega^2c|n|}\right)^{1/2} - 1\right] \tag{6.70}$$

For $|\mathbf{E}_r| < E^r$ the nonrelativistic value of the oscilation energy, (Eq. 6.52), is reproduced

$$\varepsilon_{\mathrm{osc}} = \frac{1}{2}\frac{e_2}{m_0}\frac{\mathbf{E}_r^2}{\omega^2} \qquad (|\mathbf{E}_r| < E^r) \tag{6.71}$$

E^r is the relativistic limit where, in Eq. (6.70) $\varepsilon_{\mathrm{osc}} = mc^2$ given by a relativistic threshold intensity I_r

$$E^r = \sqrt{3}m_0\omega c/e; \qquad I_r = \frac{3m_0^2\omega^2c^3}{8\pi e^2} \tag{6.72}$$

This limit for neodymium glass lasers [with a wavelength $\lambda = 2\pi c/\omega = 1.06\ \mu$m (Nd)] and for CO_2 lasers ($\lambda = 10.6\ \mu$m) amounts to

$$E^r = \begin{cases} 5.17 \times 10^{10}\ \mathrm{V/cm} & (\mathrm{Nd}) \\ 5.17 \times 10^{9}\ \mathrm{V/cm} & (\mathrm{CO_2}) \end{cases} \tag{6.73}$$

The corresponding laser intensities $I = I_r$ in vacuum are

$$I^r = \begin{cases} 3.68 \times 10^{18}\ \mathrm{W/cm^2} & (\mathrm{Nd}) \\ 3.68 \times 10^{16}\ \mathrm{W/cm^2} & (\mathrm{CO_2}) \end{cases} \tag{6.74}$$

For the pure relativistic case at very high laser intensities I and amplitudes $E_r > E^r$, Eq. (6.70) leads to

$$\varepsilon_{\mathrm{osc}} = \frac{ec|\mathbf{E}_r|}{\omega\sqrt{|n|}} \propto I^{1/2} \qquad (|\mathbf{E}_r| \gg E^r) \tag{6.75}$$

This result demonstrates that the oscillation energy of relativistic particles does not depend on their rest mass m_0 and that the increase is proportional only to the square root of the intensity I.

In Eq. (6.62) the last term on the right-hand side was neglected. The complete solution results in an expression for the oscillation energy with a correction function $A(I)$.

$$\varepsilon_{\mathrm{osc}} = m_0c^2\left\{\left[1 + \frac{A(I)e^2|\mathbf{E}_r|^2}{m^2\omega^2c^2|n|}\right]^{1/2} - 1\right\} = m_0c^2\left\{\sqrt{1 - 3A(I)\frac{I}{I_r}} - 1\right\} \tag{6.76}$$

The function $A(I)=1$ for $I \ll I^r$ and grows monotically to the limit of $\pi/2^{3/2}$ according to Max et al. [148] or $3/2^{3/2}$ according to Schwarz and Talenski [149]. The correction of ε_{osc} due to A is therefore less than 5.3%. This permits the use of Eq. (6.70) or (6.76) with $A=1$ as a good approximation in most applications.

Owing to the relativistic change of the rest mass of the electron, the plasma frequency ω_p has to be corrected to:

$$\omega_p^2 = \frac{4\pi e^2 n_e}{m_0}\sqrt{1-v^2/c^2} = \frac{4\pi e^2 n_e}{m_0}\,\frac{1}{[1+3A(I)I/I_r]^{1/2}} \tag{6.77}$$

The relativistic cutoff density is then

$$n_{ec} = \frac{\omega^2 m_0}{4\pi e^2}\,[1+3A(I)I/I_r]^{1/2} \tag{6.78}$$

The collision frequency (6.58) of the electrons in a plasma at relativistic conditions changes from the value of Eq. (2.34) if $(T \ll \varepsilon_{osc}/K)$ to become:

$$\nu = \frac{n_e Z \pi^{3/2} e^4 \ln \Lambda^4}{m_0^2 c^3 \{[1+3A(I)I/I_r]^{1/2}-1]^{3/2} 2^{5/2} [1+3A(I)I/I_r]^{1/2}} \tag{6.79}$$

or

$$\nu = \begin{cases} \dfrac{n_e Z \pi^{3/2} e^4 \ln \Lambda^*}{m_0^2 c^3 2^{5/2} A(I) I/I_r}, & \text{if } I \gg I_r \\[2ex] \dfrac{n_e Z \pi^{3/2} e^4 \ln \Lambda}{(3I/I_r)^{3/2} 2 m_0^{1/2}}, & \text{if } I \ll I_r \end{cases} \tag{6.80}$$

The expression $\ln \Lambda^*$ is the relativistic Coulomb logarithm for $\varepsilon_{osc} \gg KT$ with

$$\Lambda^* = \frac{3}{2Ze^3(\pi n_e)^{1/2}}(m_0 c)^{3/2}\{[1+3A(I)I/I_r]^{1/2}-1\}^{3/2} \tag{6.81}$$

Finally, it has to be noted that, for absorption in plasmas, the only Coulomb collision induced processes have been considered. Their extension to superdense conditions ($\omega < \omega_p$), the nonlinear extension ($\varepsilon_{osc} \gg KT$) and the relativistic extension (neglecting spatial generalizations of Eq. (6.61) leading to harmonic generation) are given here. Another type of absorption is due to instabilities, or due to the macroscopic acceleration of plasma by the nonlinear forces, described below, or due to soliton generation. The direct transfer of optical energy without collision into kinetic energy of plasma has to be distinguished from the usual Coulomb collision absorption as a purely macroscopic nonlinear absorption mechanism.

SEVEN

Waves in Inhomogeneous Plasma

Some properties of electromagnetic waves in plasmas with spatially varying refractive index (inhomogeneous plasma) will be discussed in this section before discussing the two-fluid equation of motion. Equation (6.5) and the necessary nonlinear generalization for laser-plasma interaction and some properties of electromagnetic waves in plasmas with spatially varying refractive index (inhomogeneous plasma) will be discussed in this section. It is important to note that the temporal change of the refractive index n is not taken into account. Only the wave equations are discussed,

$$\nabla^2 \mathbf{H}_r + \frac{\omega^2}{c^2} n^2 \mathbf{H}_r = 0 \tag{7.1}$$

and

$$\nabla^2 \mathbf{E}_r + \frac{\omega^2}{c^2} n^2 \mathbf{E}_r = 0 \tag{7.2}$$

These follow from Eqs. (6.24) and (6.30) if the terms with spatial derivatives of the refractive index n are neglected. The coupling between **E** and **H** is given by one of the Maxwell equations (6.10) or (6.12). The spatial dependence of the plasma frequency ω_p on the electron density n_e and the nonlinear and/or relativistic dependence on the magnitude of **E** and **H** as well as the dependence of the collision frequency ν on n_e, **E**, **H**, electron temperature T, and ion charge Z have all been described in the preceding chapter. This results in a spatial dependence of the refractive index n:

$$n^2 = 1 - \frac{\omega_p^2}{\omega^2(1 - i\nu/\omega)} \tag{7.3}$$

The solutions of the general stationary equations (6.24) and (6.29) including the spatial dependent term are neglected in this section and will be discussed in Section 11. The restriction to the simplified wave equations (7.1) and (7.2) is useful for understanding the essential behaviour of electromagnetic waves propagating in inhomogeneous media. The mathematical problem consists of the discussion of differential equations of the type

$$\frac{\partial^2}{\partial x^2} f(x) + \bar{a}(x) f(x) = 0 \tag{7.4}$$

$\bar{a}(x)$ is a given function and $f(x)$ has to be determined. This is an extensive problem for differential equations. A simple case is with $\bar{a}=\text{const}$ given by the general solutions with elementary functions.

$$\begin{aligned} f(x) &= C_1 \cos(\bar{a}^{1/2}x) + C_2 \sin(\bar{a}^{1/2}x), && \text{if } (\bar{a} \geqslant 0) \\ &= C_3 \exp(\sqrt{-\bar{a}}x) + C_4 \exp(-\sqrt{-\bar{a}}x), && \text{if } (\bar{a} \leqslant 0) \end{aligned}$$

with integration constants $C_1 \cdots C_4$. For general $\bar{a}(x)$, higher functions are determined where in most cases a very special algebraic form of $\bar{a}(x)$ defines a whole class of functions, for example, the Bessel functions, Hermite functions, confluent hypergeometric functions and others, between which, as well as with the elementary functions, relations exist and solutions in the form of series of higher functions exist. Since the computer era has been established, the solution of (7.4) for any reasonable $\bar{a}(x)$ is a relatively easy task for a direct numerical treatment (using Runge–Kutta difference schemes).

To understand the general properties of waves in inhomogeneous media, first approximate solutions of Eqs. (7.1) and (7.2) are discussed. They are called WKB solutions.

The properties of the waves can also be seen from the special case of an inhomogeneous medium with refractive index

$$n(x) = \frac{1}{1+\alpha x} \qquad (\alpha = \text{const}) \tag{7.5}$$

The corresponding differential equation (7.4) is Euler's differential equation. This has solutions with elementary functions which have been discussed by Rayleigh. The solutions for a linearly increasing $n(x)$ leads to the more complicated Airy functions.

7.1 WKB Approximation for Perpendicular Incidence

The approximate solution of a wave equation with spatially variable refractive index n [general $\bar{a}(x)$ in Eq. (7.4)] was known in the last century. It is called WKB approximation, after Wentzel, Kramers, and Brillouin, who used this approximation for the wave equation of electron waves (Schrödinger equation).

A linear polarized plane electromagnetic wave ($\mathbf{E}=\mathbf{i}_y E_y$ and $\mathbf{H}=\mathbf{i}_z H_z$) is assumed to be perpendicularly incident on a stratified plasma, whose n is only dependent on the x coordinate (the general elliptic polarized light can be considered as superposition of linear polarized waves of different $\mathbf{E}$-directions and different temporal phase). Equations (7.1) and (7.2) are then

$$\frac{\partial^2}{\partial x^2} E_y + \frac{\omega^2}{c^2} n^2(x) E_y = 0 \tag{7.6}$$

and

$$\frac{\partial^2}{\partial x^2} H_z + \frac{\omega^2}{c^2} n^2(x) H_z = 0 \tag{7.7}$$

The WKB approximation is the ansatz

$$E_y = \frac{E_v}{|n|^{1/2}} \exp\left(i \frac{\omega}{c} \int^x \mathrm{Re}(n)\, d\xi - \frac{\omega}{c} \int^x \mathrm{Im}(n)\, d\xi \right) \tag{7.8}$$

or

$$E_y = \frac{E_v}{|n|^{1/2}} \exp iF; \qquad F = i \frac{\omega}{c} \int^x n\, d\xi \tag{7.9}$$

E_v is the vacuum amplitude of the incident field. This solution is valid in Eq. (7.6) only as long as certain terms are sufficiently small.
Differentiating Eq. (7.8), for $\nu \ll \omega (\rightarrow |n| \approx n)$

$$\frac{\partial}{\partial x} E_y = -\frac{E_v}{2|n|^{3/2}} (\exp iF) \frac{\partial |n|}{\partial x} + \frac{E_v}{\sqrt{n}} (\exp iF) i \frac{\omega}{c} |n|$$

$$\frac{\partial^2}{\partial x^2} E_y = \frac{3E_v}{4|n|^{5/2}} (\exp iF) \left(\frac{\partial |n|}{\partial x} \right)^2 - \frac{E_v}{2|n|^{3/2}} \frac{\partial^2 |n|}{\partial x^2} (\exp iF)$$
$$- \frac{E_v}{|n|^{3/2}} \frac{\partial |n|}{\partial x} i \frac{\omega}{c} |n| (\exp iF) + i \frac{\omega}{c} \frac{E_v}{|n|^{1/2}} (\exp iF) \left(\frac{\partial |n|}{\partial x} \right)$$
$$- \frac{\omega^2}{c^2} |n|^2 \frac{E_v}{|n|^{1/2}} (\exp iF)$$

and inserting into Eq. (7.6)

$$-\frac{\omega^2}{c^2} |n^2| \frac{E_v}{\sqrt{|n|}} (\exp iF) + \frac{3E_v}{4|n|^{5/2}} (\exp iF) \left(\frac{\partial |n|}{\partial x} \right)^2$$
$$- \frac{E_v}{2|n|^{3/2}} \frac{\partial^2 |n|}{\partial x^2} (\exp iF) + \frac{\omega^2}{c^2} |n|^2 \frac{E_v}{\sqrt{|n|}} (\exp iF) = 0$$

shows that Eq. (7.8) is a good approximation of Eq. (7.6), if

$$\frac{\sqrt{3}}{2|n|}\left|\frac{\partial n}{\partial x}\right| \ll \frac{\omega}{c}|n| \qquad \text{or} \qquad \Theta = \frac{\sqrt{3}}{2}\frac{c}{\omega |n|^2}\left|\frac{\partial n}{\partial x}\right| \ll 1 \tag{7.10}$$

and

$$\frac{1}{2|n|}\left|\frac{\partial^2 n}{\partial x^2}\right| \ll \frac{\omega^2}{c^2}|n^2| \qquad \text{or} \qquad \psi = \frac{1}{2|n|^3}\frac{c^2}{\omega^2}\left|\frac{\partial^2 n}{\partial x^2}\right| \ll 1 \tag{7.11}$$

The second-order equation (7.11) is usually not taken into account. The WKB condition (7.10) restricts the refractive index to a relatively small spatial variation. Using the definitions (7.8) and (6.32) and the geometry for plane waves perpendicularly incident on a stratified plasma leads to

$$\mathbf{E} = \mathbf{i}_y \frac{E_v}{|n|^{1/2}} \cos\left(\frac{\omega}{c}\int^x \operatorname{Re}(n)\, d\xi - \omega t\right) \exp\left(\frac{-\bar{k}x}{2}\right) \tag{7.12}$$

$\bar{k}$ is an averaged absorption constant

$$k = 2\frac{\omega}{xc}\int^x \operatorname{Im}(n)\, d\xi \tag{7.13}$$

The real part of $\mathbf{E}$ is given for further use in calculating forces. The denominator for E_v is used approximatively by the absolute value of n. This approximation causes the neglect of terms of the third order, as will be shown below from a discussion where the complex denominator of $\mathbf{E}$ is used.

The solution of Eq. (7.7) can be derived in a similar way. To derive the correct phase between E_y and H_z, the value of H_z is obtained by using $\mathbf{E}$ from Eq. (7.12) in the Maxwell equation (6.10) and by spatial differentiation and temporal integration. The spatial differentiation produces the following term with the derivative of n.

$$H_z = -\frac{ic}{2\omega}\frac{E_v}{|n|^{3/2}}\frac{dn}{dx}(\exp iF) + E_v\sqrt{|n|}\,(\exp iF) \tag{7.14}$$

or with an approximation of n by $|n|$

$$\mathbf{H} = -\mathbf{i}_z \frac{c}{2\omega}\frac{E_v}{|n|^{3/2}}\frac{d|n|}{dx} \sin\left(\frac{\omega}{c}\int^x \operatorname{Re}(n)\, d\xi - \omega t\right)\exp(-\tfrac{1}{2}kx)$$
$$+\mathbf{i}_z E_v |n|^{1/2} \cos\left(\frac{\omega}{c}\int^x \operatorname{Re}(n)\, d\xi - \omega t\right)\exp(-\tfrac{1}{2}kx) \tag{7.15}$$

The second terms in the last two equations correspond to the usual form of the plane waves in vacuum. Decreasing n leads to decreasing $\mathbf{H}$ and increasing $\mathbf{E}$ in Eqs. (7.15) and (7.12), respectively. This increase or swelling of the amplitude of $\mathbf{E}$ in a plasma will be of essential importance in the following sections. The product $|\mathbf{E}\times\mathbf{H}|$, however, is not influenced by $|n|$. Thus, the

Poynting vector or the electromagnetic energy flow is unchanged. The first terms in Eqs. (7.14) and (7.15) correspond to the phase shift between **E** and **H**. Similar properties are shown below from the discussion of the exact mathematical solution for the Rayleigh case. The phase shift term will be essential for the generation of forces in plasma due to the electromagnetic radiation. Decreasing Re(n) in a plasma causes an increase of the wavelength in the oscillating factors in Eqs. (7.12) and (7.15).

7.2 Oblique Incidence and WKB Solution

If plane waves are obliquely incident on a stratified plasma, they have to be connected analytically with the plane waves in vacuum outside the plasma. For vacuum, $n=1$, Eq. (7.2) is

$$\frac{\partial^2}{\partial x^2}\mathbf{E}_r+\frac{\partial^2}{\partial y^2}\mathbf{E}_r+\frac{\partial^2}{\partial z^2}\mathbf{E}_r+\frac{\omega^2}{c^2}\mathbf{E}_r=0 \tag{7.16}$$

Each vector component of $\mathbf{E}_r$ can be separated, for example, from the y-component, by a product ansatz.

$$E_y(x,y,z)=E_{yx}(x)E_{yy}(y)E_{yz}(z) \tag{7.17}$$

Substituting into Eq. (7.16) and dividing by E_y leads to

$$\frac{1}{E_{yx}}\frac{\partial^2}{\partial x^2}E_{yx}+\frac{1}{E_{yy}}\frac{\partial^2}{\partial_y^2}E_{yy}+\frac{1}{E_{yz}}\frac{\partial^2}{\partial_z^2}E_{yz}+\frac{\omega^2}{c^2}=0$$

In this way, the solution of the partial differential equation (7.16) is reduced to a solution of the ordinary differential equations

$$\frac{d^2}{dx^2}E_{yx}+(-k_x^2)E_{yx}=0$$

$$\frac{d^2}{d_y^2}E_{yy}+(-k_y^2)E_{yy}=0$$

$$\frac{d^2}{dz^2}E_{yz}+(-k_z^2)E_{yz}=0$$

The constant k_x, k_y, and k_z are called eigenvalues. The eigenfunctions E_{yx}, E_{xy}, and E_{yz} can be combined by Eq. (7.17) to give a solution

$$E_y=E_v\exp(i\mathbf{k}\cdot\mathbf{r}) \tag{7.18}$$

k is the propagation vector of the plane wave, whose components

$$k_x=\frac{\omega}{c}\cos u_x;\qquad k_y=\frac{\omega}{c}\cos u_y;\qquad k_z=\frac{\omega}{c}\cos u_z \tag{7.19}$$

determine the angle **u** between the direction of **k** and the x-, y-, or z-axis of a Cartesian coordinate system, respectively. Obviously

$$k_x^2+k_y^2+k_z^2=\frac{\omega^2}{c^2}$$

Without losing generality, the plane of incidence is assumed to be the x–y plane (Fig. 7.1). The plasma is still stratified so that n is only dependent on x.

The plane waves have an angle of incidence u_0 in vacuum. The solution of the Maxwell equations for a plane wave with an electric vector $\mathbf{E}_p$ oscillating in the plane of incidence (p-polarization) are then, from the preceding steps,

$$\mathbf{E}_p=E_v(-\mathbf{i}_x\cos u_0+\mathbf{i}_y\sin u_0)\cos\left[\frac{\omega}{c}(\cos u_0)x+\frac{\omega}{c}(\sin u_0)y-\omega t\right] \tag{7.20}$$

$$\mathbf{H}_p=E_v\mathbf{i}_z\cos\left[\frac{\omega}{c}(\cos u_0x+\frac{\omega}{c}(\sin u_0)y-\omega t\right] \tag{7.21}$$

and, for the perpendicular polarization (s-polarization)

$$\mathbf{E}_s=E_v\mathbf{i}_z\cos\left[\frac{\omega}{c}(\cos u_0)x+\frac{\omega}{c}(\sin u_0)y-\omega t\right] \tag{7.22}$$

$$\mathbf{H}_s=E_v(\mathbf{i}_x\cos u_0-\mathbf{i}_y\sin u_0)\cos\left[\frac{\omega}{\tau}(\cos u_0)x+\frac{\omega}{c}(\sin u_0)y-\omega t\right] \tag{7.23}$$

The linear combination of the two cases generate the general polarization in vacuum.

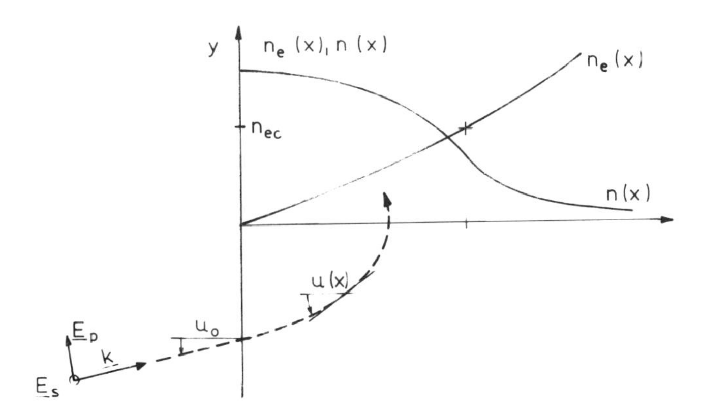

Figure 7.1 Linearly polarized plane waves with p- or s-polarization obliquely incident at an angle u_0 in vacuum onto a stratified plasma. This angle $u(x)$ varies in the plasma along the x-direction.

In this subsection, the formulation of the WKB approximation is given for the case of a collisionless plasma only ($\nu=0$), where the refractive index n is real. The substitution of the following solutions into the wave equations leads to first- and second-order restrictions of validity of the WKB approximation in a similar way to that of the preceding subsection for perpendicular incidence. These restrictions are

$$\Theta(u)=\frac{\sqrt{3}}{2}\frac{c}{\omega}\frac{1}{n^2\cos^2 u}\left|\frac{d(n\cos u)}{dx}\right|\ll 1 \tag{7.24}$$

$$\psi(u)=\frac{1}{2}\frac{c^2}{\omega^2}\frac{1}{n^2\cos^2 u}\left|\frac{d^2(n\cos u)}{dx^2}\right|\ll 1 \tag{7.25}$$

The angle $u(x)$ of the direction of wave propagation in plasma follows Snell's law:

$$n^2(x)\sin^2 u(x)=\sin^2 u_0 \tag{7.26}$$

This can be seen by analytical connection of the solution in vacuum with that in plasma.

First to be resolved is the question whether the WKB conditions (7.24) and (7.25) are met in plasmas. The general linearly polarized, obliquely incident plane wave is written as a sum of two parts, one with the oscillating **E** vector parallel to the plane of incidence (subscript p) and one with the **E** perpendicular to this plane (subscript s). It is well known that this separation is not generally possible [150]. In order to obtain the electric vector in the perpendicular case, the following equation has to be solved

$$\left(\frac{\partial^2}{\partial x^2}+\frac{\partial}{\partial y^2}+\frac{\omega^2}{c^2}n^2(x)\right)\mathbf{E}_s=0 \tag{7.27}$$

The WKB solution gives

$$\mathbf{E}_s=\mathbf{i}_z\frac{E_v\sqrt{\cos u}}{(n(x)-\sin^2 u_0)^{1/4}}\cdot\cos G \tag{7.28}$$

where

$$G=\pm\frac{\omega}{c}\left(\int_0^x\sqrt{n^2(\xi)-\sin^2(u_0)}\,d\xi+n(x)y\sin u(x)\right)+\omega t \tag{7.29}$$

The upper sign is chosen for an electromagnetic wave whose propagation direction has a component into the plasma. The lower sign is for the opposite direction. Applying the Maxwell equations, Eq. (7.28) leads to

$$\begin{aligned}\mathbf{H}_s=\pm&\frac{E_v\sqrt{n\cos u_0}}{\sqrt{\cos u(x)}}[\mathbf{i}_x\sin u(x)-\mathbf{i}_y\cos u(x)]\cos G\\ &-\mathbf{i}_y\frac{c}{\omega}\frac{E_v\sqrt{\cos u_0}}{2n^{3/2}(x)\cdot\cos^{5/2}u(x)}\frac{dn}{dx}\sin G\end{aligned} \tag{7.30}$$

In the case of E_p the electric vector is no longer perpendicular to the direction of propagation as it was for the general case, and the electric field is no longer divergenceless. The components of the electric field strength **E** are given by Eq. (6.31)

$$\left(\frac{\partial^2}{\partial x^2}+\frac{\partial^2}{\partial y^2}+\frac{\omega^2}{c^2}\,n^2(x)\right)E_{px}-\frac{\partial}{\partial x}\,\nabla\cdot\mathbf{E}_p=0 \qquad (7.31)$$

and

$$\left(\frac{\partial^2}{\partial x^2}+\frac{\partial^2}{\partial y^2}+\frac{\omega}{c^2}\,n^2(x)\right)E_{py}-\frac{\partial}{\partial y}\,\nabla\cdot\mathbf{E}_p=0 \qquad (7.32)$$

The last terms in Eqs. (7.31) and (7.32) couple all components of the electric field so that a representation as a sum of the two differently polarized components is not possible. In the case of the WKB condition, however, the last terms in Eqs. (7.31) and (7.32) are negligible, as will be shown. Using the relation

$$n(x)\nabla\mathbf{E}_p=-2E_{px}\frac{dn}{dx} \qquad (7.33)$$

which can be derived from Ginzburg [150, Eq. (19.19)] from (7.31) and (7.32)

$$\left(\frac{\partial^2}{\partial x^2}+\frac{\partial^2}{\partial y^2}+\frac{\omega^2}{c^2}\,n^2(x)\right)E_{px}+2\frac{\partial}{\partial x}\left(E_{px}\frac{d\ln n(x)}{dx}\right)=0 \qquad (7.34)$$

and

$$\left(\frac{\partial^2}{\partial x^2}+\frac{\partial^2}{\partial y^2}+\frac{\omega^2}{c^2}\,n^2(x)\right)E_{py}+2\frac{\partial}{\partial y}\left(E_{py}\frac{d\ln n(x)}{dx}\right)=0 \qquad (7.35)$$

Comparison with the WKB conditions (7.24) and (7.25) shows that the time averaged last term in Eq. (7.34) is always $(\Theta+2\psi)$ times the $n^2(\mathbf{x})$ term in the first bracket of Eq. (7.34). Therefore, neglect of the last term in Eq. (7.35), compared with n^2, is possible in the same way when

$$2\tan(x)\sin u(x)\leqslant 1 \qquad (7.36)$$

that is, when $u\leqslant 40°$. So, with the conditions given by Eqs. (7.24), (7.25), and (7.36), one can solve Eqs. (7.34) and (7.35) separately without coupling terms to get

$$\mathbf{E}_p=\frac{E_v\cos^{1/2}u_0}{(n(x)\cos u(x))^{1/2}}\left[-\mathbf{i}_x\sin u(x)+\mathbf{i}_y\cos u(x)\right]\cos G \qquad (7.37)$$

Substitution of Eq. (7.37) into Maxwell's equations gives

$$\mathbf{H}_p=\frac{E_v(n\cos u_0)^{1/2}}{\cos^{1/2}u(x)}\,\mathbf{i}_z\left[\cos G-\frac{c}{\omega}\frac{\frac{1}{2}-\sin^2 u(x)}{n\cos u(x)}\frac{dn}{dx}\sin G\right] \qquad (7.38)$$

These solutions for the *s*- and *p*-polarization will be used in the following evaluations.

7.3 The Rayleigh Profile

The Rayleigh profile of the refractive index $n(x)$ corresponds to a very special mathematically defined electron density profile of a plasma. It looks very artificial in the first instance. However, it is important for the electrodynamic forces in the plasma. Furthermore, this profile has the advantage of leading without any restriction to an exact solution of the Maxwell equations which can be described with elementary functions. This is used for a very basic study of the wave structure in inhomogeneous media and is a tool for comparison with numerical approximations.

The case of perpendicular incidence on a collisionless, stratified plasma will be considered. For a linearly polarized plane wave with

$$n = \frac{1}{1+\alpha x} \qquad \alpha \geqslant 0 \tag{7.39}$$

$$\mathbf{E}_r = i_y E_y; \qquad \mathbf{H}_r = i_z H_z \tag{7.40}$$

the complete wave equation (6.29) can be written as

$$\frac{\partial^2}{\partial x^2} E_y + \frac{\omega^2}{c^2} n^2 E_y = 0 \tag{7.41}$$

Equation (6.28) neglecting the second spatial derivative of the refractive index is

$$\frac{\partial^2}{\partial x^2} H_z - 2\left(\frac{\partial}{\partial x} \ln n\right)\frac{\partial}{\partial x} H_z + \frac{\omega^2}{c^2} n^2 H_z = 0 \tag{7.42}$$

Using the Rayleigh profile of Eq. (7.39) for the refractive index in Eq. (7.42), for real $\alpha \geqslant 0$ one obtains

$$\frac{\partial^2}{\partial x^2} E_y + \frac{\omega^2}{c^2} \frac{1}{(1+\alpha x)^2} E_y = 0 \tag{7.43}$$

The corresponding electron density, according to the definition of the plasma frequency ω_p [Eq. (2.6)] and of the cutoff density n_{ec} [Eq. (6.37)], is

$$n_e(x) = n_{ec}[1-(1+\alpha x)^{-2}] \qquad (x \geqslant 0) \tag{7.44}$$

As can be seen, a plasma with $n \leqslant 1$ is possible only for $\alpha \geqslant 0$. For $x \to \infty$, n_e is increasing from $n_e = 0$ at $x = 0$ monotonically to the cutoff density. No higher densities than the cutoff density are possible for the Rayleigh case without collisions.

For the solution of Eq. (7.43), the following substitution is used

$$\xi = 1 + \alpha x \tag{7.45}$$

An Euler differential equation is obtained

$$\xi^2 \frac{\partial^2}{\partial \xi^2} E_y + \frac{4\omega^2}{c^2\alpha^2} E_y = 0 \tag{7.46}$$

where

$$E_y = E_v \xi^{1/2 \pm 1/2[1 - 4\omega^2/(c^2\alpha^2)]^{1/2}} \tag{7.47}$$

After resubstitution of (7.45), the result is

$$E_y = (1+\alpha x)^{1/2} E_v \exp\left[\frac{i}{\mp 2} \left(\frac{4\omega^2}{c^2\alpha^2} - 1 \right)^{1/2} \ln(1+\alpha x) \right] \tag{7.48}$$

or using Eqs. (7.39), (7.40), and (6.20)

$$\mathbf{E} = \mathbf{i}_y \frac{E_y}{|n|^{1/2}} \exp\left\{ \frac{i}{\mp 2} \left[\frac{4\omega^2}{c^2\alpha^2} - 1 \right]^{1/2} \ln(1+\alpha x) - i\omega t \right\} \tag{7.49}$$

The solution for H_z can be derived from Eq. (7.42) or by substitution of E_y into a Maxwell equation and subsequent differentiation and integration exactly, for $\mu = 0$:

$$\mathbf{H} = \mp \left(\sqrt{1 - \frac{\alpha^2 c^2}{4\omega^2}} - i \frac{\alpha c}{2\omega} \right) i_z E_v n \exp\left[\mp \frac{i}{2} \sqrt{\frac{4\omega^2}{c^2\alpha^2} - 1} \ln(1+\alpha x) - i\omega \iota \right] \tag{7.50}$$

The increase of the absolute value of **E** with z due to the decreasing refractive index n in Eq. (7.49) is exactly compensated by a decrease of **H** very similar to the WKB approximation. The signs in the exponential functions indicate a wave propagating to the $+x$ and $-x$ direction, respectively. The linear combination of both solutions is the complete solution of the differential equations. The change in the purely oscillating exponential factor of Eqs. (7.49) and (7.50) into that of the case of vacuum ($\alpha = 0$) is evident by Taylor expansion of the ln function near $x = 1$:

$$\lim_{\alpha \to 0} \exp[\cdots] = \exp\left(\pm \frac{i}{2} \sqrt{\frac{4+\omega^2}{c^2\alpha^2} - 1}\; \alpha x \right) = \exp\left[\mp i \frac{\omega}{c} x \right]$$

The phase between **E** and **H**, given by the complex first factor of **H** in (7.50), is responsible for reflection. If a medium with refractive index 1 (vacuum) for $x \leqslant 0$ is connected with a Rayleigh medium with a density distribution given by Eq. (7.44), then the refractive index and the density slope are continuous. If for $x > 0$ only a wave propagating to $+x$ is permitted, necessarily a reflected wave in the vacuum ($x \leqslant 0$) is necessary to fit the phase shift of

H. This reflection is very small, as long as

$$\alpha \ll \frac{2\omega}{c} \tag{7.51}$$

but it is definitely different from zero, despite the continuous connection between vacuum and inhomogeneous plasma.

Calculation of the reflection R [151] shows that $R < 10\%$ for $\alpha = \omega/c$. However, a quick rise appears. Total reflection occurs for

$$\alpha \geqslant \frac{2\omega}{c} \tag{7.52}$$

This phenomenon is curious again. For perpendicular incidence of a wave onto a *discontinuity* from vacuum to a nonabsorbing medium of a certain refractive index n the Fresnel formula

$$R = \frac{(1-n)^2}{(1+n)^2} \tag{7.53}$$

results in $R=1$ only for $n=0$ or $n=\infty$. The *continuous* connection of vacuum with a Rayleigh plasma, with a slow increase in electron density where $\alpha = 2\omega/c$, results in $R=1$.

There was a basic discussion about the question of whether an "internal reflection" is generated in an inhomogeneous plasma or not. The treatment with ray-optical approximations or with an approximation using a sequence of stepwise homogeneous regions with exact Fresnel-type calculations of the reflectivity at each interface resulted in the standard opinion of an "internal reflection." The Rayleigh solutions, however, did not show an internal reflexion. There were simply the two penetrating waves, one to $+x$, the other to $-x$, and a reflection was produced only at a discontinuity of the derivatives of the refractive index. Osterberg [152] found the same very generally from a medium with an analytical function of $n(x)$: no internal reflection.

The solution is that the stepwise homogeneous approximation is a tool to determine the "local reflection," if the condition of a purely propagating wave should follow for $x \to +\infty$. This "Ausstrahlungsbedingung" for plane waves, in analogy to the Sommerfeld conditions for spherical waves, must be used in the analytical case as well as in the step approximation. The plane wave approximation of the two possible exact solutions is probing from point to point. This is illustrated by the following numerical examples, which, however, lead to a paradox, as shown by V. F. Lawrence [153].

The following media are considered (Fig. 7.2): vacuum ($x \leqslant 0$) is continuously connected with a Rayleigh medium ($0 \leqslant x \leqslant D$), which becomes a homogeneous plasma for $x \geqslant D$. The refractive index is real owing to the

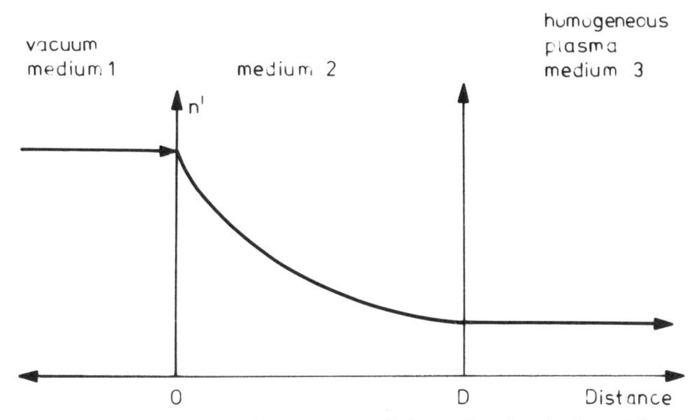

Figure 7.2 The schematic dependence on x of the refractive index n' for connections of homogeneous media 1 and 3 with an inhomogeneous Rayleigh-like density profile [153].

assumption of no absorption.

$$\begin{aligned} n_1 &= 1 = n_v && \text{medium 1} \quad x \leqslant 0 \\ n_2 &= \frac{1}{1+\alpha x} && \text{medium 2} \quad 0 \leqslant x \leqslant D \\ n_3 &= \frac{1}{1+\alpha D} = \text{const} && \text{medium 3} \quad D \leqslant x \end{aligned} \tag{7.54}$$

The components of the electric and magnetic fields can be described by plane waves

$$E_3 = C_{3+} \exp\left[i(n_3 \frac{\omega}{c} x - \omega t) \right] \tag{7.55}$$

$$H_3 = C_{3+} n_3 \exp\left[i(n_3 \frac{\omega}{c} x - \omega t) \right] - C_{3-} n_3 \exp\left[-i\left(n_3 \frac{\omega}{c} x + \omega t \right) \right] \tag{7.56}$$

The subscript denotes the medium and the constants C_+, C_-, the amplitudes of the transmitted and reflected waves, respectively. In medium 3 only the transmitted waves and no reflected waves are postulated. This is specified at $x = D$ by $C_{3+} = 1.0$, $C_{3-} = 0.0$. Then for medium 3 results at $t = 0$

$$E_3 = \exp\left(i n_3 \frac{\omega}{c} D \right) \qquad H_3 = n_3 \exp\left(i \frac{\omega}{c} n_3 D \right) \tag{7.57}$$

In the Rayleigh medium 2 from Eqs. (7.49) and (7.50) the components of the E and H fields are $(t = 0)$

$$E_{2+} = C_{2+} \sqrt{1+\alpha x} \exp\left[i \sqrt{\frac{\omega^2}{c^2 \alpha^2} - \frac{1}{4}} \ln(1+\alpha x) \right] \tag{7.58}$$

$$E_{2-}=C_{2-}\sqrt{1+\alpha x}\exp\left[-i\sqrt{\frac{\omega^2}{c^2\alpha^2}-\frac{1}{4}}\ln(1+\alpha x)\right] \tag{7.59}$$

$$H_{2+}=C_{2+}\frac{1}{\sqrt{1+\alpha x}}\left[\sqrt{1-\frac{\alpha^2}{4\omega^2c^2}}-\frac{i\alpha}{2\omega c}\right]\exp\left[i\sqrt{\frac{\omega^2}{c^2}-\frac{1}{4}}\ln(1+\alpha x)\right] \tag{7.60}$$

$$\mathbf{H}_{2-}=C_{2-}\frac{1}{\sqrt{1+\alpha x}}\left[-\sqrt{1-\frac{\alpha^2}{4\omega^2c^2}}+\frac{i\alpha}{2\omega c}\right]\exp\left[-i\sqrt{\frac{\omega^2}{c^2}-\frac{1}{4}}\ln(1+\alpha x)\right] \tag{7.61}$$

The indices y and z are dropped. The boundary conditions at the interfaces can be derived for $t=0$ at the junction of media 2 and 3, $x=D$:

$$E_{2+}+E_{2-}=E_3 \tag{7.62}$$

$$H_{2+}+H_{2-}=H_3 \tag{7.63}$$

The reflection and transmission coefficients of medium 2, C_{2+} and C_{2-}, can be derived algebraically.

The same can be written for waves in medium 1 where $n=1$

$$E_1=C_{1+}\exp\left(i\frac{\omega}{c}n_1x-i\omega t\right)+C_{1-}\exp\left(-i\frac{\omega}{c}n_1x-i\omega t\right) \tag{7.64}$$

$$H_1=C_{1+}n_1\exp\left(i\frac{\omega}{c}n_1x-i\omega t\right)-C_{1-}n_1\exp\left(-i\frac{\omega}{c}n_1x-i\omega t\right) \tag{7.65}$$

At the junction of medium 1 and medium 2 ($x=0$) at $t=0$, the continuity of E_y and H_z requires

$$E_1=C_1+C_{1-} \tag{7.66}$$

$$H_1=C_{1+}n_1-C_{1-}n_1 \tag{7.67}$$

$$E_2=C_{2+}+C_{2-} \tag{7.68}$$

$$H_2=(C_{2+}-C_{2-})\left[\left(1-\frac{\alpha^2}{4\omega^2c^2}\right)^{1/2}-\frac{i\alpha}{2\omega c}\right] \tag{7.69}$$

Equating

$$E_1=E_2;\qquad H_1=H_2 \tag{7.70}$$

the reflection coefficient R_A of the amplitudes as an exact solution depending on α and D is obtained:

$$R_A=\frac{E_2-H_2}{E_2+H_2} \tag{7.71}$$

Now the case of Fig. 7.2 is approximated by a series of homogeneous media with a stepwise decreasing refractive index n in the Rayleigh medium 2. Equations (7.55) and (7.57) remain the same for medium 3. In medium 2, however, using plane wave approximations at $t=0$, the first-order single-step approximation with $n_2=(n_1+n_3)/2$ leads to

$$E_2=C_{2+}\exp\left(in_2\frac{\omega}{c}x\right)+C_{2-}\exp\left(-in_2\frac{\omega}{c}x\right) \tag{7.72}$$

$$H_2=C_{2+}n_2\exp\left(in_2\frac{\omega}{c}x\right)-C_{2-}n_2\exp\left(in_2\frac{\omega}{c}x\right) \tag{7.73}$$

For higher approximations $n_2^i=(n_2^{i+1}+n_2^i)/2$ are used. i denotes the slab number depending on the number of steps. It is then possible to calculate the reflectivity from the conditions of medium 3 back to medium 1, where the refractive index $n_2(x)$ in medium 2 is given by the Rayleigh case.

It is necessary to note that in using step approximations the refractive index varies, depending on the number of steps used. For each step beginning at distance D, the values of C_{2+} and C_{2-} are calculated These constants are used as a basis for the next step approximation until $x=0$. The analytical result of the first approximation is derived here. Eqs. (7.63) and (7.64) can be written for $n=1$, $t=0$ as

$$E_1=C_{1+}\exp\left(i\frac{\omega}{c}x\right)+C_{1-}\exp\left(-i\frac{\omega}{c}x\right) \tag{7.74}$$

$$H_1=C_{2+}\exp\left(i\frac{\omega}{c}x\right)-C_{1-}\exp\left(-i\frac{\omega}{c}x\right) \tag{7.75}$$

At the junction of medium 1 and medium 2 ($x=0$), $n=1$, and $t=0$

$$C_{1+}+C_{1-}=C_{2+}+C_{2-} \tag{7.76}$$

$$C_{1+}-C_{1-}=n_2C_{2+}-n_2C_{2-} \tag{7.77}$$

so that

$$R=\frac{C_{1+}}{C_{1-}}=\frac{(1+n_2)C_{2+}+(1-n_2)C_{2-}}{(1-n_2)C_{2+}+(1+n_2)C_{2-}} \tag{7.78}$$

The following calculations are performed for a set of cases, where the Rayleigh parameter α is set constant (for a wavelength $\lambda=1.06$ μm corresponding to the wavelength of neodymium glass laser radiation) and the thickness D of the Rayleigh medium is varied. Figure 7.3 shows the results, where the abscissa gives D in micrometers and the ordinate the reflectivity. It is obvious that an oscillation of the reflectivity R is found with zero values at such thicknesses D, where the phases of the incident and reflected waves in the Rayleigh medium are just canceling the reflection, as known from the interference for transmission of light through parallel plates. Evidently, a higher

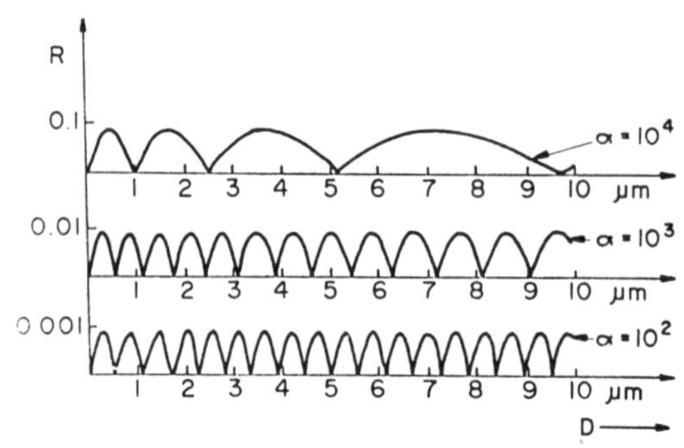

Figure 7.3 Reflectivity R depending on the distance D of the Rayleigh-like plasma interface between homogeneous plasmas described in Fig. 7.2. The parameter α determines the refractive index n from Eq. (7.39). The computed values of analytical solutions for the reflectivity, Eq. (7.7), is shown for the decreasing refractive index [153].

α leads to a higher reflectivity. At $\alpha = 2\omega/c = 1.18 \times 10^5$ cm^{-1}, total reflection at the discontinuity between the profile and the vacuum occurs, although the waves are perpendicularily incident as discussed above.

The maxima of R are of the same height. This is because no absorption is assumed and the reflection is determined only at $x=0$ and $x=D$. As known from the Rayleigh case [151], in agreement with the general result of Osterberg [152], the reflection is only determined by α. Therefore, the maxima are of constant value, although the refractive index of medium 3 is monotonically decreasing with increasing D.

The stepwise approximation gives for a small number of steps corresponding minima, but increasing maxima with medium thickness D, by orders of magnitudes larger than the exact case. This can be understood from the crude approximation of the refractive index n, which shows the insufficiency of the approximation with few steps. This is hardly surprising as large numbers of steps imply the use of small mesh sizes, which increases the accuracy of calculations.

The numerical calculations with large numbers of steps (100 or 1000) Fig. 7.4 converges to the exact case. The same reflectivity as in the exact case, the same constancy of the maxima and the corresponding distances of zero reflectivities are found. On closer observation, however, there is a slight difference: zero reflectivity distances increase for the stepwise approximation. Such a "wavelength" effect is of a basic nature. Any numerical inaccuracies or instabilities were excluded, there is a definite paradox in the difference between the exact solution and the asymptotic value of the stepwise model.

Leaving aside the aforementioned paradox, the Osterberg problem can be discussed in the following manner. It is a mathematical fact that there are two exact linearly independent solutions in the homogeneous medium,

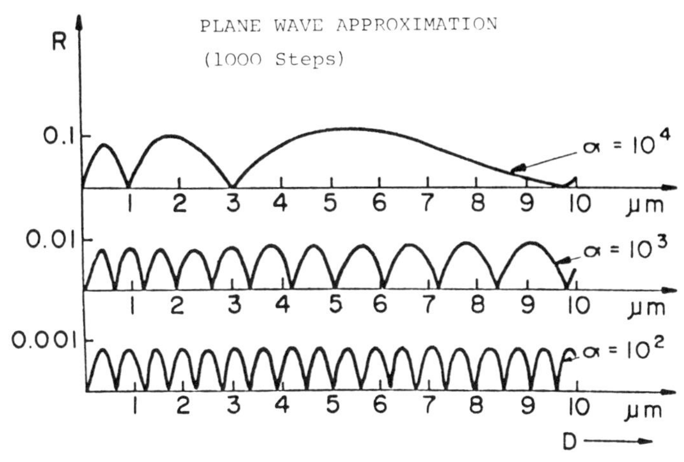

Figure 7.4 The plane wave stepwise approximation using 1000 steps for the corresponding calculations of reflectivity R as in Fig. 7.3. Both figures have varying refractive indices depending on x [153].

whose ratios are the reflectivity determined by the boundaries to the homogeneous media. The condition of only penetrating waves, that is, only transmitted waves in medium 3 (no standing waves), determines the phase between **E** and **H** in media 1 and 2. The phase depends on the thickness D and on α. This is originally reproduced by the stepwise approximation

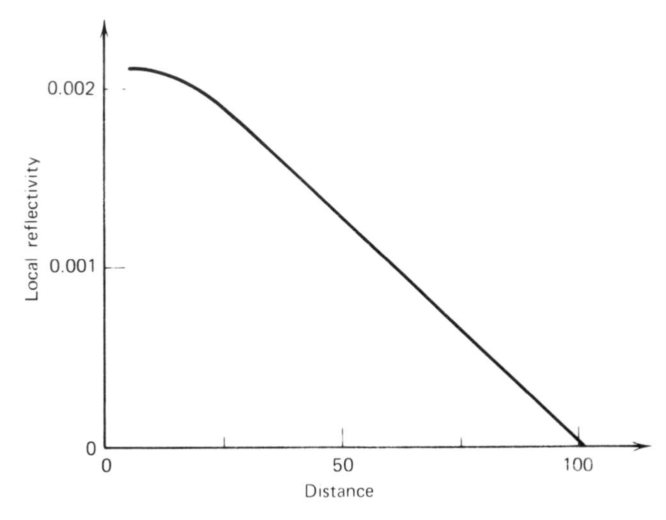

Figure 7.5 Variation of the reflection on the coefficient R of the plane wave approximation with distance in medium 2, representing a local reflection.

(apart from the paradox). However, in the case of approximation, the reflection coefficient of the plane waves within the steps decreases from medium 1 to 3, Fig. 7.5. Therefore, the results of Osterberg "nonreflectivity" and the plane waves "local reflectivity" do not contradict each other.

The problem is how the solutions for the inhomogeneous medium 2 have to be determined. The only condition is similar to that of Sommerfeld's spherical radiation condition (at large distance r, the amplitude has to decrease as $1/r$) expressed here for the plane waves: when $x \to +\infty$, the solution is approximately that of a homogeneous medium with only forward propagating waves and no standing waves, while any approximation of the exact case by fine steps for other x will produce, in effect, internal reflection or standing waves. The exact solution does not exhibit reflection properties and the stepwise approximation is, so to speak, a probe for mathematically detecting local reflectivity.

7.4 The Airy Profiles

One example of an exact solution of the wave equation for an inhomogeneous plasma with a special refractive index profile will be mentioned, which is solved by higher functions. The refractive index n should have a linear spatial dependence

$$n^2 = -ax + is \tag{7.79}$$

where

$$a = \frac{2\Theta\omega}{c}; \qquad s = \frac{\nu}{\omega} \ll 1 \tag{7.80}$$

using Θ from Eq. (7.10). The wave equation for plane waves perpendicularly incident for the E_y component is then [154]

$$\frac{\partial^2 E_y}{\partial x^2} + \frac{\omega^2}{c^2}(-ax + is)E_y = 0 \tag{7.81}$$

Using the substitution

$$\zeta = \left(\frac{\omega}{ca}\right)^{2/3}(-ax + is) = \rho^{2/3}(-ax + is); \qquad \rho = \frac{\omega}{ca} \tag{7.82}$$

Eq. (7.81) reduces to

$$\frac{d^2 E_y}{d\zeta^2} + \zeta E_y = 0 \tag{7.83}$$

The solution of this differential equation is the well-known Airy function a_i, which can be expressed by Bessel functions J and I (see Watson [155]) of the order $\frac{1}{3}$

$$E_y = 3Aa_i(-\zeta) = \begin{cases} A\zeta^{1/2}[J_{1/3}(\frac{2}{3}\xi^{3/2}) + J_{-1/3}(\frac{2}{3}\xi^{3/2})]; & \text{if Re } \xi > 0 \\ A(-\zeta)^{1/2}[I_{-1/3}(\frac{2}{3}(-\xi)^{3/2}) + I_{1/3}(\frac{2}{3}(-\xi)^{3/2})]; & \text{if Re } \xi < 0 \end{cases} \tag{7.84}$$

where

$$A = \frac{2}{3}\pi^{1/2}\left(\frac{\omega}{ca}\right)^{1/6} E_v \exp\left[\frac{i\omega}{c}\int_{-i/a}^{+is/a} n^{1/4}\,dz - \frac{i\pi}{4}\right] \tag{7.85}$$

The solution of the magnetic field is derived from the Maxwellian equations

$$H_z = i\zeta^{-1/3}\frac{dE_y}{d\zeta} \tag{7.86}$$

These solutions for E_y and H_z will be used below when the results derived by Lindl and Kaw [154] are discussed.

EIGHT

Equation of Motion

This section is devoted to the central question of the basic equations for the nonclassical treatment of laser produced plasmas, the equation of motion including its nonlinear properties. The equation of motion contains the nonlinear forces of which one part is the high-frequency (optical) extension of the ponderomotive forces, and of which another part, however, is of non-ponderomotive nature.

One should remember the steps of the preceding discussions: after the introduction to the very complex phenomena of laser-plasma interaction with fast ions, nonlinearities, and other unusual observations (Section 1), basic microscopic plasma parameters as the Debye length and the plasma and collision frequency were derived (Section 2). The properties of the kinetic theory were then discussed showing the derivation of macroscopic equations from the Boltzmann equation, the nondissipative return into equilibrium (Landau damping), but also indicating basic problems with respect to the Liouville equation and irreversibility (Section 3). After introducing and evaluating hydrodynamics (Section 4), the first global description of laser produced plasmas based on the self-similarity model was possible (Section 5) where the necessary fast thermalization for times of several nanoseconds of interaction was confirmed by agreement with experiments. Section 6 derived the basis of the macroscopic plasma theory (two-fluid model of magnetohydrodynamics) including the electrodynamics, on which basis the optical properties as refractive index and absorption (nonlinear and relativistic) were deduced. The wave-optical properties of the plasma with respect to its inhomogeneities, propagation problems, and reflection were described in Section 7.

The following discussion of the equation of motion (conservation of momentum) for plasmas including electrodynamics is essentially related to the combination of mechanical and electrodynamic phenomena, which are basically of nonlinear nature, as for example, electrostriction, magnetostriction, or the Maxwellian stress tensor, to mention the long-known phenomena. In Section 6, the plasma dynamics were based on the two Euler equations, one for ions and the other for electrons. By adding both equations, an equation of motion (first Schlüter equation) for the net plasma was achieved [136]; whereas by subtracting the Euler equations, the second Schlüter equation was found: the generalized Ohm's law, with which the Lorentz type electrodynamics of plasma was formulated and the wave properties in inhomogeneous media were derived.

This section will discuss the equation of motion, the first Schlüter equation (6.5), and its generalization as far as it is needed for the nonlinear processes of laser plasma interactions. To take it in advance, the following general equation of motion reproduces all known results of subrelativistic plasma dynamics at interactions with plane electromagnetic waves. The treatment of laser beams of finite diameter involves the solution of a basic question including the need for exactness in the presumed linear theories.

If plasma physics is based on kinetic equations, see Section 3, Spitzer derived an equation of motion given by a force density **f** [107]

$$\mathbf{f}=m_i n_i \frac{\partial \mathbf{v}}{\partial t}+m_i n_i \mathbf{v}\cdot\nabla\mathbf{v}=-\nabla p+\frac{1}{c}\mathbf{j}\times\mathbf{H} \qquad \text{(Spitzer)} \tag{8.1}$$

Several simplifications were made and terms neglected in the derivation. In Schlüter's derivation [136] of the equation of motion, (6.5), from adding the Euler equations for electrons and ions,

$$\mathbf{f}=-\nabla p+\frac{1}{c}\mathbf{j}\times\mathbf{H}-\frac{1}{4\pi}\frac{\omega_p^2}{\omega^2}\mathbf{E}\cdot\nabla\mathbf{E} \qquad \text{(Schlüter)} \tag{8.2}$$

one nonlinear term appeared, see Eq. (6.6). The most general equation of motion, satisfying all cases of plane optical wave interaction with plasma, is [138]

$$\begin{aligned}\mathbf{f}=&-\nabla p+\frac{1}{c}\mathbf{j}\times\mathbf{H}+\frac{1}{4\pi}\mathbf{E}\nabla\cdot\mathbf{E}\\&-\frac{1}{4\pi}\frac{\omega_p^2}{\omega^2+\nu^2}\left(1+\frac{\nu}{\omega}\right)\mathbf{E}\nabla\cdot\mathbf{E}-\frac{1}{4\pi}\left(1+i\frac{\nu}{\omega}\right)\mathbf{E}\cdot\nabla\mathbf{E}\\&-\frac{1}{4\pi}\mathbf{E}\mathbf{E}\cdot\nabla\frac{{\omega_p}^2}{\omega^2+\nu^2}\left(1+i\frac{\nu}{\omega}\right)\end{aligned} \tag{8.3}$$

where the derivation is based on the exact connection with the Maxwellian stress tensor, by which way the problems of dispersion and dissipation were solved automatically (to be described). It should be mentioned that Dragila

[156] derived an additional term in Eq. (8.3), $(\partial\varepsilon/\partial t)\mathbf{E}\times\mathbf{B}/(4\pi c)$, which, however, only contributes to relativistic plasma dynamics. While we shall include microscopic relativistic effects, for example, relativistic electron quivering (Section 6.5), we shall not consider effects where the net macroscopic plasma velocities or currents are becoming relativistic. We can drop therefore the Dragila term and other terms of relativistic dynamics.

In Eq. (8.3) "nonlinear terms" follow the thermokinetic force

$$\mathbf{f}_{\text{th}} = -\nabla p \tag{8.4}$$

These remaining terms in Eqs. (8.2) and (8.3) can be seen for the case of electromagnetic waves, to introduce quantities **j**, **E**, and **H** which are fast oscillating at least with the frequency ω, or as a result of a product, with higher harmonics. As binary products of these quantities appear, oscillations with 2ω are produced by these "nonlinear terms."

Historically, forces are called ponderomotive forces if they are generated by electric or magnetic fields. It will be shown by a special example that, apart from the purely field generated ponderomotive terms, mixed terms exist. These nonponderomotive terms are characterized by dissipative effects and are proportional to the collision frequency. Due to these mixed terms the terminology "nonlinear force" f_{NL} will be preferred:

$$\mathbf{f}_{\text{NL}} = \mathbf{f} - \mathbf{f}_{\text{th}} \tag{8.5}$$

This terminology was accepted by several authors [156–157].

8.1 Equivalence to Maxwellian Stress Tensor

The derivation of the most general formulation up to the present stage of the equation of motion (8.3) or of the nonlinear force (8.5)

$$\mathbf{f}_{\text{NL}} = \frac{1}{c}\mathbf{j}\times\mathbf{H} + \frac{1}{4\pi}\mathbf{E}\nabla\cdot\mathbf{E} + \frac{1}{4\pi}\nabla\cdot(n^2-1)\mathbf{E}\mathbf{E} \tag{8.6}$$

is connected with a physical criterion: the forces due to obliquely incident (plane wave) laser radiation on a stratified collisionless plasma must produce a net force only toward or against the plasma surface, because no momentum transfer is possible perpendicular to the surface. This criterion of oblique incidence is a tool to check whether the nonlinear force (8.6) is complete, whether it contains all and not too many terms. This was shown in 1969 [138]. In the meantime, many authors could not suppress their excitement in discovering several nonlinear terms, for which only one measure exists: these terms are those that exist only since 1969 (the rederivation nevertheless is appreciable and can lead to new insights) or they are incomplete or overdetermining and therefore wrong.

The direct derivation of Eq. (8.6) from the two-fluid theory is given in Appendix C. With regard to the assumptions involved there, especially on the interpretation of (oscillating) space charges and high frequency dielectric displacement, one should be skeptical about their correctness; but we have the tool for correctness, the criterion of oblique incidence. This can be shown by algebraic rewriting of Eq. (8.6) into an expression using the Maxwellian stress tensor that we discuss in this subsection.

Historically, the steps leading to this formulation were quite complicated and required prior steps. For perpendicular incidence, the Schlüter formulation, Eq. (8.2), was sufficient, arriving at a nonlinear force of laser plasma interaction which indicated the basic mechanism of the gradient of the refractive index [158]. The formula agreed at low density approximation with the force-formula derived at microwave interaction with plasmas [159–161], but in cases of absorption, first discussed 1969 [138], interpretation of the basic formulation (8.6) led to interpretations of nonponderomotive terms of the nonlinear force [this has nothing to do with the thermokinetic force (8.4)] [162–164] and higher order terms [165].

The motivation to use the formulation of the Maxwellian stress tensor [138] came from the expression of Landau and Lifshitz [166] where the problems of variable electromagnetic fields had not yet been solved, however. These authors underlined that the time averaged stress tensor for generating forces in matter in a variable electromagnetic field had not been derived then. Landau and Lifshitz used only nondispersive and nonabsorbing media. Pitaevski found a generalization to dispersive fluids with high frequency electric fields and only static magnetic fields [167]. Our formal use of this very restricted Landau–Lifshitz formula [166] as equation (6) in reference [138] was therefore for the completely different conditions of high frequency electromagnetic waves in dispersive plasma with absorption. This was justified later only by algebraic identity to the general plasma equation (8.3) derived from the two-fluid model and was proved only by the previously mentioned criterion of oblique incidence [138].

Following these outlines, the mathematical transformation of the general equation of motion, Eq. (8.3), including all nonlinear terms, into one, using the Maxwellian stress tensor is given. For this derivation, vector algebraic and vector analytic identities as well as the Maxwell equations are used, where, finally, the Poynting vector **S** will be included.

$$\frac{\partial}{\partial t}\left(\frac{\mathbf{E}\times\mathbf{H}}{4\pi c}\right)=\frac{\partial}{\partial t}\mathbf{S} \tag{8.7}$$

Compared with quantities of the nonlinear force, this is less by the ratio

$2\pi/\omega\tau_R$. τ_R is the rise time of the laser pulse. For neodymium glass lasers $\tau_R \simeq 10^{-12}$ sec and $\omega/2\pi = 2.8 \times 10^{14}$ sec^{-1}. A similar ratio is valid for CO_2 lasers. In the following sections, the Poynting term can be neglected, but the derivation here is without any neglect. Rewriting Eq. (8.5) from Eq. (8.3) as

$$\mathbf{f}_{\mathrm{NL}} = \mathbf{A} + \mathbf{B} \tag{8.8}$$

where

$$\mathbf{A} = \frac{1}{c}\mathbf{j} \times \mathbf{H} \tag{8.9}$$

$$\mathbf{B} = \frac{1}{4\pi}\mathbf{E}\nabla \cdot \mathbf{E} + \frac{1}{4\pi}\nabla \cdot (n^2 - 1)\mathbf{E}\mathbf{E} \tag{8.10}$$

The refractive index n is defined by Eq. (6.27). From Eq. (6.18) $\mathbf{j}$ is eliminated

$$\mathbf{j} = \frac{c}{4\pi}(\nabla \times \mathbf{H}) - \frac{1}{4\pi}\frac{\partial}{\partial t}\mathbf{E} \tag{8.11}$$

and inserted in Eq. (8.9)

$$\begin{aligned} \mathbf{A} &= \frac{1}{4\pi}(\nabla \times \mathbf{H}) \times \mathbf{H} - \frac{1}{4\pi c}\left(\frac{\partial}{\partial t}\mathbf{E}\right) \times \mathbf{H} \\ &= -\frac{1}{4\pi}\mathbf{H} \times (\nabla \times \mathbf{H}) - \frac{1}{4\pi c}\left(\frac{\partial}{\partial t}\mathbf{E}\right) \times \mathbf{H} \end{aligned} \tag{8.12}$$

Using the vector identity

$$\mathbf{H} \times (\nabla \times \mathbf{H}) = \tfrac{1}{2}\nabla \mathbf{H}^2 - \mathbf{H} \cdot \nabla \mathbf{H} \tag{8.13}$$

$\mathbf{A}$ can be written as

$$\mathbf{A} = -\frac{1}{4\pi}\left[\frac{1}{2}\nabla \mathbf{H}^2 - \mathbf{H} \cdot \nabla \mathbf{H}\right] - \frac{1}{4\pi c}\left(\frac{\partial}{\partial t}\mathbf{E}\right) \times \mathbf{H} \tag{8.14}$$

Turning to $\mathbf{B}$, Eq. (8.10)

$$\mathbf{B} = \frac{1}{4\pi}\mathbf{E}\nabla \cdot \mathbf{E} + \frac{1}{4\pi}\nabla \cdot n^2\mathbf{E}\mathbf{E} - \frac{1}{4\pi}\mathbf{E}\nabla \cdot \mathbf{E} - \frac{1}{4\pi}\mathbf{E} \cdot \nabla \mathbf{E} \tag{8.15}$$

can be changed to

$$\mathbf{B} = \frac{1}{4\pi}n^2\mathbf{E} \cdot \nabla \mathbf{E} + \frac{1}{4\pi}\mathbf{E}\nabla \cdot (n^2\mathbf{E}) - \frac{1}{4\pi}\mathbf{E} \cdot \nabla \mathbf{E} \tag{8.16}$$

From Eq. (6.27),

$$\nabla \cdot (n^2\mathbf{E}) = 4\pi\rho_e \tag{8.17}$$

the time average of which is zero in a space charge neutral plasma [136]. We note that the space charge density ρ_e cannot be neglected here, as oscillating space charges of the frequency ω can be generated by the electromagnetic waves. This process is the driving of electrostatic (Langmuir) waves in a plasma and is possible in a plasma with incident electromagnetic waves of frequency ω. These electrostatic waves will be induced even at plasma densities n_e where no reaonance exists with the electrostatic oscillation at the plasma frequency, see Fig. 2.1.

Adding zero by the following additional two last terms and combining the first terms in Eq. (8.16)

$$\mathbf{B}=\frac{1}{4\pi}[\nabla\cdot n^2\mathbf{EE}-\mathbf{E}\cdot\nabla\mathbf{E}+\tfrac{1}{2}\nabla\mathbf{E}^2-\tfrac{1}{2}\nabla\mathbf{E}^2] \tag{8.18}$$

the second and the third term are combined to

$$\frac{1}{4\pi}\mathbf{E}\times(\nabla\times\mathbf{E})=\frac{1}{8\pi}\nabla\mathbf{E}^2-\frac{1}{4\pi}\mathbf{E}\cdot\nabla\mathbf{E}. \tag{8.19}$$

Using the Maxwell equation (6.18) in Eq. (8.19) leads to

$$\mathbf{B}=\frac{1}{4\pi}[\nabla\cdot n^2\mathbf{EE}-\tfrac{1}{2}\nabla\mathbf{E}^2]-\frac{1}{4\pi c}\mathbf{E}\times\frac{\partial}{\partial t}\mathbf{H} \tag{8.20}$$

Writing **B** from Eq. (8.20) and **A** from Eq. (8.14), Eq. (8.8) is

$$\mathbf{f}_{\mathrm{NL}}=\frac{1}{4\pi}\nabla\cdot[\mathbf{EE}+\mathbf{HH}-\tfrac{1}{2}(\mathbf{E}^2+\mathbf{H}^2)\mathbf{1}+(n^2-1)\mathbf{EE}]-\frac{1}{4\pi c}\frac{\partial}{\partial t}\mathbf{E}\times\mathbf{H} \tag{8.21}$$

where use was made of

$$\mathbf{H}\cdot\nabla\mathbf{H}=\nabla\cdot\mathbf{HH}-\mathbf{H}\nabla\cdot\mathbf{H} \tag{8.22}$$

with a negligible last term because of Eq. (6.16). The unity tensor **1** is in Cartesian coordinates defined by

$$\mathbf{1}=\mathbf{i}_x\mathbf{i}_x+\mathbf{i}_y\mathbf{i}_y+\mathbf{i}_z\mathbf{i}_z \tag{8.23}$$

Eq. (8.20) can be written in the form of

$$\mathbf{f}_{\mathrm{NL}}=\nabla\cdot\left[\mathbf{T}+\frac{n^2-1}{4\pi}\mathbf{EE}\right]-\frac{1}{4\pi c}\frac{\partial}{\partial t}\mathbf{E}\times\mathbf{H}. \tag{8.24}$$

T is the Maxwellian stress tensor

$$\mathbf{T}=\frac{1}{4\pi}[\mathbf{EE}+\mathbf{HH}-\tfrac{1}{2}(\mathbf{E}^2+\mathbf{H}^2)\mathbf{1}] \tag{8.25}$$

The components **T** are given by the scalar components of **E** and **H**

$$4\pi\mathbf{T}=\begin{pmatrix}0.5(E_x^2-E_y^2-E_z^2+H_x^2-H_y^2-H_z^2) & E_xE_y+H_xH_y & E_zE_z+H_xH_z\\ E_xE_y+H_xH_y & 0.5(-E_x^2+E_y^2-E_z^2-H_x^2+H_y^2-H_z^2) & E_yE_z+H_yH_z\\ E_xE_z+H_xH_z & E_yE_z+H_yH_z & 0.5(-E_x^2-E_y^2+E_z^2-H_x^2-H_y^2+H_z^2)\end{pmatrix} \tag{8.26}$$

Formula (8.24) is very similar for the formulation of the force density in a dielectric medium without dispersion following Landau and Lifshitz [166]. The only problem is that a plasma is a medium with dispersion, and an appropriate interpretation of the factor of **EE** in the brackets in Eq. (8.24) is necessary [162]. If no absorption is present, the force is purely of a ponderomotive nature, while collisions cause an extension (see the following Section 8.3) to nonponderomotive terms of the nonlinear force.

For the very special case of plane waves of linear polarization (**E** in y-direction), perpendicularly incident on a stratified, inhomogeneous plasma with a depth along the x-axis, is from Eq. (8.24), given by

$$\mathbf{f}_{\mathrm{NL}} = -\frac{\mathbf{i}_x}{8\pi}\frac{\partial}{\partial x}(E_y^2 + H_z^2) \tag{8.27}$$

where the Poynting term has been neglected for the reasons mentioned before Eq. (8.8). This formula will be used as the most general expression for describing nonlinear forces in stratified plasma with perpendicularly incident waves. One curiosity should be pointed out at this stage. When deriving the nonlinear force for perpendicularly incident plane waves from the $(\mathbf{j}\times\mathbf{H})$ expression, Eq. (8.3), without collisions [158], the term with $\partial n/\partial x$ in the WKB solution of **H**, Eq. (7.15), was essentially necessary. When evaluating Eq. (8.27), this term is not necessary to arrive at the same formula of the nonlinear force for the collisionless case. A refractive index n, following the WKB condition, is used, Eq. (7.10),

$$\Theta = \frac{c}{2\omega}\frac{\sqrt{3}}{|n|^2}\frac{\partial |n|}{\partial x} \ll 1 \tag{8.28}$$

The complex electric and magnetic field strengths are then

$$\mathbf{E} = \mathbf{i}_2 \frac{E_v}{|n|^{1/2}} \exp iF_0 \exp[\mp \bar{k}(x)x/2] \tag{8.29}$$

and approximately)

$$\mathbf{H} = \mathbf{i}_3 E_v |n|^{1/2} \exp iF_0 \exp[\mp \bar{k}(x)x/2] \tag{8.30}$$

where

$$F_0 = \omega\left(t \mp \int^x (\mathrm{Re}\, n(\xi)/c)\, d\xi\right) \tag{8.31}$$

Terms with derivatives of n in space are neglected here, for these only give terms of second and higher order in the following time-averaged expressions. By selecting only the real parts, from Eq. (8.29) is found

$$E_y^2 = E_v^2 \left[\left(\frac{1}{\mathrm{Re}|n|^{1/2}}\right)^2 \cos^2 F_0 + \left(\mathrm{Im}\frac{1}{|n|^{1/2}}\right)^2 \sin^2 F_0 \right. \tag{8.32}$$
$$\left. + \tfrac{1}{2}\mathrm{Re}|n|^{-1/2}\, \mathrm{Im}|n|^{-1/2} \sin 2F_0\right] \exp(\bar{k}(x)x/2)$$

This leads to the time-averaged value

$$\overline{E_y^2}=\frac{E_v^2}{2|n|}\exp(\mp\bar{k}(x)x/2) \tag{8.33}$$

In the same way, from Eq. (8.30) is obtained

$$\overline{H_z^2}=\pm\tfrac{1}{2}E_v^2|n|\exp(\mp\bar{k}(x)x/2) \tag{8.34}$$

The time-averaged nonlinear force density (8.27) is then

$$\overline{\mathbf{f}_{\mathrm{NL}}}=-\mathbf{i}_x\frac{E_v^2}{16\pi}\frac{\partial}{\partial x}\left(\frac{1}{|n|}+|n|\right)\exp(\mp\bar{k}(x)x/2) \tag{8.35}$$

or

$$\begin{aligned}\overline{\mathbf{f}_{\mathrm{NL}}}=&\mathbf{i}_x\frac{E_v^2}{16\pi}\frac{1-|n|^2}{|n|^2}\exp(\mp\bar{k}(x)x/2)\frac{\partial|n|}{\partial x}\\ &\pm\mathbf{i}_x\frac{E_v^2}{16\pi}\frac{1+|n|^2}{|n|^2}\frac{2\omega}{c}\mathrm{Im}(n)\exp(\mp\bar{k}(x)x/2)\end{aligned} \tag{8.36}$$

In the collisonless case with $k=0$ and n taken from Eq. (6.36), the second term of Eq. (8.36) vanishes and the following result is obtained

$$\overline{\mathbf{f}_{\mathrm{NL}}}=i_x\frac{E_v^2}{16\pi}\frac{\omega_p^2}{\omega^2 n^2}\frac{\partial n}{\partial x} \tag{8.37}$$

It is important to note that this result, as it was first derived in 1969 [138], produces a collisional term and gives the basis of the approximation for the nonlinear force in the WKB condition by

$$\overline{\mathbf{f}_{\mathrm{NL}}}=-\mathbf{i}_x\frac{E_v^2}{16\pi}\frac{\partial}{\partial x}\left(\frac{1}{|n|}+|n|\right)\exp(\mp kx) \tag{8.38}$$

The exponential factor in Eq. (8.36), with the absorption constant $\bar{k}$, is close to one in the ranges where WKB is valid and where $\omega_p \leqslant \omega$.

8.2 Obliquely Incident Plane Waves

The insufficiency of a nonlinear force in Eq. (8.2) is seen when the force density in a stratified plasma for obliquely incident plane waves is calculated. Using only Eq. (8.2), net forces along the plasma surface appear. This cannot be accepted, because there is no recoil possible for the necessary momentum transfer. Therefore, the nonlinear force from the equation (8.3) is used

$$\mathbf{f}_{\mathrm{NL}}=\frac{1}{c}\mathbf{j}\times\mathbf{H}+\frac{1}{4\pi}\mathbf{E}\nabla\cdot\mathbf{E}+\frac{1}{4\pi}\nabla\cdot(n^2-1)\mathbf{E}\mathbf{E} \tag{8.39}$$

or from the equivalent Eq. (8.24)

$$\mathbf{f}_{\mathrm{NL}}=\nabla\cdot\left[(\mathbf{EE}+\mathbf{HH}-(\mathbf{E}^2+\mathbf{H}^2)\mathbf{1}/2)4\pi+\frac{n^2-1}{4\pi}\mathbf{EE}\right]-\frac{1}{4\pi c}\frac{\partial}{\partial t}\mathbf{E}\times\mathbf{H} \quad (8.40)$$

Both are reconfirmed by the following procedure, valid for plane electromagnetic waves.

The calculation of the time-averaged net force for obliquely incident plane waves will be done for a collisionless plasma $\nu=0$, for a WKB approximation, see Section 7. With the same geometry of the p-polarized plane waves, Eqs. (7.37) to (7.38), this results in time-averaged values of

$$\frac{\partial}{\partial y}\mathbf{EE}=\frac{\partial}{\partial z}\mathbf{EE}=\frac{\partial}{\partial y}\mathbf{HH}=\frac{\partial}{\partial z}\mathbf{HH}=0. \quad (8.41)$$

The time-averaged nonlinear force density (equation of motion) from Eq. (8.40) for a collisionless plasma ($\nu=0$) is then

$$\begin{aligned}\overline{\mathbf{f}_{\mathrm{NL}}}=&\frac{1}{8\pi}\mathbf{i}_x\frac{\partial}{\partial x}[(2n^2-1)(\overline{E_{px}^2}-\overline{E_{py}^2})\cos\beta-\overline{E_{sx}^2}\sin^2\beta\\&+(\overline{H_{sx}^2}-\overline{H_{sy}^2})\sin^2\beta+\overline{H_{px}^2}\cos^2\beta]\\&+\frac{1}{4\pi}\mathbf{i}_y\frac{\partial}{\partial x}[n^2\overline{E_{px}E_{py}}\cos^2\beta+\overline{H_{sx}H_{sy}}\sin^2\beta]\\&+\frac{1}{8\pi}\mathbf{i}_z\frac{\partial}{\partial x}[n^2\overline{E_{px}E_{sx}}+\overline{H_{px}H_{sx}}]\sin 2\beta\end{aligned} \quad (8.42)$$

where β is the angle between the electrical vector $\mathbf{E}$ and the plane of incidence. The $\mathbf{i}_z$ component shows a coupling between parallel and perpendicular polarization. To evaluate this component, one finds from Eqs. (7.28), (7.30), (7.37), and (7.38)

$$n^2E_{px}E_{sx}=-\frac{nE_v^2\cos\alpha_0}{\cos\alpha(x)}\sin\alpha\sin^2G \quad (8.43a)$$

$$H_{sx}H_{px}=\frac{nE_v^2\cos\alpha_0}{\cos\alpha(x)}\sin\alpha\cos^2G \quad (8.43b)$$

As a result, the time-averaged bracket in the $\mathbf{i}_z$ component of Eq. (8.42) vanishes identically.

Because of the last result, one can construct an expression for the case of a general β by a simple addition of the two expressions valid for each polarization. For $E_s(\beta=\pi/2)$ the expression of the time-averaged perpendicular component $\overline{\mathbf{f}}_s$ of the force density is

$$\overline{\mathbf{f}}_s=\frac{1}{8\pi}\mathbf{i}_x\frac{\partial}{\partial x}(-E_{sz}^2+H_{sx}^2-H_{sy}^2)+\frac{1}{4\pi}\mathbf{i}_y\frac{\partial}{\partial x}(H_{sx}H_{sy}) \quad (8.44)$$

From (7.30) is found

$$H_{sx}H_{sy} = -\mathbf{n}E_v \cos\alpha_0 \sin\alpha \cos^2 G$$
$$-\frac{c}{+k\omega}\frac{E_v^2 \cos\alpha_0 \sin\alpha}{|n|^{1/2}\cos^3\alpha}\frac{dn}{dx}\sin 2G \tag{8.45}$$

From Eq. (7.29) the time averaged value of Eq. (8.45) is

$$\overline{H_{sx}H_{sy}} = -\tfrac{1}{2}E_v^2 \cos\alpha_0 \sin\alpha_0 = \text{const} \tag{8.46}$$

and as a result, the $\mathbf{i}_y$ component of $\mathbf{f}_s$ vanishes in (8.44). From (7.28) and (7.30) one finds

$$\overline{-E_{sz}+H_{sx}-H_{sy}} = -E_v^2 \cos\alpha_0 \frac{\cos^2\alpha_0 + n^2\cos\alpha}{2n\cos\alpha} + \frac{\tilde{A}}{2} \tag{8.47}$$

No second-order term was found, but a third-order term was given by

$$\tilde{A} = \frac{c^2}{4\omega^2}\frac{1}{n^3\cos^5\alpha}\left(\frac{dn}{dx}\right)^2 \tag{8.48}$$

The force density from (8.45) is then

$$\mathbf{f}_s = \mathbf{i}_x \frac{E_v^2\cos\alpha_0}{16\pi}\left(\frac{1}{\cos^3\alpha}\frac{\omega_p^2}{\omega^2}\frac{1}{n^2}\frac{dn}{dx} - \frac{d}{dx}\tilde{A}\right) \tag{8.49}$$

In the case of $E_p(\beta=0)$ one finds in (8.42)

$$\mathbf{f}_p = \mathbf{i}_x \frac{1}{8\pi}\frac{\partial}{\partial x}\left[\overline{E_{px}^2\left(1-2\frac{\omega_p^2}{\omega^2}\right) - E_{py}^2 - H_{pz}^2}\right] + \mathbf{i}_y \frac{1}{4\pi}\frac{\partial}{\partial x}\overline{n^2 E_{px}E_{py}} \tag{8.50}$$

The last term vanishes because of the constance in space of

$$\overline{n^2 E_{px}E_{py}} = \frac{E_v^2}{2}\cos\alpha_0 \sin\alpha_0 \tag{8.51}$$

Only the general formulas (8.39) or (8.40) give the correct result of a vanishing force in the plane of the plasma surface. Thus, the importance of the nonlinear terms of Eq. (8.39) is demonstrated. Using Eq. (8.48) and n with $\nu=0$ leads to

$$\overline{(2n^2-1)E_{px}^2 - E_{py}^2 - H_{pz}^2} = -E_v^2\cos\alpha_0 + \frac{\cos^2\alpha_0 + n^2\cos^2\alpha}{2n\cos\alpha}$$
$$+\frac{\tilde{A}}{2}\cos^2\alpha(1-2\sin 2\alpha)^2 \tag{8.52}$$

and finally from Eqs. (8.50) and (8.57)

$$\bar{\mathbf{f}}_p=\mathbf{i}_x \frac{E_v^2 \cos\alpha}{16\pi}\left[\frac{1}{\cos^3\alpha}\frac{\omega_p^2}{\omega^2}\frac{1}{n^2}\frac{dn}{dx}-\frac{d\tilde{A}}{dx}\cos^2\alpha(1-2\sin^2\alpha)^2 + 2\tilde{A}\frac{dn}{dx}\frac{\sin^3\alpha}{\sin\alpha_0}(2\sin^2\alpha-1)(3\sin 2\alpha-2\cos^4\alpha-1)\right] \tag{8.53}$$

A comparison with the collisionless case of perpendicular incidence, Eq. (8.37), shows in the first order from Eqs. (8.49) and (8.53) that

$$\bar{\mathbf{f}}(\alpha)=\overline{\mathbf{f}(0)}\frac{\cos\alpha_0}{\cos^3\alpha} \tag{8.54}$$

The force density of this analysis is only in the negative direction, that is, toward decreasing electron density, up to the third-order in the spatial variation of the refractive index in all polarizations and is independent of the propagation direction of the light. The magnitude of the force density, however, is weakly dependent on the polarization of the light in the third-order terms. The consideration of these third-order terms is also justified in the WKB approximation, because it can be shown that deviations of the WKB approximation from the exact solution are exponentially small [168].

8.3 Nonponderomotive Collisional Term of the Nonlinear Force

Perpendicular incidence of linearly polarized electromagnetic waves on an inhomogeneous plasma along the x-direction is assumed within this subsection.

$$\mathbf{E}=\mathbf{i}_y E_y; \qquad \mathbf{H}=\mathbf{i}_z H_z \tag{8.55}$$

The generated force density in the irradiated inhomogeneous plasma can be calculated from Eq. (8.1). Because all other nonlinear terms in (8.3) vanish ($\mathbf{E}$ is perpendicular to ∇n_e and $\nabla\cdot\mathbf{E}=0$) the following equation is obtained:

$$\mathbf{f}_{\mathrm{NL}}=\frac{1}{c}\mathbf{j}\times\mathbf{H} \tag{8.56}$$

If the complex refractive index n of the plasma is given by Eq. (6.27) and if the inhomogeneous plasma of a certain temperature T has an electron density profile fitting the WKB condition Eq. (7.10), the (complex) electric field is

$$E_y'=\frac{E}{\sqrt{n}}\exp iF \tag{8.57}$$

where

$$F=\frac{\omega}{c}\int^{x} n(\xi)\, d\xi+\omega t \tag{8.58}$$

$\bar{k}$ is given by Eq. (7.13). The (complex) magnetic field is

$$H_z'=E_v|n|^{1/2}\exp(-\bar{k}x/2)\exp iF-\frac{icE_v\exp(\bar{k}x/2)\exp iF}{2\omega n^{3/2}}\frac{dn}{dx} \tag{8.59}$$

The temporal integration constants are neglected since f_{NL} will be time averaged later.

For calculating the time-averaged force, real quantities are required:

$$\mathbf{H}=\mathrm{Re}(\mathbf{H}')=\tfrac{1}{2}(\mathbf{H}'+\mathbf{H}'^*) \tag{8.60}$$

Thus

$$\begin{aligned}\mathbf{H}=\mathbf{i}_y\Bigg\{&E_v\exp(-\bar{k}x/2)[\mathrm{Re}(n)^{1/2}\cos F-\mathrm{Im}(n)^{1/2}\sin F]\\ &-\frac{cE_v\exp(-\bar{k}x/2)}{2\omega}\left[Im\left(\frac{1}{|n|^{3/2}}\frac{dn}{dx}\right)\cos F+\mathrm{Re}\left(\frac{1}{|n|^{3/2}}\frac{dn}{dx}\right)\sin F\right]\Bigg\}\end{aligned} \tag{8.61}$$

The crucial point for deriving the collisional term is to use the complete diffusion equation with respect to collisions has to be used to find $\mathbf{j}$ for the plasma, still with neglect of the Hall and Lorentz terms

$$\frac{\partial \mathbf{j}'}{\partial t}+\nu\mathbf{j}'=\frac{\omega_p^2}{4\pi}\mathbf{E}' \tag{8.62}$$

The importance of ν was suggested [127]. Both ν and ω_p are assumed to be constant in time. In the numerical evaluations the nonlinear generalization [146] of Spitzer's collision frequency, Eq. (6.59) is used. This shows that heating of plasma by electromagnetic waves and other dynamic effects have to be neglected. Using the WKB solution for $\mathbf{E}$, Eq. (8.57) in Eq. (8.62), the following equation is found

$$\mathbf{j}'=\mathbf{i}_y\left[\frac{iG'F_0}{|n|^{1/2}(\nu-i\omega)}\right] \tag{8.63}$$

where

$$G'=\frac{\omega_p^2E_v^2}{4\pi} \tag{8.64}$$

is real and Eq. (8.58) has to be used. Again the temporal integration constant has been neglected for the same reason as in earlier papers.

The real part of $\mathbf{j}$ [remembering that $\exp(iF_0)=\exp(-\bar{k}x/2)\exp iF$],

(8.63) can be written as

$$\mathbf{j}=\mathbf{i}_y\left\{\frac{G'\exp(-\bar{k}x/2)}{|n|(\nu^2+\omega^2)}[(\mathrm{Re}|n|\nu+\mathrm{Im}|n|\omega)\cos F\right.$$
$$\left.-(\mathrm{Re}|n|^{1/2}\omega-\mathrm{Im}|n|^{1/2}\nu)\sin F]\right\} \tag{8.65}$$

Now, the time-averaged general nonlinear force f_{NL} in Eq. (8.56), is calculated for the considered plane waves. Eq. (8.56) and the expressions in Eqs. (8.61) and (8.65) for $\mathbf{j}$ and $\mathbf{H}$ are used. The only time-dependent terms in these equations are cos F and sin F terms since noting $\langle\cos^2 F\rangle=\langle\sin^2 F\rangle=\frac{1}{2}$ and $\langle\cos F\sin F\rangle=\langle\cos F\rangle=\langle\sin F\rangle=0$ the following equation is obtained for the nonlinear force

$$\overline{\mathbf{f}_{\mathrm{NL}}}=\mathbf{i}_x\frac{\omega_p^2E_v^2\exp(-\bar{k}x)}{8\pi c(\nu^2+\omega^2)|n|}[(\mathrm{Re}^2|n|-\mathrm{Im}^2|n|^{1/2})\nu+2\omega\,\mathrm{Im}|n|\mathrm{Re}|n|^{1/2}]$$
$$-\mathbf{i}_x\frac{\omega_p^2E_v^2\exp(-\bar{k}x)}{16\pi\omega|n|(\omega^2+\nu^2)}\left[\mathrm{Im}\left(\frac{1}{|n|^{3/2}}\frac{dn}{dx}\right)\left(\nu\,\mathrm{Re}|n|^{1/2}+\omega\,\mathrm{Im}|n|^{1/2}\right)\right. \tag{8.66}$$
$$\left.-\mathrm{Re}\left(\frac{1}{|n|^{3/2}}\right)\left(\omega\,\mathrm{Re}|n|^{1/2}-\nu\,\mathrm{Im}|n|^{1/2}\right)\right]$$

This equation holds under the following assumptions are made. A plane electromagnetic wave travelling in the x-direction is incident perpendicularly on a plasma with an inhomogeneity in the x-direction only. The WKB condition is satisfied, and ν and ω_p are approximately constant over the time of interaction. Eq. (8.66) consists basically of two terms each with a square bracket. To arrive at a more transparent discussion, one reduces Eq. (8.66) for $\bar{k}\approx 0$ and

$$\overline{\mathbf{f}}_{\mathrm{NL}}=\mathbf{i}_x\left(\frac{\omega_p^2E_v^2}{8\pi c\omega^2}\nu+\frac{\omega_p^2E_v^2}{16\pi\omega^2n^2}\frac{dn}{dx}\right) \tag{8.67}$$

where now $n^2=1-(\omega_p^2/\omega^2)$. The second term of Eq. (8.67) is recognized as the usual ponderomotive force [158],

$$\overline{\mathbf{f}}_{\mathrm{NL}}=-i_x\frac{E_v^2\omega_p^2}{16\pi\omega^2}\frac{d}{dx}\frac{1}{|n|} \tag{8.68}$$

which, for $\nu=0$ and for the WKB formulation of the electric field $\mathbf{E}$, Eqs. (8.55) and (8.57), is

$$\overline{\mathbf{f}}_{\mathrm{NL}}=-\frac{\omega_p^2}{8\pi\omega^2}\overline{\frac{\partial}{\partial x}\mathbf{E}^2} \tag{8.69}$$

The factor $\frac{1}{2}$ is generated by temporal avaraging of $\mathbf{E}$. This formula (8.69) is formally identical with that derived by Boot [159], Gapunov and Miller

[160], and by Weibel [161] for the forces generated by microwaves in plasmas, if the motion of low-density collisionless plasma towards the nodes of a standing electromagnetic wave is considered. The difference to the derivation in the case of lasers was the force in high-density plasma, when a propagating wave is incident. In this case the spatial variation of the dielectric properties, given by $(d/dx)(1/n)$ in Eq. (8.68) [158], is the essential driving mechanism. The general case of high-plasma densities includes the low-density case of microwaves discussed earlier. A further generalization was to include collisions [162] and to generalize the equation of motion for oblique incidence, as was discussed in Section 8.2.

The description of the collision property by a separate term, namely the first in Eq. (8.67) (in contrast to the use of a correction factor only [162]) was given by Stamper [163], and, in the more general form of Eq. (8.66) by Miller et al. [164]. For relating the first term of Eq. (8.67) to Stamper's term his expression is discussed.

$$\mathbf{f}_{NL} = \frac{\mathrm{Re}(\varepsilon) - 1}{8\pi} \nabla \overline{\mathbf{E}^2} + \left(\frac{\omega_p}{\omega}\right)^2 \frac{\nu I}{c^2} \frac{\mathbf{k}}{|\mathbf{k}|} \tag{8.70}$$

The first term is the usual nonlinear (ponderomotive) force, and the second term is referred by Stamper to the collisional force. This can be seen from the energy flux density

$$I = \frac{c}{4\pi} |\mathbf{E} \times \mathbf{H}|$$

Absorption in the WKB solutions for $\mathbf{E}$ and $\mathbf{H}$ is neglected.

$$\mathbf{E} = \mathbf{i}_y \frac{E_v}{|n|^{1/2}} \cos F_N$$

$$\mathbf{H} = \mathbf{i}_x \left\{ E_v |n|^{1/2} \cos F_0 - \frac{c}{2\omega} \frac{E_v}{|n|^{3/2}} \frac{d|n|}{dx} \sin F_0 \right\}$$

Inserting these last equations in the second term for the collisional force $\mathbf{f}_c$ in Eq. (8.70) leads to the nonponderomotive part of the nonlinear force

$$\mathbf{f}_c = \mathbf{i}_x \frac{\omega_p^2 E_v^2 \nu}{8\pi c \omega^2} \tag{8.71}$$

This is the first term of Eq. (8.67). The first term of Eq. (8.66) can be considered as a generalization of the collisional force.

The importance of Stamper's term in the special formulation of Eq. (8.67) can be seen in the following examples of a parabolic electron density

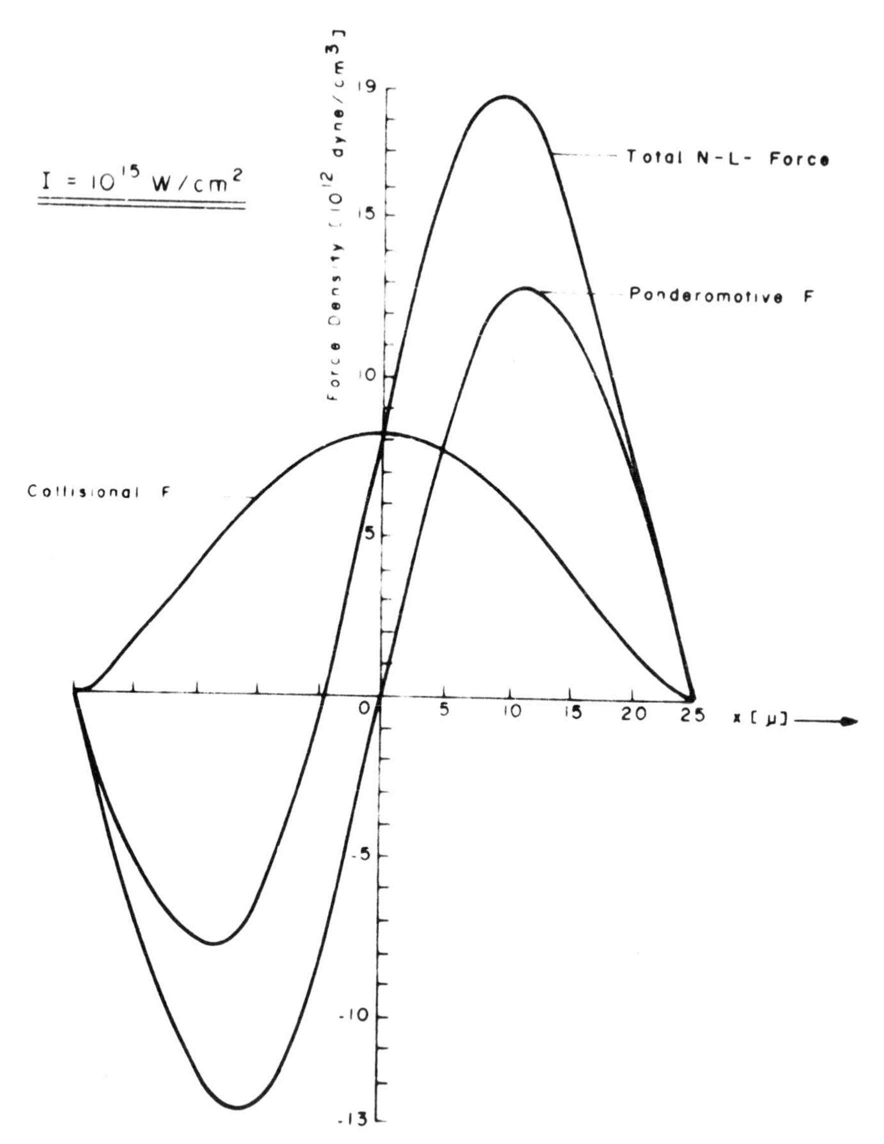

Figure 8.1 Total nonlinear force. collisional (Stamper) term and usual ponderomotive force for a 50 μm thick parabolic density profile [Eq. (22)] with a maximum of $n_e = n_{ec}/2$ for hydrogen plasma of 100 eV temperature at plane wave perpendicular neodymium glass laser irradiation of 10^{15} W/cm^2 intensity [164].

profile of a plasma:

$$n_e(x) = \begin{cases} n_0\left(1 - \dfrac{x^2}{b^2}\right) & \text{if } |x| < b \\ 0, & \text{elsewhere} \end{cases} \tag{8.72}$$

n_0 is chosen to be $n_0 = n_{ec}/2$ to see a strong effect. For higher densities, the dielectric swelling $(1/|n|)$ will be more dominant over the Stamper term, and the restriction to the WKB condition may be lost. b is chosen to be $b = 25\ \mu$m, to allow for some reasonable conditions for irradiation by neodymium glass laser radiation ($\lambda = 1.06\ \mu$m; $n_{ec} = 10^{21}$ cm). The plasma is a Hydrogen plasma ($Z = 1$) at a temperature of 100 eV. Two vacuum intensities of 10^{15} W/cm^2 and 10^{16} W/cm^2 are used. The light (plane waves) is incident from the left in the Figs. 8.1, 8.2 and 8.3.

The profile of Eq. (8.72) satisfies the WKB condition, Eq. (7.10), resulting in

$$\Theta \leqslant 1.9 \times 10^{-3} \tag{8.73}$$

In Figs. 8.1 and 8.2 the ponderomotive force and collisional force of Eq. (8.18)

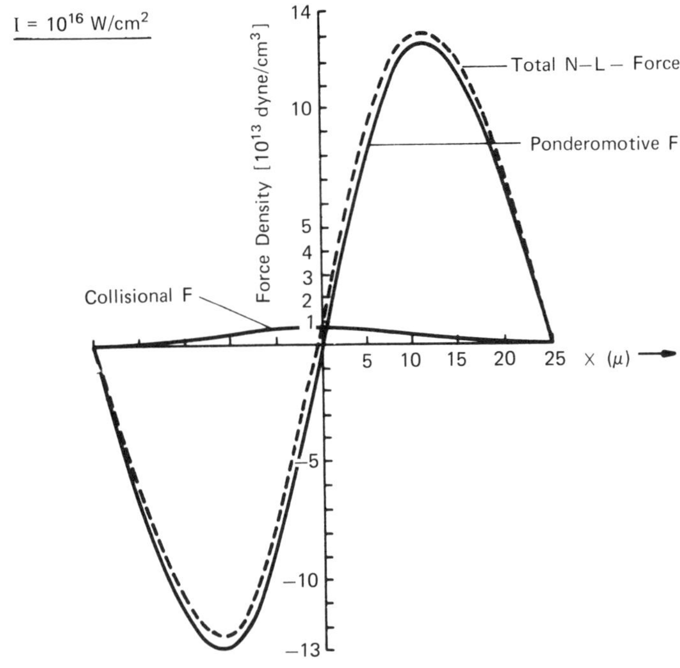

Figure 8.2 Same as Fig. 8.1 for 10^{16} W/cm^2.

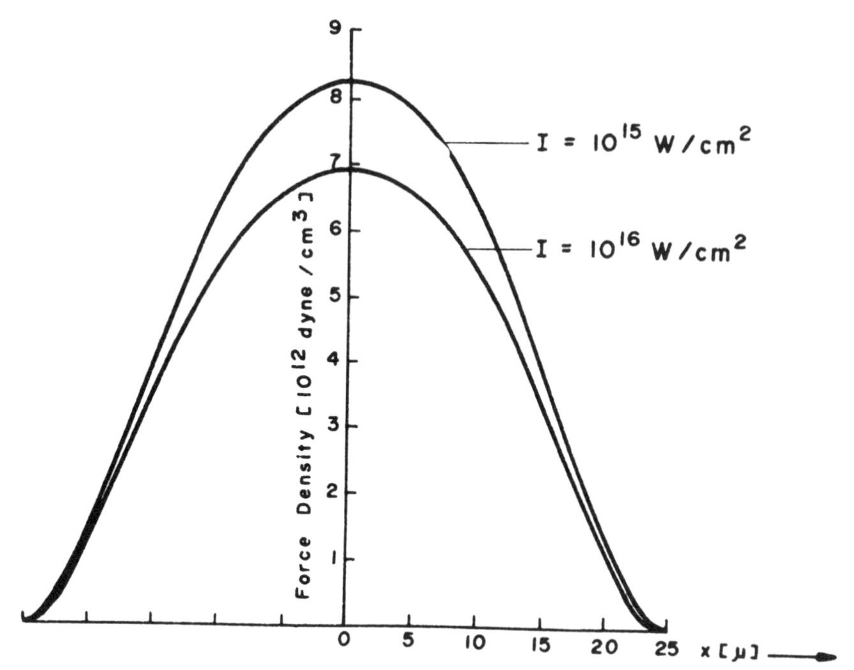

Figure 8.3 The collisional force for the case of Figs. 8.1 and 8.2 decreases for higher intensity due to the nonlinear intensity of the collision frequency.

(second and first term, respectively) are plotted for intensities of 10^{15} and 10^{16} W/cm^2, which are added to obtain the overall shape of the nonlinear force $f_{\rm NL}$. Fig. 8.3 shows the Stamper's collisional force for both intensities. At an intensity of 10^{15} W/cm^2, which is very close to the threshold for the predominance of the nonlinear force, the collisional force is relatively large. At higher intensities, the nonlinear force is much less influenced by the collisional force. Their intensity dependence, Fig. 8.3, corresponds to the fully nonlinear description of the intensity dependence of the collision frequency.

The collisional part diminishes with increasing temperature and intensity, when the fully nonlinear dependence of the collision frequency (6.59) is used. It is less than about 5% of the total nonlinear force for $T=100$ eV at neodymium glass laser intensities of 10^{16} W/cm^2 for a maximum density of one-half of the cutoff density.

8.4 Additional Third-Order Terms for Perpendicular Incidence

It is remarkable that higher order terms appear if the generally valid formula (8.27) for perpendicularly incident plane waves is evaluated for the WKB approximation. Using E_y, Eq. (8.57), and the first terms of Eq. (8.61) for H_z (neglecting then the phase between E_y and H_z), the nonlinear force without collisions (8.68) and the nonponderomotive Stamper term [first in bracket of (8.67)] had been derived in 1969 [138], while the phase was essential in Section 8.3 for the derivation from the $\mathbf{j}\times\mathbf{H}/c$ expression. If the phase is used in the formula (8.27) we should find additional third-order terms as will be shown now.

Using the same expression for the WKB approximation (8.57) and (8.61) as before, from Eq. (8.27), we arrive at the nonlinear force [169, 170]

$$\operatorname{Re}\overline{\mathbf{f}_{\mathrm{NL}}} = -i\frac{1}{4\pi}\left\{\overline{\operatorname{Re}(E_y)\operatorname{Re}\left(\frac{\partial E_y}{\partial x}\right)} - \overline{\operatorname{Re}(H_z)\operatorname{Re}\left(\frac{\partial H_z}{\partial x}\right)}\right\} \tag{8.74}$$

From Eqs. (8.57) and (8.61) we find

$$\operatorname{Re}(E_y) = \frac{E_v}{|n|^{1/2}}\exp(-\bar{k}\omega x/2)\left[\operatorname{Re}\frac{1}{n^{1/2}}\cos F - \operatorname{Im}\left(\frac{1}{n^{1/2}}\right)\sin F\right] \tag{8.75}$$

and

$$\operatorname{Re}\left(\frac{\partial E_y}{\partial x}\right) = -\frac{E_v}{2}\exp(-k(x)x/2)\left[\operatorname{Re}\left(\frac{1}{n^{3/2}}\frac{dn}{dx}\right)\cos F - \operatorname{Im}\left(\frac{1}{n^{3/2}}\frac{dn}{dx}\right)\cos F\right]$$
$$-\frac{E_v\omega}{c}\exp(-\bar{k}(x)x/2)\left[\operatorname{Re}(n^{1/2})\sin F + \operatorname{Im}(n^{1/2})\cos F\right], \tag{8.76}$$

also

$$\operatorname{Re}(H_z) = E_v\exp(-\bar{k}(x)x/2)[\operatorname{Re}(n^{3/2}\cos F - \operatorname{Im}(n^{3/2})\sin F]$$
$$-\frac{E_v\omega}{c}\exp(-\bar{k}(x)x/2)\left[\operatorname{Im}\left(\frac{1}{n^{3/2}}\frac{dn}{dx}\right)\cos F + \operatorname{Re}\left(\frac{1}{n^{3/2}}\frac{dn}{dx}\right)\sin F\right] \tag{8.77}$$

and

$$\operatorname{Re}\left(\frac{\partial H_z}{\partial x}\right) = -\frac{E_v\omega}{c}\exp(-\bar{k}(x)x/2)[\operatorname{Re}(n^{2/3})\sin F + \operatorname{Im}(n^{3/2})\cos F]$$
$$-\frac{E_v c}{2\omega}\exp(-\bar{k}(x)x/2)\left\{\left[\operatorname{Re}\frac{1}{n^{3/2}}\frac{d^2n}{dx^2} - \frac{3}{2}\frac{1}{n^{3/2}}\left(\frac{dn}{dx}\right)^2\right]\sin F\right.$$
$$\left.+\operatorname{Im}\left[\frac{1}{n^{3/2}}\frac{d^2n}{dx^2} - \frac{3}{2}\frac{1}{n^{5/2}}\left(\frac{dn}{dx}\right)^2\right]\cos F\right\} \tag{8.78}$$

Hence, Eq. (8.74) becomes

$$\operatorname{Re}\{\bar{f}_{\mathrm{NL}}\} = \frac{E_v^2}{16\pi}\exp(-\bar{k}(x)x)\left[\operatorname{Re}\left(\frac{1}{n^{1/2}}\right)\operatorname{Re}\left(\frac{1}{n^{3/2}}\frac{dn}{dx}\right)\right.$$

$$\left.+\operatorname{Im}\left(\frac{1}{n^{1/2}}\right)\operatorname{Im}\left(\frac{1}{n^{3/2}}\frac{dn}{dx}\right)\right]$$

$$-\frac{E_v^2}{16\pi}\exp(-\bar{k}(x)x)\left[\operatorname{Re}\left(\frac{1}{n^{3/2}}\frac{dn}{dx}\right)\operatorname{Re}(n^{3/2})+\operatorname{Im}\left(\frac{1}{n^{3/2}}\frac{dn}{dx}\right)\operatorname{Im}(n^{3/2})\right]$$

$$-\frac{E_v^2\omega}{8\pi c}\exp(-\bar{k}(x)x)\left[\operatorname{Im}\left(\frac{1}{n^{1/2}}\right)\operatorname{Re}(n^{1/2})-\operatorname{Im}(n^{1/2})\operatorname{Re}\left(\frac{1}{n^{1/2}}\right)\right]$$

$$+\frac{E_v^2\omega}{8\pi c}\exp(-k(x)x)[\operatorname{Re}(n^{1/2})\operatorname{Im}(n^{3/2})-\operatorname{Im}(n^{1/2})\operatorname{Re}(n^{3/2})]$$

$$-\frac{E_v^2 c}{16\pi\omega}\exp(-\bar{k}(x)x)\left\{\operatorname{Re}\left[\frac{1}{n^{3/2}}\left(\frac{d^2n}{dx^2}\right)-\frac{3}{2n^{5/2}}\left(\frac{dn}{dx}\right)^2\right]\operatorname{Im}(n^{1/2})\right.$$

$$\left.-\operatorname{Im}\left[\frac{1}{n^{3/2}}\left(\frac{d^2n}{dx^2}\right)-\frac{3}{2n^{5/2}}\left(\frac{dn}{dx}\right)^2\right]\operatorname{Re}(n^{1/2})\right\}$$

$$-\frac{E_v^2 c}{32\pi c}\exp(-\bar{k}(x)x)\left\{\operatorname{Re}\left[\frac{3}{n^{3/2}}\left(\frac{d^2n}{dx^2}\right)-\frac{3}{2n^{5/2}}\left(\frac{dn}{dx}\right)^2\right]\operatorname{Re}\left(\frac{1}{n^{3/2}}\frac{dn}{dx}\right)\right.$$

$$\left.+\operatorname{Im}\left[\frac{1}{n^{3/2}}\left(\frac{d^2n}{dx^2}\right)-\frac{3}{2n^{5/2}}\left(\frac{dn}{dx}\right)^2\right]\operatorname{Im}\left(\frac{1}{n^{3/2}}\frac{dn}{dx}\right)\right\} \tag{8.79}$$

This is the hitherto most extensive evaluation of the general nonlinear force [Eq. (8.3)] or (8.24) for plane waves perpendicularly incident on a stratified plasma for the WKB approximation. If we consider the case in which collisions are neglected, then $\nu=0$ and the imaginary terms vanish, and the above expression then yields

$$\bar{\mathbf{f}}_{\mathrm{NL}} = \frac{E_v^2\omega_p^2}{16\pi\omega^2 n^2}\frac{dn}{dx} - \frac{E_v^2 c^2}{32\pi\omega^2}\left[\frac{1}{n^3}\left(\frac{d^2n}{dx^2}\right)\left(\frac{dn}{dx}\right)\right] + \frac{3E_v^2c^2}{64\pi\omega^2}\frac{1}{n^4}\left(\frac{dn}{dx}\right)^3 \tag{8.80}$$

The first term in Eq. (8.80) is the one obtained before [138], Eq. (8.68) neglecting the phase term between E_y and H_z. It would be reasonable to assume that the higher order terms in Eq. (8.80) are due to the phase terms. This is not as can be seen by spatially differentiating our expression (8.57) for the magnetic field strength **H**. When the phase term is differentiated, one of its terms cancel with one of the derivatives of the first term of Eq. (8.57). Hence the nonlinear force arises from the phase term alone. This is in agreement with the model of quivering motion [54]. As described before [138], the first term of Eq. (8.80) indicates a deconfining collisionless acceleration, because the direction of the force density is toward decreasing plasma densities, which is clearly independent of the polarization of the incident laser light from symmetry considerations. The first of the higher order terms is a confining one, while the other is a deconfining force term. These third-order terms may contribute to the momentum transferred to the homogeneous interior [138]. With collisions, the other terms can be interpreted as collision produced radiation pressure of the light, within the inhomogeneous plasma an analogy with the usual radiation pressure of homogeneous media, where, however, the very complex influence of the refractive index is included now.

8.5 Summary

The most general equation of motion of a plasma consists of a thermokinetic term, Eq. (8.5), and the following nonlinear terms derived 1969 [138]

$$f_{NL} = \frac{1}{c}\,\mathbf{j}\times\mathbf{H} + \frac{1}{4\pi}\,\mathbf{E}\nabla\cdot\mathbf{E} + \frac{1}{4\pi}\,\nabla\cdot(n^2-1)\mathbf{E}\mathbf{E} \tag{8.81}$$

which is identical with the formulation using the Maxwellian stress tensor $\mathbf{T}$, Eq. (8.25)

$$f_{NL} = \nabla\cdot\left(\mathbf{E}\mathbf{E} + \mathbf{H}\mathbf{H} - \tfrac{1}{2}(\mathbf{E}^2 + \mathbf{H}^2)\mathbf{1} + (n^2-1)\mathbf{E}\mathbf{E}\right)/4\pi - \frac{1}{4\pi c}\frac{\partial}{\partial t}\,\mathbf{E}\times\mathbf{H} \tag{8.82}$$

This formulation is formally identical with the nondispersive Landau–Lifshitz expression [166] if a special formulation of the density is used, where our derivation of the algebraic identity of Eq. (8.82) with Eq. (8.81) proves the general validity for dispersive media of a plasma with dissipation. Only expressions (8.81) or (8.82) result in nonlinear forces for oblique incidence of plane wave on stratified collisionless plasma which have no wrong time-averaged component in the plasma surface.

For perpendicular incidence (x-direction) of plane waves, the forces in the plasma are especially (from 8.81)

$$f_{NL} = \frac{1}{c}\,\mathbf{j}\times\mathbf{H} \tag{8.83}$$

or (from 8.82)

$$f_{NL} = \frac{\partial}{\partial x}\,\frac{\mathbf{E}^2 + \mathbf{H}^2}{8\pi} \tag{8.84}$$

These equations are valid for any general density profiles (differences to the WKB approximation have been discussed by Lindl and Kaw [154]).

For the simplification of the WKB approximation, the special result is for a plasma with collisions

$$f_{\mathrm{NL}} = \mathbf{i}_x \frac{E_v^2}{16\pi} \frac{\omega_p^2}{\omega^2} \frac{1}{|n|^2} \frac{d|n|}{dx} + \mathbf{i}_x \frac{E_v^2}{16\pi} \frac{\omega_p^2}{\omega^2} \frac{2\omega}{c} \frac{v}{\omega} \tag{8.85}$$

where the second term is a non ponderomotive dissipative part of the nonlinear force, first derived 1969 [138] and derived later in another way by Stamper [163].

For nondissipative (collisionless) plasmas, the nonlinear force (for the very special case of plane waves perpendicularly incident on a WKB-like plasma) reduces to

$$\mathbf{f}_{\mathrm{NL}} = -\mathbf{i}_x \frac{E_v^2}{16\pi} \frac{\omega_p^2}{\omega^2} \frac{\partial}{\partial x} \frac{1}{n} = -\mathbf{i}_x \frac{1}{16\pi} \frac{\omega_p^2}{\omega^2} \frac{\partial}{\partial x} \mathbf{E}^2 \tag{8.86}$$

The last relation can simply be seen from the fact that the WKB approximation results in $\bar{E} = E_v/(n)^{1/2}$.

The terminology for $\mathbf{f}_{\mathrm{NL}}$ is not unique. While the expression "nonlinear force" was used [3, 156, 163, 171, 172, 173, 174], the expressions "nonlinear radiation force" [175], electrostrictive force" [176], "$j \times B$-force" [177, 178], or an unspecified "ponderomotive force" have also been used. The fact that additional (nonponderomotive) terms appear in the nonlinear force apart from those of ponderomotive forces, may be a convincing reason for remaining with the name "nonlinear force."

NINE

Momentum and Instabilities by the Nonlinear Forces

This section considers some general properties of plasmas due to the nonlinear forces at incident laser radiation—or equivalent microwave interaction. One long-known property is the usual radiation pressure which turns out to be as a trivial case from the following general treatment. Another well-known property of low-density plasma within a standing electromagnetic wave is the pushing of the plasma toward the nodes of the standing wave just by the gradients of $\mathbf{E}^2$ of the wave field [see Eq. (8.86)] [159 to 161]. What was discussed at laser plasma interaction first [158] was the reaction to high-density plasma due to dielectric effects. If a propagating electromagnetic wave runs into a high-density plasma, the spatial derivative of the refractive index is essential. The dielectric increase of the $\mathbf{E}^2+\mathbf{H}^2$ value for perpendicular incidence of the electromagnetic wave due to the dielectric swelling is the driving process. Dielectric swelling means the increase of

$$\mathbf{E}^2+\mathbf{H}^2=E_v^2\left(\frac{1}{|n|}+|n|\right)\exp(-\bar{k}x) \tag{9.1}$$

above its vacuum value, due to the decreasing refractive index n; this causes forces in the plasma. The corona is driven toward lower density (as long as the integral absorption constant $\bar{k}$ is small enough) causing an appropriately strong recoil to the plasma interior as an increased radiation pressure due to the swelling. The swelling factor is

$$S=\frac{1}{|n|} \tag{9.2}$$

Figure 9.1 schematically describes the explosion process of the plasma corona due to the nonlinear forces because of dielectric swelling.

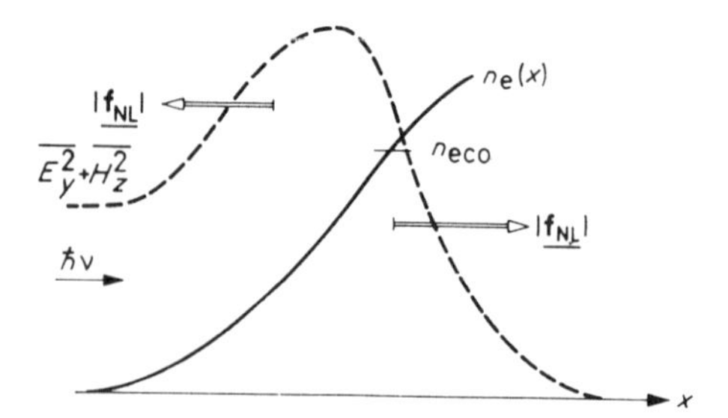

Figure 9.1 Schematic description of the nonlinear force in a plasma corona of an electron density profile $n_e(x)$ where the swelling of $\mathbf{E}^2+\mathbf{H}^2$ gives rise to the nonlinear force $\mathbf{f}_{NL}$.

This section first evaluates under what conditions the nonlinear forces are larger than the thermokinetic forces. The transfer of the momentum to the plasma corona will then be demonstrated generally, as well as the transfer of kinetic energy to the plasma. From that result, a conclusion about the momentum of photons or of the electromagnetic energy in plasma can be drawn. This question touches the 70-year-old Abraham–Minkowski controversy. From the aspect of energy density in blackbody radiation, a short interpretation of the energy density in a plasma will be given, though a general solution of the Abraham–Minkowski controversy—at least for nonabsorbing or little absorbing media—has been evaluated now by M. M. Novak [179]. This general result supports indirectly the validity of the special formulations of the nonlinear force, as described in the preceding subsection as appropriate formulation for plane waves. A further general property of the nonlinear forces at perpendicular incidence on plasmas is the generation of parametric instabilities about which the last subsection will report. The details of the numerical evaluation of the nonlinear force for perpendicular incidence and a comparison with various experiments will be presented in the following section.

9.1 Range of Predominance of the Nonlinear Force

The nonlinear force $\mathbf{f}_{\mathrm{NL}}$ will be observed only when the thermokinetic force $\mathbf{f}_{\mathrm{th}}$, contained in the total force $\mathbf{f}$ [Eqs. (8.4) and (8.5)], is comparable to or less than the nonlinear force. The formula (8.84) for plane waves at perpendicular incidence is rewritten

$$\mathbf{f} = -\nabla p + \nabla(\mathbf{E}^2 + \mathbf{H}^2)/8\pi \tag{9.3}$$

The comparison of the nonlinear force with the thermokinetic force would be simply the comparison of the potentials $n_e kT$ with that value of the nonlinear force before spatial differentiation in Eq. (9.3) where, however, an appropriate integration constant has to be added. This integration constant is determined by the fact that for a constant dielectric constant there has to be no net nonlinear force if the electromagnetic waves have a temporally constant amplitude. The value of $n_e KT$ must be compared, therefore, with the excess of $\mathbf{E}^2 + \mathbf{H}^2$ over its vacuum value. E^* is defined as the electric field amplitude in vacuum, where the nonlinear force equals the thermokinetic force.

$$n_e\left(1+\frac{1}{Z}\right)KT_{\text{th}}=\frac{E_v^{*2}}{16\pi}\left[\left(\frac{T^{3/4}}{a}+\frac{a}{T^{3/4}}\right)\exp_0-1\right] \tag{9.4}$$

The minimum absolute value of the refractive index n has been used from Eq. (6.49) with Eq. (6.50). The exponential function in Eq. (9.4) is abbreviated by $\exp_0$. The temperature T for the nonlinear refractive index consists of the temperature T_{th} of random motion and the energy of coherent motion, thus T^* in Eq. (6.57). The electron density n_e in Eq. (9.4) is then the cutoff density. Equation (9.4) is then a high-order equation between the plasma temperature T_{th} and the critical laser field E^* for the predominance of the nonlinear force. The solution is found by an iteration, where the first-order approximation is $T = T_{\text{th}}^{(1)}$. This is used for the calculation of the second-order solution $T_{\text{th}}^{(2)}$ including Eqs. (6.55) to (6.58)

$$\begin{aligned} n_e\left(1+\frac{1}{Z}\right)KT_{\text{th}}^{(2)}=\frac{E_v^{*2}}{16\pi}\Bigg\{\Bigg[\left(T_{\text{th}}^{(1)}-\frac{E_v^{*2}T_{\text{th}}^{(1)3/4}}{16\pi a n_{ec}K}\right)^{3/4}\frac{1}{a} \\ +a\Bigg/\left(T_{\text{th}}^{(1)}-\frac{E_v^{*2}T_{\text{th}}^{(1)3/4}}{16 a n_{ec}K}\right)^{3/4}\Bigg]\exp_0-1\Bigg\} \end{aligned} \tag{9.5}$$

The iteration proceeds until the difference between the last two iterations is less than 1%. The threshold electric field strength depends on the plasma temperature, on $\exp_0$, and on laser radiation; see Fig. 9.2, in which the electric field strength is expressed in laser intensities, Eq. (6.56). The threshold for the predominance of the nonlinear force over the thermokinetic force is a little above 10^{14} W/cm^2. For CO_2 lasers this threshold is more than 100 times less.

Apart from the mentioned numerical calculation of the nonlinear threshold intensities I^* [146], a more direct estimation can be used by an extension of the calculation of Steinhauer and Ahlstrom [180]. The pressures compared were those generated by the thermalization of the radiation and those governed by the nonlinear effects. The ratio

$$\frac{|\mathbf{f}_{\text{th}}|}{|\mathbf{f}_{\text{NL}}|}=R\geqslant\frac{1.14\times10^5}{T^{54}}\left(\frac{Z}{\lambda_0}\right)^{1/2} \tag{9.6}$$

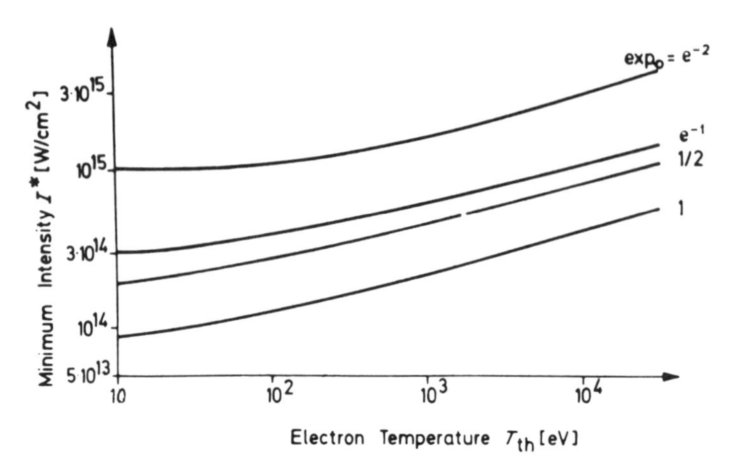

Figure 9.2 Minimum intensity I^* for neodymium glass laser radiation to create larger nonlinear forces $\mathbf{f}_{NL}$ than thermokinetic forces with an undefined exponential function for the collision induced attenuation [146].

was a realistic upper bound, because the maximum values of temperatures T were used. An increase due to thermal conductivity was neglected. The laser wavelength λ_0 is given in micrometers and T in electron volts. The result is that even at intensities less than the above-given threshold I^*, the nonlinear force should predominate, if the kinetic temperature T of the plasma is larger than 10 keV for neodymium glass laser wavelengths. The connection with the above-mentioned result for I^* can be seen in the following way. Because Steinhauer and Ahlstrom's temperature determines the optical constants, Eq. (6.57), we can use [146]

$$T = T_{\text{th}} + \frac{\varepsilon_{\text{osc}}}{2K} = T_{\text{th}} + \frac{\omega_p^2 E_v^2}{16\pi\omega^2 n_e K|n|} \approx \frac{E_v^8}{(16\pi a n_{ec} K^{3/4})^4} \tag{9.7}$$

The predominance of the nonlinear force claims from Eq. (9.6), that R is less than unity, therefore

$$1 > (7.2 \times 10^8 / E_v)^{10} \tag{9.8}$$

with E_v in V/cm and for the neodymium glass laser wavelength of 1.06 μm. This immediately shows that there is a strong increase of the nonlinear force, as soon as I_v exceeds 7.2×10^8 V/cm, which corresponds to a laser intensity of about 10^{14} W/cm^2. The nonlinear intensity dependence of the refractive index was not used by Steinhauer and Ahlstrom but their result could be rectified later [146]. The then following result (9.8) shows the strong, resonance like increase of the nonlinear force, as soon as the threshold is exceeded. This resonance like behavior of the nonlinear force is a basic property which could be seen from the mentioned interaction process in a more precise way. The use of the constant R in Eq. (9.6) to be one is somehow arbitrary and can give the right order of magnitude only.

9.2 Momentum Transfer to the Plasma Corona and Compression

Considering again plane electromagnetic waves perpendicularly incident on a stratified plasma with a depth given by the x-coordinate, the momentum transferred by the nonlinear force—if it is predominant under the conditions given in the preceding subsection—following the scheme of Fig. 9.1, will be evaluated. A laser pulse is incident between the times t_1 and t_2. If the time variation of the amplitude E_v of the electric laser field strength in vacuum is slow enough, the Poynting vector can be neglected. It can be shown that even a picosecond substructure of the laser pulses does not change these assumptions [181, 182]. The total energy ε_L of the laser pulse is then

$$\varepsilon_L = c \int_{\tilde{K}} dy\, dz \int_{t_1}^{t_2} dt\, \frac{E_v^2(y, z, t)}{8\pi} \tag{9.9}$$

Here the integration is performed across the entire cross section $\tilde{K}$ (coordinates y and z), where light interacts with the plasma. The cross section is assumed to be sufficiently large to use a plane wave description. The momentum of all photons in the vacuum is

$$P_0 = \frac{\varepsilon_L}{c} \tag{9.10}$$

Under the condition for applying the WKB approximation (7.10), the nonlinear force produces a total momentum P_{inh} transferred to the inhomogeneous plasma between the depths x_1 and x_2 in the direction of $\mathbf{f}$ given by

$$P_{\text{inh}} = \int_{K} dy\, dz \int_{x_1}^{x_2} dx \int_{t_1}^{t_2} \mathbf{f}_{\text{NL}}\, dt \tag{9.11}$$

Using Eq. (8.72) for the nearly collisionless case is found

$$\begin{aligned} P_{\text{inh}} &= -\int_{K} dy\, dz \int_{x_1}^{t_2} dt \int_{x_1(t)}^{x_2(t)} dx \left[\frac{\partial}{\partial x} \frac{E_v^2(y, z, t)}{16\pi} \frac{1}{|n|} - |n| \right] \\ &= -\frac{P_0}{2|n_2|} (1 - |n_2|)^2 \end{aligned} \tag{9.12}$$

n_2 equals $n(x_2)$ by definition. x_1 is assumed to be outside the plasma in the vacuum. Equation (9.12) expresses the momentum of the accelerated inhomogeneous plasma layer in terms of the momentum P_0 of the laser pulse in the vacuum. To get a high momentum one needs low values of n_2, which means a high swelling S according to Eq. (9.2). The magnitude of the momentum is limited by damping; this causes an effective collision frequency. The momentum [146] transferred to the plasma can be much more than the usual radiation pressure, which is given by the photon momentum, because the momentum at the front of the plasma can be compensated by the momentum transferred to the plasma interior. Figure 9.3 shows the ablative acceleration of the plasma corona towards negative x if the swelling is sufficient, to produce the negative momentum P_{inh}. If x_2 exceeds the value for the maximum swelling, the net momentum transferred to the plasma between x_1 and the then reached depth x_3 is then increasing again. It reaches positive values up to the value of the usual radiation pressure P_0. The conservation of momentum gives

$$P_0 = P_{\text{int}} + P_{\text{inh}} \tag{9.13}$$

When all photons are absorbed in the plasma interior, then the total momentum transferred to the plasma interior, that is, to the plasma below the critical density, is

$$P_{\text{int}} = P_0 + P_{\text{inh}} = \frac{P_0}{2}\left(\frac{1}{|n|} + |n|\right) \tag{9.14}$$

In terms of P_0 can be written

$$\frac{P_{\text{int}}}{P_0} = 1 + \frac{P_{\text{inh}}}{P_0} \tag{9.15}$$

The momentum for compressing the interior of the plasma by the nonlinear force can be multiples of the radiation pressure, if the swelling provides a sufficiently large momentum to the inhomogeneous surface.

In order to see the maximum values of the swelling and of the explosion by

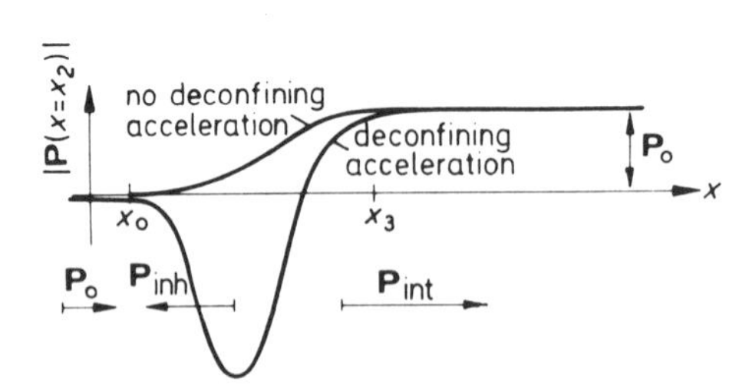

Figure 9.3 Momentum P transferred to the plasma between the vacuum and the interior of the inhomogeneous plasma. At sufficient swelling, the momentum can be negative P_{inh}, showing an ablation of the corona up to the minimum value of the refractive index [146].

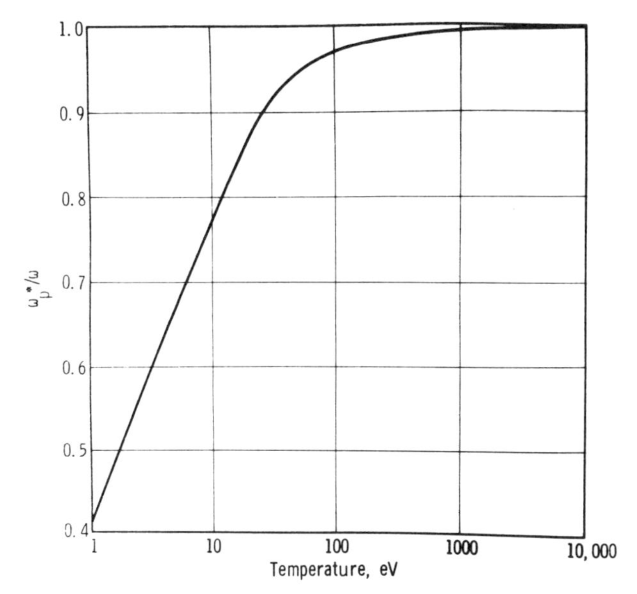

Figure 9.4 Electron densities corresponding to a plasma frequency ω_p* below which the collision produced absorption (the exponential factors in Eq. (8.72) for the nonlinear force can be neglected at WKB conditions) [138].

the nonlinear forces, it has been evaluated [138] numerically up to plasma densities (expressed by a plasma frequency ε_p^*) for which the exponential damping factor in the formulation of Eq. (8.36) can be neglected. A pessimistic estimate of absorption is determined considering Coulomb collisions without nonlinear effects if the WKB approximation is used with a value of $\theta < 0.25$ (Eq. 7.10). The results are given in Fig. 9.4. Using these values, the ratio of P_{inh}/P_0 or the nonlinear increase of the radiation pressure by swelling corresponding to the nonlinear force is given as dependent on the plasma temperature in Fig. 9.5. Note that these are optimum values, while the practical cases can provide conditions of lower swelling. On the other hand, the evaluations of the Figs. 9.4 and 9.5 were done for a linear behavior of the collision frequency. The resonancelike increase for the nonlinear collision processes can then increase the values given in Figs. 9.4 and 9.5.

9.3 Energy Transfer by Integration of the Nonlinear Force

In order to discuss some experimental cases, in which the conditions for the predominance of the nonlinear force hold, an exact integration of the equa-

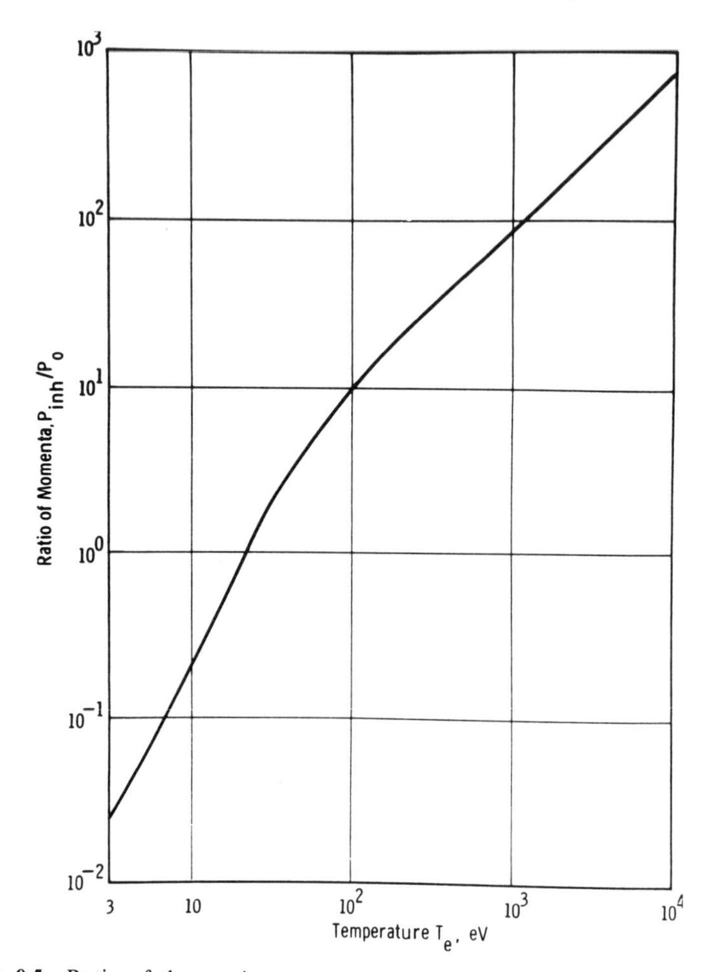

Figure 9.5 Ratio of the maximum momentum transferred to a WKB-like plasma corona by the nonlinear force depending on the temperature, if the densities of Fig. 9.4 are used and if the collision frequency is linear [138].

tion of the nonlinear force is possible if the resulting plasma profiles vary sufficiently slowly in time. The same geometry of perpendicularly incident plane wave and the WKB approximation are considered. The equation of motion is used in formulation

$$\mathbf{f}_{\mathrm{NL}} = -\mathbf{i}_x \frac{E_v^2}{16\pi} \frac{n_0}{n_{ec}} \frac{\partial}{\partial x} \frac{1}{|n|} \tag{9.16}$$

We recall that the velocity v_N gained by a body falling along a distance x

with an acceleration dv_i/dt is

$$v_0=[2(dv_i/dt)x]^{1/2} \tag{9.17}$$

The differential gain of the square of the velocity for varying acceleration is

$$\delta v_0^2=2(dv_i/dt)\Delta x \tag{9.18}$$

The increase of the kinetic energy of the ions due to the force (9.16) is then (using $n_e=Zn_i$, with the ion charge Z, and avoiding the restriction of the collisionless case)

$$d(\tfrac{1}{2}m_i v_i^2)=m_i\frac{dv_i}{dt}\,dx=\left|\frac{E_v^2}{16\pi}\frac{Z}{n_{ec}}\frac{\partial}{\partial x}\frac{\exp(-\bar{k}x/2)}{|n|}\right|dx \tag{9.19}$$

The exact integral for the ion energy is then

$$\begin{aligned}\frac{m_i}{2}v_i^2&=\frac{E_v^2}{16\pi}\frac{Z}{n_{ec}}\int_{x_1}^{x_2}\frac{\partial}{\partial x}\left[\frac{\exp(-\bar{k}x/2)}{|n|}\right]dx\\ &=\frac{E_v^2}{16\pi}\frac{Z}{n_{ec}}\left[\frac{\exp(-\bar{k}x_2/2)}{|n(x_2)|}-\frac{\exp(-\bar{k}x_1/2)}{|n(x_1)|}\right]\end{aligned} \tag{9.20}$$

The integration is performed between the depth x_1, a point in vacuum ($n=1$; $\bar{k}=0$), and x_2, the point in the plasma corresponding to minimum refractive index $|n|$. $\bar{k}$ is assumed to be sufficiently small as shown in Fig. 9.4; see also Marhic [183]. The maximum energy of the ions in the nonlinear-force driven ablation region is then

$$\varepsilon_i^{\text{trans}}=\frac{E_v^2}{16\pi}\frac{Z}{n_{ec}}\left(\frac{1}{|n|_{\min}}-1\right) \tag{9.21}$$

The ion energy of the plasma interior corresponding to a compression is given by

$$\varepsilon_i^{\text{compr}}=\frac{E_v^2}{16\pi}\frac{Z}{n_{ec}}\frac{1}{|n|_{\min}} \tag{9.22}$$

The limits of the integration in Eq. (9.20) are now those for $|n|_{\min}$ and for the interior of the superdense plasma with $\bar{k}\gg 2/x$ a nonvanishing refractive index. The acceleration of the plasma parallel to the laser beam corresponds to a (dynamical macroscopic) nonlinear absorption process.

The results (9,21) and (9.22) can be interpreted very easily by taking into account the fact that the average kinetic energy of the oscillation, $\varepsilon_{\text{osc}}^{\text{kin}}$ of the electrons in the laser field, Eq. (6.56), can be expressed for a WKB case with dielectric swelling by

$$\varepsilon_{\text{osc}}^{\text{kin}}=\frac{E_v^2}{16\pi n_{ec}|n|}\exp(-\bar{k}x/2) \tag{9.23}$$

$\bar{k}$ is assumed to be much less than one at the maximum of the swelling in agreement with the evaluation of Fig. 9.4, but a damping mechanism is included to prevent an infinite resonancelike process. Finally, the maximum energy of the ablated ions is found to be

$$\varepsilon_i^{\text{transl}} = \frac{Z(\varepsilon_{\text{osc,max}} - \varepsilon_{\text{osc,vac}})}{2} \tag{9.24}$$

$\varepsilon_{\text{osc,max}}$ and $\varepsilon_{\text{osc,vac}}$ are the oscillation energy of the electrons in the plasma and in vacuum, respectively. In cases where the swelling of the oscillation energy

$$\varepsilon_{\text{osc,max}} = S\varepsilon_{\text{osc,vac}} = \frac{\varepsilon_{\text{osc}}}{|n|_{\text{min}}} \tag{9.25}$$

reaches a factor $S = 100$, the oscillation energy of the vacuum can be neglected. This correction represents the radiation pressure in the vacuum. For strong swelling is found

$$\varepsilon_i^{\text{transl}} = Z\varepsilon_{\text{osc,vac}}/2 \tag{9.26}$$

The ion energy and the ion velocity are predominant for the motion of the plasma. This is used in the above consideration. The nonlinear force is acting on the plasma electrons, and the driven electrons are electrostatically attached to the ions. Therefore, the whole model works for plasma dimensions only.

9.4 Photon Momentum in Plasma (Abraham–Minkowski Problem)

The results of the subsection before last can be used to discuss the momentum of the electromagnetic energy and the momentum of photons in plasma. It has to be pointed out, that it is not evident from the beginning that the momentum of the electromagnetic energy is the same as that of the photons, as plasmons are involved too. Considering the momentum of the plasma material in the corona, Eq. (9.12), this is a recoil against the propagation of the laser photons. Therefore, it can be concluded that the momentum of the electromagnetic energy in the plasma has been increased by this recoil. The momentum of the electromagnetic energy is then

$$P = |P_{\text{inh}}| + P_0 \tag{9.27}$$

and from (9.10) and (9.12),

$$P = \frac{P_0(1 + |n|^2)}{2|n|} \tag{9.28}$$

Transferring this to the energy per photon in vacuum

$$p_\phi = \frac{\hbar\omega}{c} \tag{9.29}$$

the photon momentum (in the sense of electromagnetic energy) in the plasma is obtained

$$p_{\phi,\mathrm{pl}}=\frac{\hbar\omega}{2|n|c}(1+|n|^2) \tag{9.30}$$

This is the result of consideration along the lines of the last but one subsections [138], which has been reproduced completely and confirmed by Lindl and Kaw [154]. Furthermore, it has been derived in a completely different way for the electromagnetic energy density of a wave packet in a homogeneous nonabsorbing plasma by Klima and Petrzilka [184]. It has been pointed out by the latter authors that this momentum (9.30) differs from the momentum

$$p_A=\frac{\hbar\omega}{nc} \tag{9.31}$$

derived by Abraham, and

$$p_M=\frac{\hbar\omega}{c}n \tag{9.32}$$

by Minkowski, respectively. Formally, the energy density (9.30) would correspond to a photon momentum which is (half that of Abraham and half that of Minkowski, the average of the two momenta) [185]

$$p_{\phi,\mathrm{pl}}=\frac{p_A+p_M}{2} \tag{9.33}$$

The fact that the momentum (9.33) is reasonable for the plasma (confirming in this way the whole derivation of the preceding section) will be seen in the following evidence [185] that a photon momentum of the kind in Eq. (9.33) in a plasma results in the Fresnel formulas for the reflection of light, when passing from vacuum discontinuously to a homogeneous plasma of refractive index n without collisions. If $\tilde{R}$ is the fraction of reflected photons and $\tilde{T}$ that of transmitted photons, the conservation of energy requires

$$1-\tilde{R}=\tilde{T} \tag{9.34}$$

The conservation of photon momentum with a correct sign for the reflected photons at the interface results in

$$p_\phi+\tilde{R}p_\phi=\tilde{T}p_{\phi,\mathrm{pe}} \tag{9.35}$$

Together with (9,34), the reflection R and transmission T can be eliminated simply

$$\tilde{R}=\left(\frac{1-n}{1+n}\right)^2; \qquad \tilde{T}=\frac{4n}{(1+n)^2} \tag{9.36}$$

resulting in the Fresnel formulas.

The conclusion is that a photon in the aforementioned sense of the radiation energy density, has a higher momentum inside the plasma according to the formula (9.30). In order to push such a photon into a plasma, a recoil is necessary to produce this increase of momentum over the vacuum value. This recoil is that of the reflected photons according to the Fresnel formulas for a discontinuous transition. For a reflectionless, WKB-like, inhomogeneous interface between vacuum and plasma, the inhomogeneous layer has to take the momentum difference as a mechanical recoil directed against the incident photons. If the photons with their increased momentum are absorbed in the plasma, their total momentum of Eq. (9.33) [identical to (9.14) per photon] is then transferred to the absorbing region in the plasma. This causes the recoil, which is increased by the nonlinear swelling of the radiation pressure. For $|n| \ll 1$, this increased radiation pressure is

$$P_{\text{int}} = \frac{P_0}{2|n|} = \frac{S}{2} P_0 \tag{9.37}$$

One should recall the factor $\frac{1}{2}$ at this point in Eq. (9.37).

As has been pointed out [185], the formal rewriting of Eq. (9.30), using (6.47) for $\nu = 0$,

$$p_{\phi,\text{pl}} = \frac{\hbar\omega}{nc}\left(1 - \frac{\omega_p^2}{2\omega^2}\right) = p_A - \frac{\omega_p^2}{2\omega^2} p_A \tag{9.38}$$

indicates that the photon propagates with an Abraham momentum. However, it is reduced by an exchange process of the radiation energy with the oscillating photons. Formally, Eq. (9.38) can be written as

$$p_{\phi,\text{pl}} = \frac{\hbar\omega}{c}\left(n + \frac{1-n^2}{2n}\right) = p_M + \frac{\omega_p^2}{2\omega^2} p_A \tag{9.39}$$

This indicates that the photon behaves like a Minkowski photon with an additional momentum due to the plasma oscillations.

Under the latter aspect of the Minkowski momentum, the photoelectric interaction of the blackbody radiation is understandable. If the blackbody radiation density $U(\omega, T)$, depending on frequency ω and temperature T is calculated from the quantum electrodynamic interaction of this radiation with particles by spontaneous emission and stimulated emission and absorption, one arrives for nonabsorbing media with refractive index n at [186]

$$U(\omega, T) = U_p(\omega, T) n^3 \left(1 + \frac{\omega}{n}\frac{\partial n}{\partial \omega}\right) \tag{9.40}$$

where

$$U_p(\omega, T) = \frac{8\pi\hbar\omega^3}{c^3[\exp(\hbar\omega/KT) - 1]} \tag{9.41}$$

is Planck's spectral distribution for vacuum. Using the refractive index n for a plasma without collisions ($\nu=0$), Eq. (6.46) in Eq. (9.40), the result is

$$U=U_p n \tag{9.42}$$

This result agrees with that derived by Bekefi [187] and Dawson [188] for plasmas. The only problem is the restriction $\omega_p \leqslant \omega$; otherwise formula (9.42) can become purely imaginary. The result (9.42) confirms the Minkowski picture, as U can be connected only to the part of the electromagnetic energy, which consists of "photons in vacuum." The additional photon part of the total electromagnetic energy is then due to the electron oscillations only, see Eq. (9.39). These plasmon oscillations do not contribute to the photoelectric excitation and de-excitation, which was the basis of Einstein's derivation of Eq. (9.41) [189].

The neglect of $\exp(-\bar{k}_x/2)$ for the energy density in a plasma is possible, as emission and absorption are in equilibrium. The total electromagnetic energy density of blackbody radiation in a plasma is

$$U_{\text{tot}}=U_p\left(|n|+\frac{\omega_p^2}{2\omega^2|n|}\right) \tag{9.43}$$

Using the absolute value of the refractive index, formula (9.43) is valid for plasmas with collisions and even for frequencies ω at or below ω_p. The photon contribution to photoelectric interaction is then

$$U(\omega, T)=U_p(\omega, T)|n| \tag{9.44}$$

This is a radiation energy density in contrast to the electromagnetic energy density of Eq. (9.43). In Figs. 9.6 and 9.7, examples of the function U for temperatures $T=10^{6\circ}$K and densities $n_e=3.14\times10^{24}$ cm^{-3} and 1.26×10^{25} cm^{-3} are given in comparison with the Planck function U_p in vacuum (Badertscher [190]).

The spectral integration of U

$$U_{\text{tot}}=\int_0^{\infty} U\,d\omega=\sigma_{\text{op}}T^4 \tag{9.45}$$

arrives at the Stefan Boltzmann constant in the usual magnitude for vacuum

$$\sigma_{\text{op}}=\sigma_0=5.67\times10^{-5}\ \text{erg/cm}^2\ \text{sec}\ {}^\circ\text{K}^4 \tag{9.46}$$

For plasma, however, σ_{op} has a very slight dependence on temperature and depends on plasma density. For the cases in Figs. 9.6 and 9.7, the values are [190]

$$\sigma_{\text{op}}=5.393\times10^{-5}\ \text{cgs} \quad \text{for } T=10^{6\circ}\text{K}; \quad n_e=3.14\times10^{24}\ \text{cm}^{-3}$$

$$\sigma_{\text{op}}=5.074\times10^{-5}\ \text{cgs} \quad \text{for } T=10^{6\circ}\text{K}; \quad n_e=1.26\times10^{25}\ \text{cm}^{-3}$$

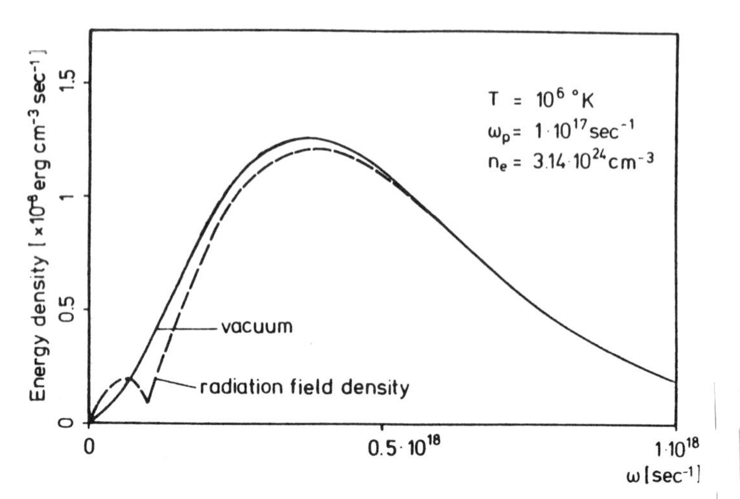

Figure 9.6 Planck's radiation formula (vacuum) for $T=10^{6\circ}$K compared with the spectral distribution of the radiation energy density for a plasma of the same temperature and an electron density of 3.14×10^{24} cm^{-3} [190].

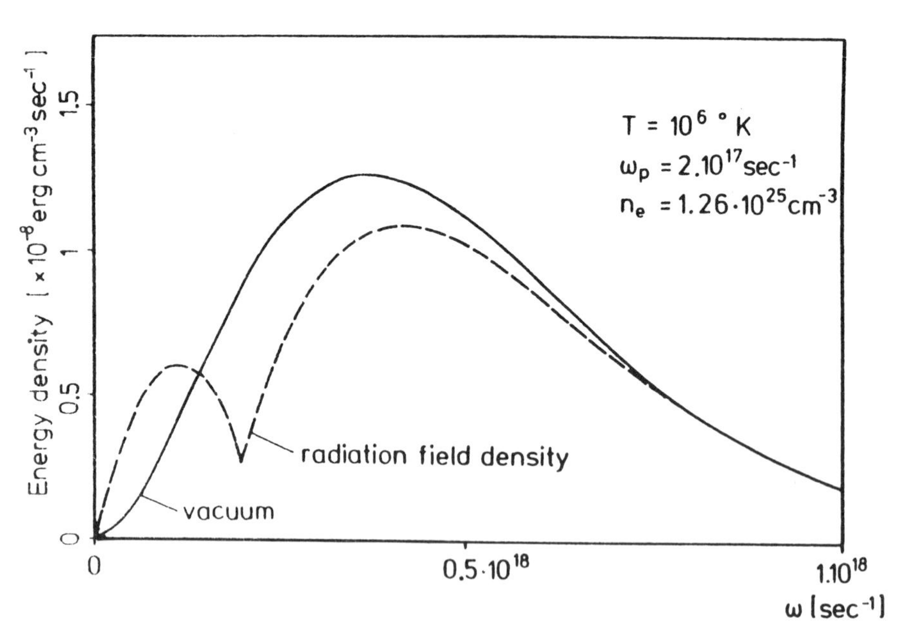

Figure 9.7 Planck's radiation formula as in Fig. 9.6 compared with the plasma of an electron density $n_e=1.26\times10^{25}$ cm^{-3} [190].

Evaluation for higher densities can formally produce σ_{op} values higher than ω_0, but this happens for densities that are above the limit of degeneracy, for which the above theory of optical constants has to be revised.

While we were considering the total energy density (determined by half of the Abraham and half of the Minkowski momentum), only the energy density of the vacuum field (bare photons = photoelectrically acting) excluding the energy exchanged with the electrons (dressed photons), led to the adequate description of the blackbody radiation in a plasma. This result was the starting point for Novak [179] to derive a basically new concept for the discussion of the Abraham–Minkowski problem.

We begin with the assumption (to be justified below) that the Minkowski momentum refers to a spineless photon while that of the Abraham describes a photon of spin unity. As is well known, the concept of spin occurs as a low-frequency phenomenon only, since its contribution at the other end of the spectrum may be neglected.

The equilibrium between the radiation and a medium is studied along the lines of Einstein [189]. Only the Minkowski form of the momentum density satisfied the equilibrium conditions, Eq. (9.40). In order to obtain this result, one imposes a condition $h\nu \gg kT$, effectively, we may say that it holds in the geometrical optics approximation.

In order to study the whole range of frequencies, we include the similarity between the dispersion relation

$$\omega^2 = \omega_p^2 + \mathbf{k}^2 c^2 \tag{9.47}$$

and the relativistic form of the energy E and momentum $\mathbf{p}$ of the particle

$$E^2 = m_0^2 c^4 + \mathbf{p}^2 c^2 \tag{9.48}$$

In view of the relations

$$E = \hbar\omega; \qquad \mathbf{p} = \hbar\mathbf{k} \tag{9.49}$$

We assert that a photon, which before entering a plasma had a zero rest mass, acquired, while inside it, an effective rest mass $\hbar\omega_p/c^2$. As a consequence, the Maxwell field in a medium can now be replaced by the Proca field, describing vector mesons. We now utilize the methods of the Proca electrodynamics and obtain the canonical energy-momentum tensor

$$T^{\mu\nu} = \frac{\partial \mathscr{L}}{\partial(\partial_\mu A^k)} \quad \partial^\nu A^k - g^{\mu\nu}\mathscr{L} \tag{9.50}$$

where the Proca Lagrangian $\mathscr{L}_p$ is

$$\mathscr{L}_p = -\frac{1}{16\pi} F_{\alpha\beta}F^{\alpha\beta} - \frac{1}{c} j_\alpha A^\alpha + \frac{\mu^2}{8\pi} A_\alpha A^\alpha \tag{9.51}$$

with $\mu = m_0 c/h$ being the effective photon's rest mass in units of the universal length. This tensor is not generally accepted. In order to satisfy the conservation laws, we require, following Belinfante [191] that

$$\Theta^{\mu\nu} = T^{\mu\nu} + T_s^{\mu\nu} \tag{9.52}$$

where $T_s^{\mu\nu}$ is interpreted as the spin energy-momentum tensor. It is then found that the total field tensor is symmetric and consists of an "orbital" part, which determines the energy and momentum, and a "spin" part, which does not contribute to either but is important in calculations involving the total angular momentum.

Hence, in accord with the promise of this work, the same conclusion must apply also to the tensor describing the electromagnetic field in a plasma. It is agreed that the Minkowski tensor readily adjusts itself to the canonical formalism. Then, in analogy to Eq. (9.52), the tensor

$$S_{\mathrm{sp}}^{\mu\nu} = S_A^{\mu\nu} - S_M^{\mu\nu} \tag{9.53}$$

ought to be interpreted as the spin energy-momentum tensor. In the component form, this tensor has the form

$$S_{\mathrm{sp}}^{oo} = 0; \qquad S_{\mathrm{sp}}^{oi} = 0; \qquad S_{\mathrm{sp}}^{io} = \frac{n-1}{4\pi c}\,\mathbf{E}\times\mathbf{H} \tag{9.54}$$

Then, it follows from the above that the force corresponding to the spin in

$$f_{\mathrm{sp}} = \frac{n-1}{4\pi c}\frac{\partial}{\partial t}\mathbf{E}\times\mathbf{H} \tag{9.55}$$

At high frequencies, the refractive index approaches unity and the (spin-) force vanishes, while its value increases toward the low-frequency end. One should remember that the spin was derived by the relativistic generalization of quantum mechanics in the correct way (Dirac equation) of spin $\frac{1}{2}$ for particles as electrons. Here we have a derivation of the spin one of photons without quantization.

Thus, from the interchangeability between the radiation in a medium and a neutral vector meson in vacuo, we conclude that the correct form for the field energy-momentum tensor is given by Abraham. However, in the limit of high frequencies, where the spin contribution may be neglected, the asymmetric Minkowski tensor represents a valid approximation. A very one of photons qualitative picture is then understanding why at low frequencies the photo effect (with exchange of spin to the electron) dominates and at high frequencies the Compton effect with no change of the electron spin. It has been derived from general principles that the electromagnetic energy exchanged between the blackbody field and the electrons is mainly classical by quivering motion even at relativistically high temperatures. The exchange by quantum processes does not exceed $5/\pi^5 \simeq 5\%$ [192].

9.5 Parametric Instabilities

Parametric instabilities are a large number of transfer processes of the laser energy into oscillation or wave modes of the plasma, especially of electrostatic (Langmuir) waves, ion-acoustic waves, and others where an increase of these oscillations or density fluctuations is due to the laser at certain conditions of frequencies (energy) and wave vector (momentum) of the involved waves. Francis Chen gets the credit for discussing and analyzing these phenomena consequently on the bases of the nonlinear force [193, 194] after Oraevski and Sagdeev [195] had introduced this type of wave interaction into plasma physics and a more consequent treatment followed by Silin [196], Dubois and Goldman [197], and Nishikawa [198]. The following broad discussion of this topic was reviewed by Cap and co-workers [199].

The basic phenomena "parametric instability" was called "parametric resonance" when Landau and Lifshitz [200] discussed an oscillating mechanical system of which one parameter was influenced by another oscillation. One example is a mathematical pendulum of length l, mass m whose origin is oscillating in the vertical direction y by a frequency ω_0, and an amplitude A ($y=A\cos\omega_N t$), see Fig. 9.8. Using angle ϕ as generalized coordinate, the Lagrangian (kinetic energy minus potential energy) is then

$$L=\frac{ml^2}{2}\phi^2+mla\omega_0^2\cos\omega_0 t\cos\phi+mgl\cos\phi \tag{9.56}$$

from the Lagrange equation of the second kind

$$\frac{\partial}{\partial t}\frac{\partial}{\partial\phi}L-\frac{\partial}{\partial\phi}L=0 \tag{9.57}$$

the following equation of motion for small amplitudes $\sin\phi\simeq\phi\ll 1$) is achieved

$$\frac{\partial^2\phi}{\partial t^2}+[a+q\cos(\omega_0 t)]\phi=0 \tag{9.58}$$

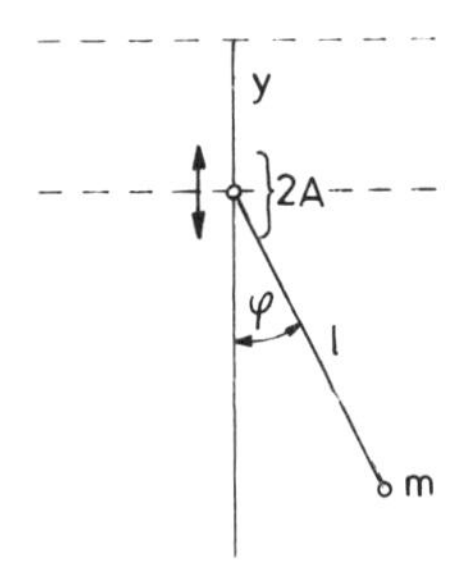

Figure 9.8 Mathematical pendulum whose origin A is oscillating in the vertical direction y.

where

$$a = \omega^2 \tag{9.59}$$

is the radian frequency of the undisturbed pendulum ($A = 0$) and

$$q = \frac{4\omega^2 A}{l} \tag{9.60}$$

Equation (9.58) is Mathieu's differential equation which has the special property of quasi-periodic (stable) and nonperiodic (unstable) solution as shown in Fig. 9.9. A quasi-periodic solution is achieved if

$$\omega_0 = 2\omega + \varepsilon' \tag{9.61}$$

where $\varepsilon \ll \omega$. In this case, the solution of $\phi(t)$ is an oscillation with temporally increasing amplitude

$$\phi \sim t \sin c_1 t + \text{oscillating terms} \tag{9.62}$$

($c_1 \sim$ const), if $q \ll a$. This can be seen from (9.58) using Eqs. (9.59) to (9.61)

$$\frac{\partial^2}{\partial t^2}\phi + \omega^2[1 + \eta \cos \omega_0 t]\phi = 0 \tag{9.63}$$

by the iterative solution

$$\phi^{(m)} = \phi_{n-1} + \phi_n \qquad (n = 2, 3, \ldots) \tag{9.64}$$

where ϕ_1 is the undisturbed solution ($q = 0$, $\eta = 0$). The second-order correction is then from

$$\frac{\partial^2}{\partial t^2}\phi_2 + \omega^2\phi_2 = -\omega^2\eta \cos(\omega_0 t) \cos(\omega t) \tag{9.65}$$

The solutions are oscillating [202], but if Eq. (9.61) is fulfilled, the solutions are

$$\phi \sim t \sin(c_1 t). \tag{9.66}$$

as was to be shown for increasing amplitude (=parametric instability).

The meaning of the result of Eq. (9.62) in the case of a plasma is that ω (pendulum frequency) is an eigenfrequency of the plasma, for example, that of electrostatic oscillations ω_p, while the external disturbance ω_0 is that of the incident laser. An instability will then occur, see Eq. (9.61), if $\omega_p \simeq \omega/2$ corresponding to an electron density of one-quarter of the cutoff density. This has been seen immediately from single-particle simulation codes [203] or from the fact that the light reflected from a plasma contains a small component of half of the frequency of the incident laser beam.

The study of other instability ranges (other shaded areas in Fig. 9.9), the

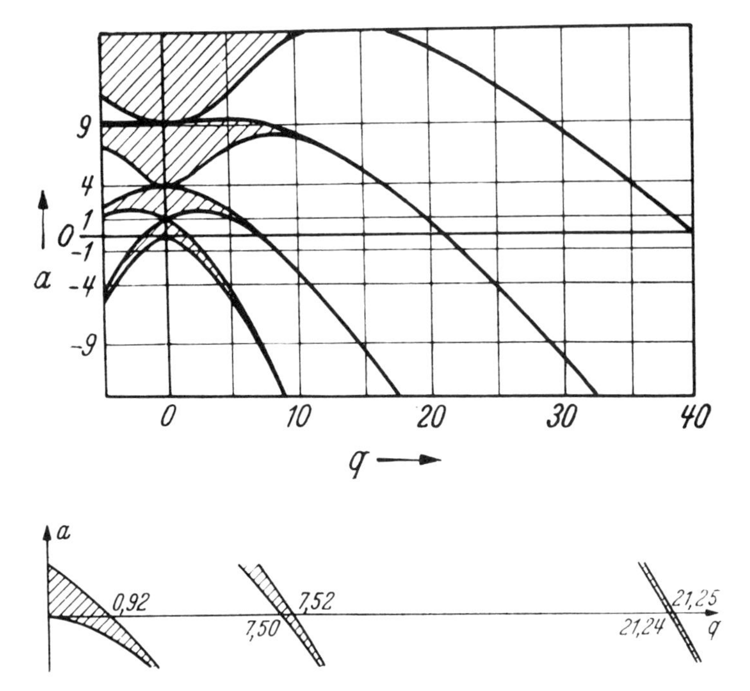

Figure 9.9 Ranges of stable (shaded) and unstable solutions of Mathieu's differential equations (9.58), after Paul and Raether [201].

possible detuning by ε', Eq. (9.61), or the complete solution [$n=\infty$ in Eq. (9.64)] are only some of the further mathematical tasks. Physics is important in following up where the energy transfer from the laser light is introduced, which needs the use of a more general Lagrangian equation (9.57), where the generalized forces depend on velocities and cannot be related to potentials.

Chen's analysis of the parametric instabilites in plasmas at laser irradiation [193] starts from the nonlinear force for perpendicular incidence of infinite plane waves on plasmas whose density profile permits a WKB approximation for the solution of the wave equation Eq. (8.86),

$$\mathbf{f}_{\mathrm{NL}} = -\mathbf{i}_\alpha \frac{1}{16\pi} \frac{\omega_p^2}{\omega^2} \frac{\partial}{\partial x} \overline{\mathbf{E}^2} \tag{9.67}$$

or

$$\mathbf{f}_{\mathrm{NL}} = \frac{\omega_p^2}{8\pi\omega^2} \{\mathbf{E}\cdot\nabla\mathbf{E} - \mathbf{E}\times(\nabla\times\mathbf{E})\} \tag{9.68}$$

where use was made of Eq. (8.19). Chen derived this formula from the quivering motion description found in most textbooks, where the difficulties due to the phase differences between **H** and **E** and **j** [54] were not followed up (which question was cleared recently by Kentwell [169]). Chen's formulation (9.68) has the merit that the last term results in forces along the propagation of the laser light, while the first term works on any deviation from the striated structure in the direction perpendicular to that of the propagation. This permits a distinguishing between *backscattering instabilities*, due to $\mathbf{E}\times(\nabla\times\mathbf{E})$, and the *electrostatic parametric instabilities*, due to $\mathbf{E}\cdot\nabla\mathbf{E}$ which is equivalent to the $\mathbf{v}\cdot\nabla\mathbf{v}$ convection term in the equation of motion (8.1). This nonlinear force in the lateral direction of the perpendicularly incident plane wave may indeed need more analysis with respect to the Maxwellian stress tensor (see Section 12). Chen [193] was aware of some of these problems, and certain limitations of the following results may be necessary.

The *electrostatic parametric instabilities* are the result of the interaction of perpendicularly incident plane waves of laser radiation with the lateral deviations n, of the electron density from its equilibrium value n, due to the $\mathbf{v}\cdot\nabla\mathbf{v}$ term in Eq. (9.67), Fig. 9.10. For the electrostatic oscillations the plasma frequency ω_p, Eq. (2.6), was due to the deviation of the electrons from their equilibrium whose oscillation is attenuated by Landau damping, Eq. (3.68). Bohm and Gross [204] studied the generated electrostatic waves which have the following frequency ω_c

$$\omega_e^2=\omega_p^2+(3/2)\mathbf{k}_3^2 v_{\mathrm{th}}^2 \tag{9.69}$$

(Bohm–Gross frequency) where the thermal electron velocity $v_{\mathrm{th}}=2KT_e/m$ determines the transport of a signal by this wave (Langmuir wave). The wave vector $\mathbf{k}_s$ is given by the phase velocity v_ϕ of the wave, which can be very large

$$|\mathbf{k}_s|=\omega/v_\phi \tag{9.70}$$

It has to be distinguished where the laser frequency ω is a little less than

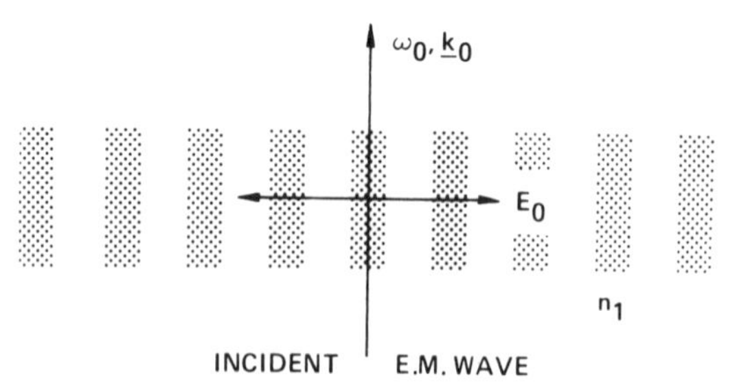

Figure 9.10 Direction of the k vector of the incident plane laser wave of frequency ω and an electric field E_0 in a plasma with some lateral (stochastic) deviation n_1 of its equilibrium electron density n resulting in electrostatic parametric instabilities. After Chen [193].

ω_e in which case we have an *oscillating two-stream instability*,

$$\omega < \omega_e \tag{9.71}$$

Density ripples in the direction of the electric laser field E_0 (Fig. 9.11) will then grow without propagation. The laterally uniform laser field E_0 will then interact with the space charge field E_1 of the density rippling by the nonlinear force $\mathbf{f}_{NL}$

$$8\pi \frac{\omega^2}{\omega_p^2} \mathbf{f}_{NL} = -2E_0 \frac{\partial E_1}{\partial x} \tag{9.72}$$

causing an increase of the ripple, Fig. 9.11.

If the laser frequency is a little larger than the Bohm–Gross frequency,

$$\omega > \omega_e \tag{9.73}$$

the *parametric decay instability* is generated. The oscillating two-stream instability does not work, and the incident wave decays into an electron wave ω_e and an ion acoustic wave ω_1. Following Fig. 9.12, the nonlinear force acts

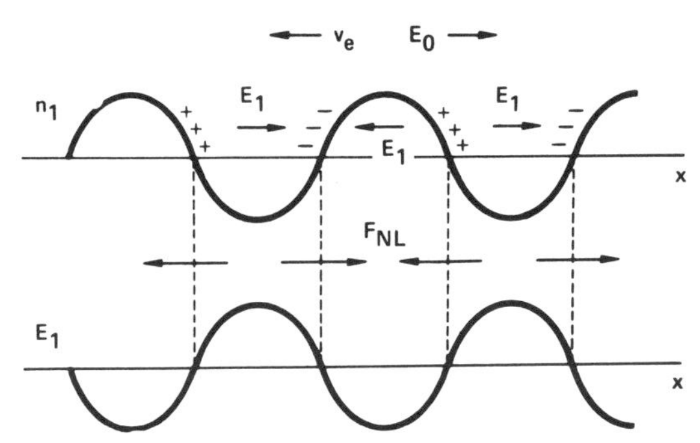

Figure 9.11 Oscillating two-stream instability. Due to Eq. (9.71), a laser frequency less than the Bohm–Gross frequency, a lateral density ripple is at rest. The electrostatic field $\mathbf{E}_1$ of the ripple interacts with the laser field $\mathbf{E}_0$ by the nonlinear force Eq. (9.72), increasing the ripple. After Chen [193].

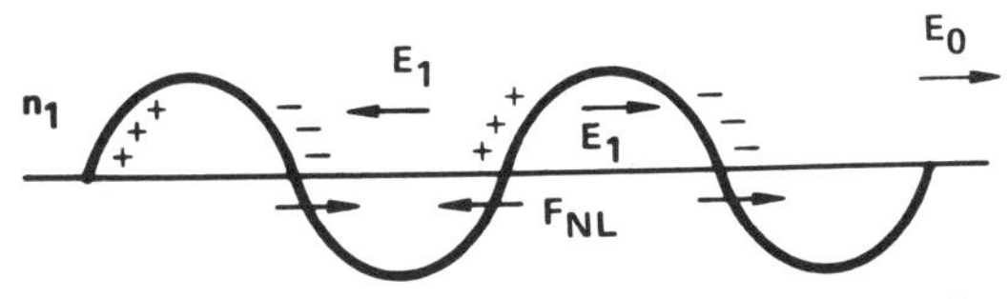

Figure 9.12 Parametric decay instability where the larger laser frequency than the Bohm–Gross frequency can be complemented by an ion wave only. Despite the decay of a density ripple by the nonlinear force in the rest frame, the Doppler shift in the frame moving with the wave results in a similar increase of the ripple in the same way as at the oscillating two-stream instability in Fig. 9.11. After Chen [193].

so as to destroy the density perturbation n_1. However, in the frame of the moving ion wave, the density perturbation would be at rest as in the oscillating two-stream case. Its mechanisms can again operate based on quasi-neutrality and a Doppler shift, causing a growing ion wave.

The lateral effects of the parametric decay instabilities will cause a *filamentation* or self-focusing of the laser beam in the plasma. This is based on the balance of the lateral nonlinear force with the gasdynamic pressure first proposed by Askaryan [205]

$$\mathbf{f}_{NL} = \nabla nKT_e \tag{9.74}$$

as shown in Fig. 9.13. Based on a Boltzmann-like density profile

$$n = n_0 \exp\left(-\frac{\omega_p^2}{\omega^2}\frac{\overline{\mathbf{E}^2}}{8\pi n_0 KT_e}\right) \tag{9.75}$$

a threshold for self-focusing at the laser power P_0 [193]

$$P_0 = 8800\left(\frac{\omega_e}{\omega_p}\right)^2 T\ \mathrm{W} \tag{9.76}$$

where $[T] = \mathrm{eV}$, was found [193]. This is the same value that we had derived before [206], where the mechanisms of total reflection and diffraction had been added to the balance of the forces Eq. (9.74) only (see Section 12).

The *backscattering instabilities* occur from nonlinear forces parallel to the direction of the wave vector **k** of the laser light, where, however, the details of the quiver motion and the phases of the initial and the induced electric and magnetic field have to be included. Again currents and velocities are produced perpendicular to **k**, reacting then with the **E**-field of the laser. While the electrostatic parametric instabilities are transverting the laser energy into electrostatic waves, the backscatter instabilities result in a transversion into an electromagnetic wave of frequency ω nearly in or against the direction

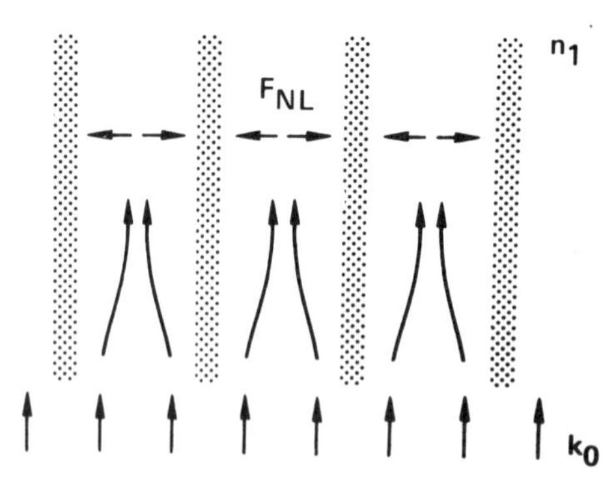

Figure 9.13 Filamentation or self-focusing due to the nonlinear force. After Chen [193].

of **k**. The lateral wave ω for this coupling can be equal to ω_e (an electron plasma wave); then we have *stimulated Raman scattering* (SRS). If ω is equal to that of the ion acoustic wave, we have *stimulated Brillouin scattering* (SBS). If ω is not perpendicular to **k**, the density perturbation can still exist if the laser field **E** is sufficiently large to maintain it against diffusion. This is then called *resistive quasi-mode scattering*. If $\omega_1 = k_1 v_e$ or $k_1 v_i$ where v_e and v_i are the thermal velocities of electrons and ions, interaction with resonant particles can cause an instability. This is the *induced Compton scattering* or *nonlinear Landau growth*.

Without looking into the detailed derivation, the following result of the threshold and growth rate in homogeneous plasma are given

	Threshold	Growth Rate	
SBS	$\frac{v_0^2}{v_e^2} = \frac{8\gamma_i \nu_{ei}}{\omega_i \omega_0}$	$\gamma_0 \simeq \frac{1}{2}\frac{v_0}{c}\left(\frac{\omega_0}{\omega_i}\right)^{1/2} \omega_{pi}$	(9.77)
SRS	$\frac{v^2}{c^2} = \frac{2\omega_p^2}{\omega_0^2}\frac{\gamma_e}{\omega_p}\frac{\nu_{ei}}{\omega_0}$	$\gamma_0 \simeq \frac{1}{2}\frac{v_0}{c}(\omega_0 \omega_p)^{1/2}$	(9.78)

where v_e are the thermal electron velocity, γ_e and γ_c the electron or ion wave damping rates, and $(\omega_p^2/\omega_0^2)(\nu_{ei}/2)$ is the damping rate of the electromagnetic waves. The following thresholds [207] are for Nd-glass and CO_2 lasers for intensities in W/cm^2

	Nd	CO_2
SBS	10^{13}	10^{10}
SRS	10^{13}	10^{9}
Oscillillation two-stream instability	10^{13}	10^{9}
Parametrametric decay instability	10^{13}	10^{10}

The action of the backscatter instabilities can be seen immediately from the electromagnetic waves reflected from the laser produced plasma. The intensity of the reflected light with half of the laser frequency or the higher harmonics is much less than the incident light, which is the most direct indication that the instabilities do not grow to infinity but are limited by saturation [208]. The use for diagnostics is very valuable. Watteau et al. [209] were able to demonstrate the change of the blue shift of the critical plasma density to a red shift, if the Nd glass laser intensity passes 10^{15} W/cm^2. Wong [210] used the fact of the small backscatter intensity as an obvious argument that the dynamics of laser-plasma interaction will not be influenced by the instabilities. This result has been confirmed theoretically by Bobin et al. [211], who emphasized that the parametric instabilities can work for Nd glass lasers for intensities between 10^{14} and 10^{16} W/cm^2 only. Above these intensities, the nonlinear force is predominant for plasma dynamics. Similar conclusions were drawn by Balescu [212]. A more detailed analysis was given by Liu et al. [213]. The nonlinear force disturbs the resonance conditions for the parametric decays. At the interesting intensities near 10^{15} W/cm^2 for Nd glass lasers, only 1% of the absorbed radiation can go into decay modes [214]. Under very artificial conditions, this contribution may grow to 10%. Experiments by Ng et al. [215] confirm these conclusions. In the discussion of the nonlinear plasma dynamics, therefore, instabilities can be neglected.

9.6 Summary

If laser radiation with an intensity above 10^{15} W/cm^2 for neodymium glass (10^{13} W/cm^2 for CO_2) is incident on an inhomogeneous plasma with densities up to cutoff, the nonlinear force will be predominant over the thermokinetic forces. If the swelling

$$S = \frac{1}{|n|} \tag{9.79}$$

is larger than 10, the corona of the plasma receives a deconfining or ablating momentum, which is

$$P_{\text{inh}} \simeq \frac{P_0}{2} S \tag{9.80}$$

P_0 is the usual radiation pressure (Eq. 9.10). The plasma interior beyond the cutoff density (if its thickness is larger than the absorption length) receives a confining momentum of compression (9.15)

$$P_{\text{int}} \simeq \frac{P_0}{2} S \tag{9.81}$$

For stationary density profiles, the translation energy, transferred to the plasma ions of the ablating corona or to the compressed interior, is

$$\varepsilon_i^{\text{transl}} \simeq \frac{Z}{2} \varepsilon_{\text{ocs}} = Z \varepsilon_{\text{osc}}^{\text{kin}} \tag{9.82}$$

Z is the ion charge, ε_{osc} is the maximum oscillation energy of the electrons near the cutoff density, and $\varepsilon_{\text{osc}}^{\text{kin}}$ is the average kinetic energy of this oscillation energy, as long as the macroscopic plasma theory is valid (Debye length $\ll$

density slope). The momentum of the electromagnetic energy per photon $p_{\phi,pl}$ in the plasma is increased over its vacuum value $p_\phi = h\omega/c$, where $S > 10$, to

$$p_{\phi,pl} \simeq \frac{p_\phi S}{2} \tag{9.83}$$

The exact value for a collisionless plasma is given by

$$p_{\phi,pl} = p_M + \frac{\omega_p^2}{2\omega^2} p_A \tag{9.84}$$

where p_M is the Minkowski momentum (effective in the photoelectric action of the radiation) and p_A is the Abraham momentum including the spin and the plasma property of electrostatic oscillations. The nonlinear force explains the different types of parametric instabilities. Its consequent inclusion in the dynamics proved the fact that these instabilities do not contribute remarkably to the absorption processes.

TEN

Numerical and Experimental Examples–Solitons

This section is devoted to plane waves perpendicularly incident on inhomogeneous plasmas. The forces in the plasma will be calculated numerically and compared with experiments. First, calculations without the nonlinear forces are considered, where the plasma dynamics are determined by the thermal pressure of the plasma after heating by the laser radiation. Subsequently, nonlinear forces are included. One of the early results was the discovery of the generation of a density minimum (caviton) by Shearer, Kidder, and Zink [171]. These minima can never be produced by thermokinetic pressures and are therefore typical for the plasma dynamics with nonlinear forces. The observation of these minima and the subsequent steepening of density profiles is the first tool for checking the action of nonlinear forces in experiments. The development of the nonlinear-force driven plasma dynamics is then shown to be typical for the generation of solitons. Macroscopic nonlinear absorption, caused by the net transfer of optical energy into kinetic energy of plasma without heating, occurs and leads to the mentioned soliton.

10.1 Thermokinetic Forces

The numerical study of laser-plasma interaction without nonlinear forces was done simultaneously by Mulser [88] and Rehm [215] for the one-dimensional case of plane electromagnetic waves perpendicularly incident on a stratified plasma. The basic hydrodynamic equations of conservation [the equations of continuity, Eq. (4.17), of motion, (4.6) (or (8.3) without the quantities **E** and **H**), and of energy conservation, Eq. (4.39)] are used to calculate the plasma density $\rho(x, t)$, given by the ion density $n_i(x, t) = \rho(x, t)/m_i$, the plasma temperature $T(x, t)$, and the plasma velocity in the x-direction $v(x, t)$. The initial conditions $\rho(x, 0)$, $T(x, 0)$, and $v(x, 0)$ are given. The boundary condition is the time dependence of the incident laser radiation. The problem is the formulation of the power generation term $W(x, t)$ in the energy equation. The question of heating the initially condensed material, its ionizaation, and Saha-equilibrium turns out not to be important for laser intensities above $10^9 \mathrm{W/cm^2}$ for neodymium glass lasers. For $W(x, t)$, the solution of the Maxwell equations (6.17) and (6.18) is necessary for the incident and the reflected wave (generated by the inhomogeneous plasma) for each instant. It was assumed that the transfer of absorbed energy in the plasma, according to the optical constants, is without any delay. This is correct if the time of interaction exceeds 1 nsec. It is interesting that Mulser [88] used a Lagrangian-type numerical code (localizing the intervals of computation to the moving mass density of the plasma). Rehm [215] used an Eulerian code (fixing spatial intervals to the coordinates). Both calculations arrived at the same result.

Figure 10.1 shows the result of a stepwise neodymium glass laser pulse

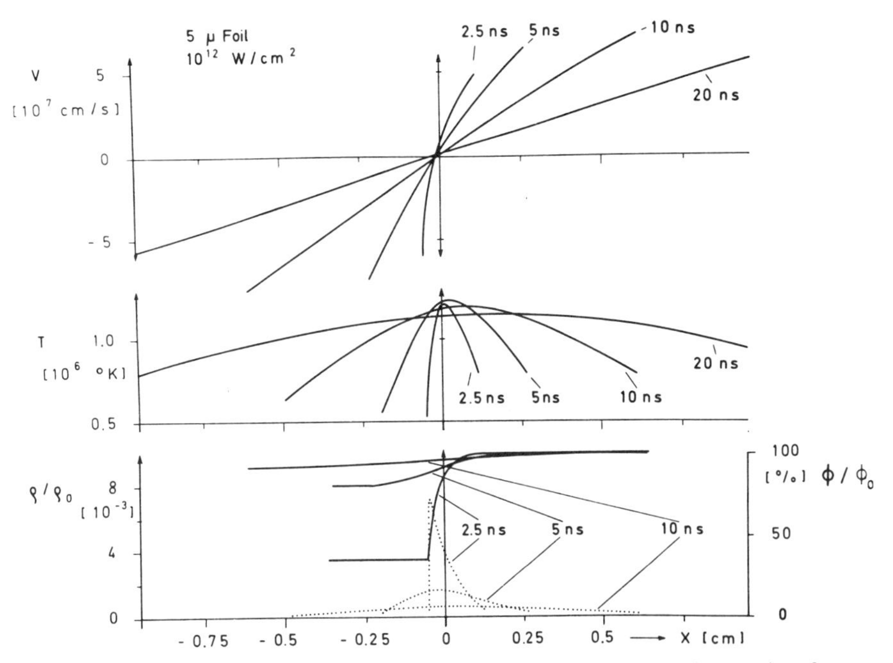

Figure 10.1 One-dimensional numerical solution of laser plasma interaction for a hydrogen foil of 5 μ thickness. A linear velocity profile and Gaussian density profile resulted at 5 nsec and later. After Mulser [88].

irradiating a solid hydrogen foil of 5 μm thickness. At and after 5 nsec, the velocity profile is nearly linear and the density profile is nearly Gaussian. The dynamics change approximately into that of the self-similarity model (Section 5).

A thick hydrogen block shows a quite different behavior (Fig. 10.2). The light is absorbed in a plasma density, which is 60 times less than the solid density. The ablation of the corona causes a compression of the plasma interior. This is shown by the negative velocity v in Fig. 10.2. In this way, a compression of the interior to multiples of the initial density results. Plasma densities of up to 250 times that of the corona are calculated.

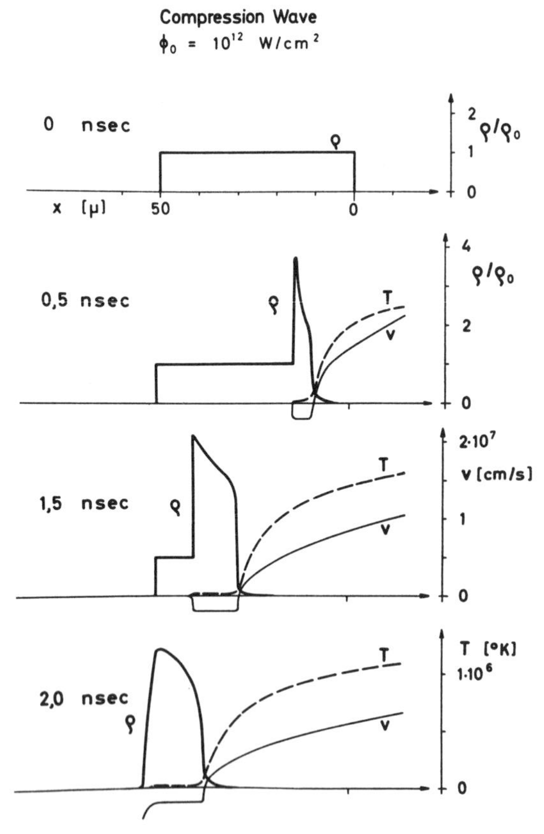

Figure 10.2 One-dimensional numerical solution of the hydrodynamic equations of conservation for laser light with a steplike intensity of 10^{12} W/cm^2 incident on a 50 μm thick slab of solid hydrogen (density ρ_0). The resulting density $\rho = n_i m_i$, velocity v, and temperature T are shown for times $t=0$, 0.5, 1.5, and 2.0 nsec. After Mulser [88].

This shock process happens only if there is no self-focusing, which could cause a more homogeneous heating of the plasma and a self-similarity expansion of thick foils also. The shock process was also shown by some analytical studies, based on hydrodynamic similarity laws by Krokhin and Afanasyev [216], Caruso and Gratton [217], or in a more general derivation, by Perth [218]. The most straightforward result, however, is that of the numerical calculations [88, 215].

A much more general hydrodynamic compression was calculated by Nuckolls [219], where use was made of a temporally increasing laser intensity. The compression process was then a growth of density similar to the concept of Guderley [220], where the addition of increasing shocks, following an appropriate sequence for meeting at one point, can produce densities of 10^4 times that of a solid [221] for a spherical geometry.

This successive compression by gasdynamic ablation requires a sufficiently short equipartition time (time to equilibrate the electron and ion temperatures). For Mulser's calculations with nanosecond pulses and no nonlinear derivations of the collision frequency, the instantaneous equipartition could be assumed. This is not the case for the calculations of Nuckolls [219]. To demonstrate the long duration of collisional equilibration, the electron collision time (6.58a)

$$\tau_{\mathrm{col}} = 1/\nu \qquad \text{where } T = T_{\mathrm{th}} + \varepsilon_{\mathrm{osc}}/2K \tag{10.1}$$

is plotted in Fig. 10.3 for various lasers and varying intensity.

In order to correlate the collision time with the interacting laser pulse and its thermokinetically caused mechanical pulse, it will be assumed for simplicity that the pulse of mechanical power density I_{th}, arising from the thermalizing interaction of the radiation with the plasma, has the form

$$I_{\mathrm{th}} = I_0 \sin^2\left(\frac{\pi t}{\tau_0}\right) \qquad \text{for } 0 \leqslant t \leqslant \tau_0 \tag{10.2}$$

τ_0 is the half-width of the pulse. A generalization to a more complicated pulse shape does not substantially change the following results. The laser pulse must then arrive earlier (see Fig. 10.4) by a precursion time, which depends on the laser intensity, and can be identified with the collision time τ_{col}. Thus, for short pulses, the laser light behaves in the plasma like a light beam in transparent glass and will produce no thermalizing coupling and no remarkable thermalizing energy transfer (except by nonthermalizing nonlinear force). The relation between the slope angles α and α' of the pulses (Fig. 10.4) can be used to find the greatest possible increase of a laser pulse. This is the instantaneous increase, corresponding to $\alpha' = \pi/2$, the highest possible increase I_{th}, which limits the gasdynamic compression models [219, 221]. Quantitatively, the maximum increase of I_{th} for a pulse of the

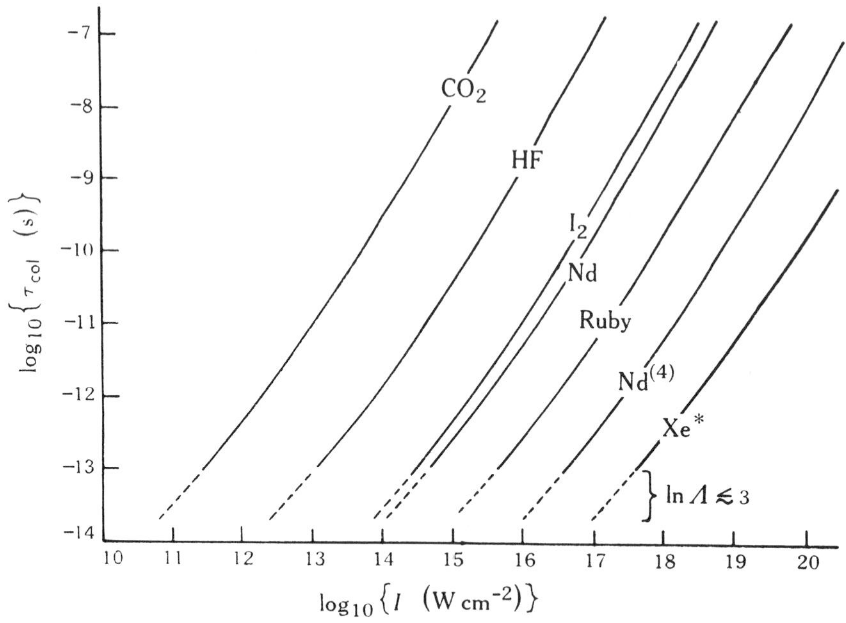

Figure 10.3 Plot of the minimum time for thermalization, as given by the collision time ω_{col} of the electrons, Eq. (10.1), as a function of intensity by the indicated lasers ($\varepsilon_{osc} \gg kT_{th}$; $|n|=1$; $n_e<1/2n_{ec}$). The values of n_{ec} in cm^{-3} are: CO_2, 10^{19}; HF, 8.6×10^{19}; I_2, 6.6×10^{20}; Nd–glass, 10^{21}; ruby, 2.3×10^{2}; fourth harmonic Nd–glass, 1.6×10^{22}; Xe*, 3.7×10^{22}.

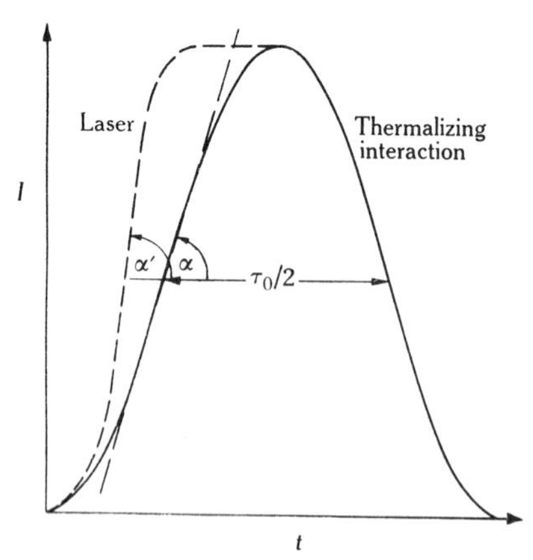

Figure 10.4 The laser pulse (dashed curve) must precede the thermalizing interaction pulse (continous curve) by an irradiance-dependent precursion time τ^* [identified with τ_{col} of Eq. (10.1)], in order to drive a gas dynamic ablation-compression process. The limitation on the thermalization is reached for $\alpha=\pi/2$ [222].

shape (10.2) is given by

$$\frac{\partial I_{\text{th}}}{\partial t}=\frac{I_0\pi}{\tau_0}\sin\left(\frac{2\pi t}{\tau_0}\right) \qquad \text{at } t=(1/4)\tau_0 \tag{10.3}$$

or

$$\left.\frac{\partial I_{\text{th}}}{\partial t}\right|_{\max}=\frac{I_0\pi}{\tau_0} \tag{10.4}$$

Inserting $\tau^*=\tau_{\text{col}}$, leads to

$$\tau^*=1.23\times10^{18}\left(\frac{1}{n_e \ln\Lambda}\right)\left(\frac{I_v}{|n|n_{ec}}\right)^{3/2} \tag{10.5}$$

$I_v=I_{0'}$ as is evident from Fig. 10.4. From the geometry of Fig. 10.4, the following condition, corresponding to $\alpha\leqslant\pi/2$, is obtained

$$\frac{\partial I_{\text{th}}}{\partial t}\leqslant\left(\frac{\partial\tau^*}{\partial I_v}\right)^{-1} \tag{10.6}$$

With Eqs. (10.4) and (10.5) and the inequality, Eq. (10.6), it is found that the laser pulse duration τ_0 has to be longer than the precursion time τ^* or

$$\frac{I_0}{\tau_0}\leqslant\frac{(|n|n_{ec})^{3/2}n_e\ln\Lambda}{(3/2)\pi 1.23\times10^{18}I_v^{1/2}} \tag{10.7}$$

Once more identifying I_0 with $I_{v'}$, the restriction for gasdynamic compression is

$$\frac{I_v^{3/2}}{\tau_0}\leqslant1.72\times10^{-19}n_e(n_{ec}|n|)^{3/2}\ln\Lambda \tag{10.8}$$

The units of I_v are W/cm^2, of n_e, and n_{ec} are cm^{-3}, and τ_0 is in seconds.

From the inequality (10.8) and with $|n|=1$ (requiring that $n_e=\frac{1}{2}n_{ec}$), $\ln\Lambda=8.1$ (for $I=10^{16}$ W/cm^2), and for neodymium glass laser radiation

$$\frac{I_v^{3/2}}{\tau_0}\leqslant2.1\times10^{34}(\text{W/cm}^2)^{3/2}/\text{sec} \tag{10.9}$$

Therefore, for $\tau_0=100$ psec, I_v has to be less than 1.7×10^{16} W/cm^2. The calculation by Nuckolls [219] has exceeded the limit (10.9). This indicates that some revision is necessary in the application for nuclear fusion calculations.

10.2 Static Case with Nonlinear Forces

Before going to fully dynamic calculations of plasma motion at laser irradiation including the nonlinear forces, some static cases will be considered.

Assuming a certain plasma density and temperature profile at a certain time, the resulting forces in the plasma can be calculated.

The most simplified case is that of a collisionless plasma with a linear density profile, as studied by Lindl and Kaw [154]. The solutions of the Maxwell equations for the electromagnetic waves are given by the Airy case, Section 7, and the resulting force is

$$\overline{\mathbf{f}_{\mathrm{NL}}} = \mathbf{i}_x \frac{\omega}{c} E_v^2 \exp(-2\rho s)\{(a_{iR}a'_{iR} + a_{iI}a'_{iI}) + \rho^{-2/3}[a'_{iR}(\zeta_R a_{iR} - \zeta_I a_{iI}) + a'_{iI}(\zeta_I a_{iR} + \zeta a_{iI})]\} \tag{10.10}$$

where the real parts and the imaginary parts of the Airy functions a_i are denoted by the indices R and I. The prime means differentiation with respect to the arguments. Figure 10.5 reports on the numerical result of the evaluation of f_{NL} from Eq. (10.1) (oscillating curve). Because of the collisionless plasma, total reflection occurs. The oscillation of the nonlinear force shows the acceleration of the plasma toward the nodes of the standing wave. The

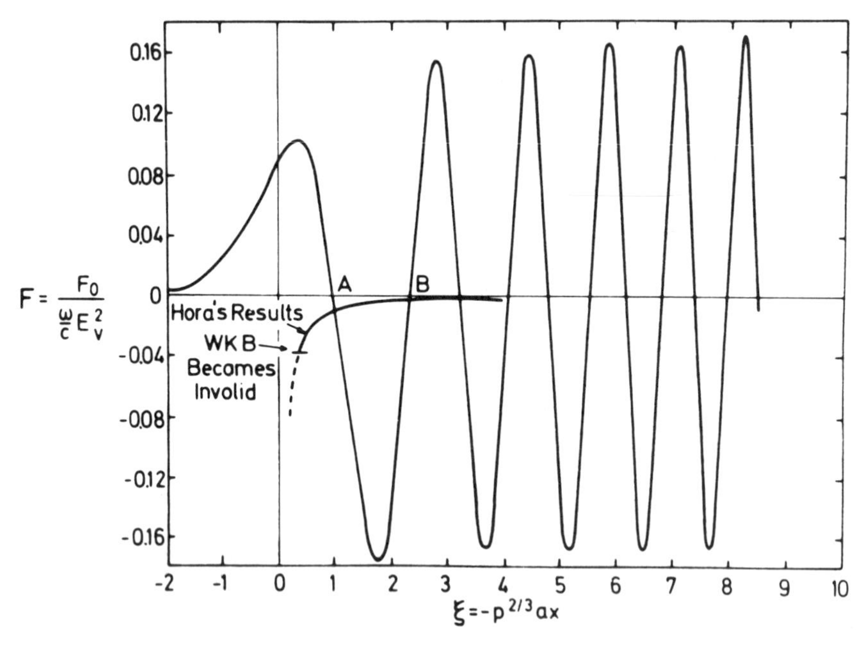

Figure 10.5 The nonlinear force for a collisionless plasma of linear density gradient is given. The light (incident from the right) creates a standing electromagnetic wave and the locally oscillating force. The net acceleration of the plasma is given by the double value of the monotonic force close to the abscissa. It is compared with the WKB approximation (Lindl and Kaw [154]).

spatially averaged net forces are drawn and show agreement of the exact results [154] with those where the WKB approximation, Eq. (8.50), is applicable.

Another static example uses a plasma with collisions given by a constant temperature $T_e = T_i$ of 1 keV and a density profile, where the WKB approximation is applicable [$\theta \leqslant 0.1$ in Eq. (7.10)]. In these calculations, n_e is derived from n_e and x, using $\Delta|n|$, Eq. (6.48), with a linear collision frequency [$T = T_{th}$, Eq. (6.58)]. The approximation $T = T_{th}$ causes a lower bound of the final forces, because the implication of the electron oscillation causes a higher temperature, hence lower values $|n|$ and then stronger force densities. The result in Fig. 10.6 shows a very slow increase of $n_e(x)$ around the cutoff density, which for neodymium glass laser radiation is around 10^{21} cm^{-3}. The second step is to calculate the intensity

$$I = \frac{E^2}{8\pi c} = \frac{E_v^2}{8\pi|n|c} \exp\left(\frac{-\bar{k}x}{2}\right) \tag{10.11}$$

The reflected wave can be neglected because of the WKB condition. The

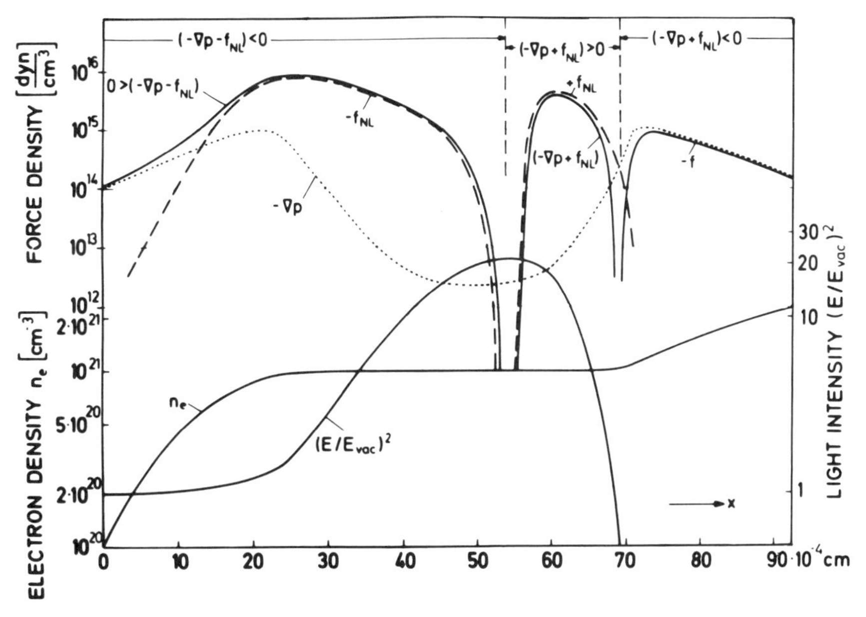

Figure 10.6 Numerical calculation of WKB-like density profile $n_e(x)$ of a plasma of temperature $T = 1$ keV. The radiation of 10^{15} W/cm^2, incident from the left side, produces an amplitude E of the electric field of the penetrating light wave compared with the amplitude E_v in vacuum before incidence. The resulting thermokinetic force f_{th} and the nonlinear force f_{NL} are calculated on the basis of the linear collision frequency [223].

strong increase of $(E/E_v)^2$, up to values beyond 20, proves that the decrease of $|n|$ occurs much earlier than the decrease of I, influenced by the integral absorption constant $\bar{k}(x)$. Therefore, the maximum values of $|n|^{-1}$ are higher than 20. This results in an increase of the effective wavelength λ^{eff} compared to the vacuum wavelength λ by more than 20.

$$\lambda_{\text{eff}} = \lambda S = \frac{\lambda}{|n|} \tag{10.12}$$

From this point of view, the nearly pathologically flat density profile from 25 to 65 μm (Fig. 10.6) is only over about four effective wavelengths.

The next step is to calculate the thermokinetic force density $\mathbf{f}_{\text{th}} = -\nabla p = -kT\nabla n_e$, which is given by the pointed curve in Fig. 10.6. The nonlinear force $\mathbf{f}_{\text{NL}}$ is negative up to a depth of 52 μm. It is larger than the thermokinetic force for a depth larger than 13 μm and can be nine times as large as the maximum of the thermokinetic force. The total force $\mathbf{f} = \mathbf{f}_{\text{NL}} + \mathbf{f}_{\text{th}}$ and its components have a negative sign, which means an acceleration of the plasma in the negative x-direction. For depths larger than 53 μm the nonlinear force f_{NL} is positive (directed in the $+x$-direction), and, because its absolute value is very high, the total force $\mathbf{f}$ is positive up to a depth from a little more than 53 to 68 μm which is nearly the skin depth, which is so large because the effective wavelength is at some depth larger than 20 vacuum wavelengths. For greater depth, n_e is increased monotonically in an arbitrary manner. A detailed dynamic calculation may also result in a decreasing n_e within the highly shocked material below the radiation zone, but this question is irrelevant at this point.

10.3 Approximative Dynamic Cases

The examples of the preceding subsection are of static nature, as these show the magnitude of the nonlinear force at a certain plasma configuration only. In real cases, the whole time-dependent dynamics should be calculated. The first calculations of this kind were performed by Shearer, Kidder, and Zink [171], using the Wazer code, which treats electrons and ions separately with respect to temperature. The only critical point of the model is that the laser radiation is approximated by an incident and a reflected wave only without a general solution of the wave field in the inhomogeneous corona. This results in too optimistic properties. The standing wave process (pushing plasma toward nodes) causes a fast growing density rippling, which leads to macroscopic Brillouin instability, which cannot be shown. Nevertheless, this approximation is useful as it shows a long time interaction process with a significant effect. The nonlinear force produces a density minimum in the initially monotonic density profile. Beyond the minimum, a very sharp increase of the density is produced, a shock wave compressing the plasma to high densities. The density minimum is called a caviton and is a characteristic of the nonlinear force. In a plasma with a uniform temperature distribution, any thermal pressure can produce monotonic density profiles only. This can be seen from the fact that the temperature of the laser irradiated plasma is proportional to the collision frequency.

$$T \sim \nu \sim c_1 \frac{n_e}{T^{3/2}}$$

For the second proportionality see Eq. (2.35). Therefore, for the monotonic thermokinetic force $\partial(n_e kT)/\partial x > 0$ is found a positive value for which

$$T\left(1 + \frac{c_1}{T^{5/2}}\right)\frac{\partial}{\partial x} n_e - \frac{3}{2}\frac{n_e c_1}{T^{5/2}}\frac{\partial T}{\partial x} > 0$$

because T, n_e, and c_1 are positive (if there is no stimulated emission). The thermokinetic force could only be nonmonotonic if the laser light produced a higher temperature at a larger depth than at a lower depth, but this is not possible in chemically uniform plasmas.

The results of Shearer, Kidder, and Zink are shown in Figs. 10.7 to 10.9. The time dependence of the laser pulse is given in the upper part of Fig. 10.7 with a maximum of 10^{16} W/cm^2. The electron temperature is much higher, Fig. 10.9, than the ion temperature because of the very long equipartition time. The strong ion temperature maximum in Fig. 10.9 is due to the adiabatic heating of the compressed plasma. One should note the first derived density minimum (caviton) in Fig. 10.7. The calculation was a first step of approximation with respect to the Maxwellian equations. This was improved in the next following case where the nonlinear intensity dependence of the optical constants was included.

The complete solution of the Maxwell equation for the whole plasma corona is used with nonlinear optical constants (see Section 6). The critical point, however, is the assumption of a stationary solution of the Maxwell equations, for which the following examples of subpicosecond interaction will not be correct. However, the basic properties of a very general nonlinear behavior can be seen [225, 226].

The equation of motion for the plasma in a one-dimensional geometry is given by the force density, distinguishing between electrons and ions. Besides the equation of continuity [Eq. (4.17) for both components], the energy equation (4.39) contains internal and external thermodynamic energies, thermal conductivity, and the incident laser radiation absorbed by collisions or by nonlinear electrodynamic motion as a source term.

The calculation begins at a time $t = 0$ with a distribution of plasma density,

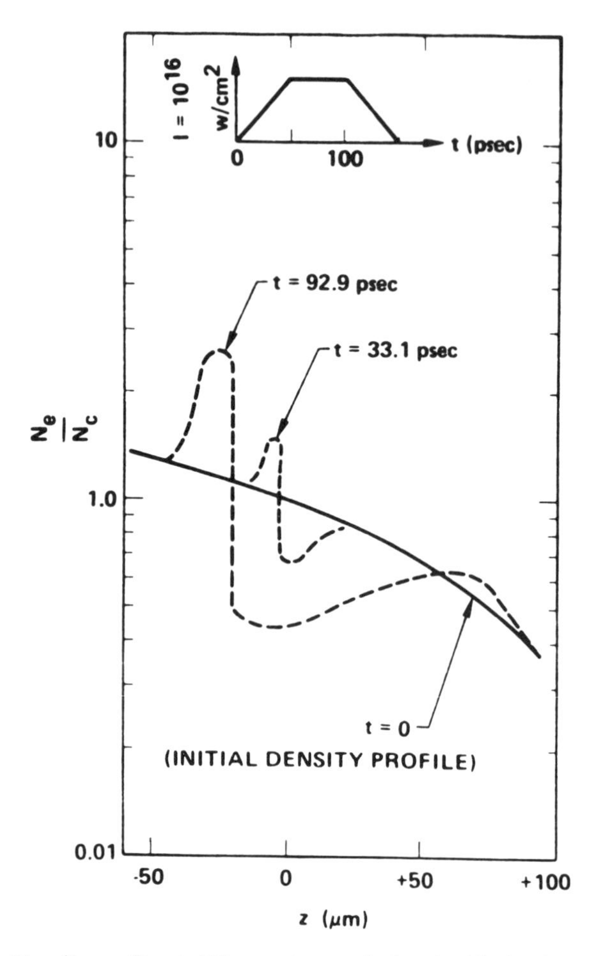

Figure 10.7 Density profile at different times calculated with the dynamic computer programme WAZER by Shearer [171, 224]. The low-density maximum is caused by the nonlinear force. The assumed laser pulse is given in the upper part. This was the first numerically discovered density minimum (caviton) and profile steepening due to nonlinear forces. After Shearer, Kidder, and Zink [171].

starting from solid-state density and a temperature profile. Without any laser field, there would be gasdynamic expansion and conservation of the total energy. At the successive time steps, a time-dependent incident laser intensity is prescribed, for which at each time step an exact stationary solution of the Maxwell equations is calculated including the actual plasma density and its refractive index; the retardation of the waves, the switching-on mechanism, and the development of the reflected wave are neglected.

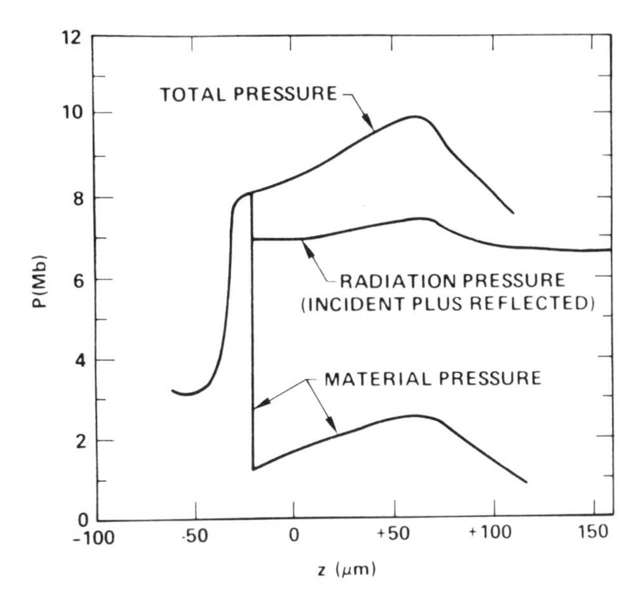

Figure 10.8 Pressure versus distance at $t = 90.9$ psec for the same WAZER computer run [224], see Fig. 10.7. After Shearer, Kidder, and Zink [171].

The motion of the plasma within the next time step (with appropriately varying step size) is described with the gasdynamic force and the nonlinear electrodynamic force. The following two examples from extensive series of computer runs illustrate characteristic results. In both cases, the initial temperature of the electrons and ions (100 eV) is constant for the whole plasma (this could be produced by a prepulse from the laser). The laser pulse increases within 10^{-13} sec up to an intensity of 2×10^{16} W/cm^2. The increase is linear for 5×10^{-14} sec and then follows smoothly a Gaussian profile. After reaching 2×10^{16} W/cm^2, the laser intensity remains constant.

The initial plasma density increases (for $x < 0$) quadratically above the cutoff density up to the solid-state density of LiD. It then changes smoothly in Fig. 10.10*a* into a linear decrease up to the length $x = 50$ μm. The resulting electromagnetic momentum flux density $(E^2 + H^2)/8\pi$ is given in Fig. 10.10*a*. Curve A is taken at an early time, when the laser intensity is 2×10^{14} W/cm^2. In a very thin plasma ($x \simeq 50$ μm), the value of $E^2 + H^2$ is constant. At about 20 μm an oscillation of $E^2 + H^2$ is found, which increases in amplitude and wavelength between 0 and 8 μm. This behavior is well known from the analytical work for the same linear density profile [154] (dealing, however, with temperature $T = 0$ and collisionless plasma). Near $x = 0$, the $|\mathbf{E}^2 + \mathbf{H}^2|$

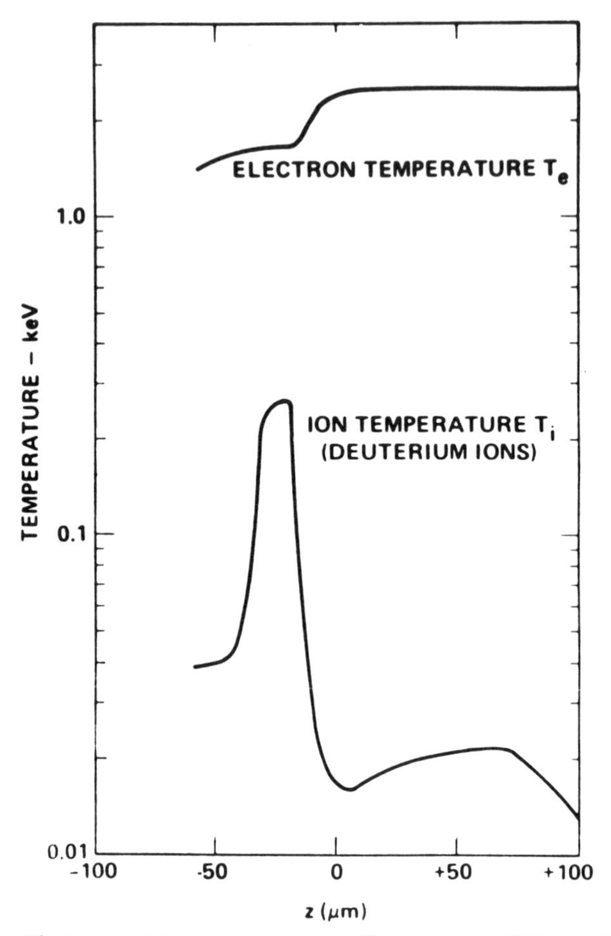

Figure 10.9 Electron and ion temperature vs distance at $t = 92.9$ psec for WAZER computer run [224] of Fig. 10.7 [171].

value has increased due to the dielectric properties of the plasma.

The increase of the laser field **E** and of the wavelength (10.12) over its vacuum values E_v and λ_0 is given by a swelling factor $S = |n|^{-1} > 1$

$$|\mathbf{E}| = E_v S = \frac{E_0}{|n|} \tag{10.13}$$

n is the optical (complex) refractive index of the plasma, provided the electromagnetic field can be described by the WKB approximation. In the case of curve A in Fig. 10.10a, the plasma fits the WKB conditions at 1.3 μm quite

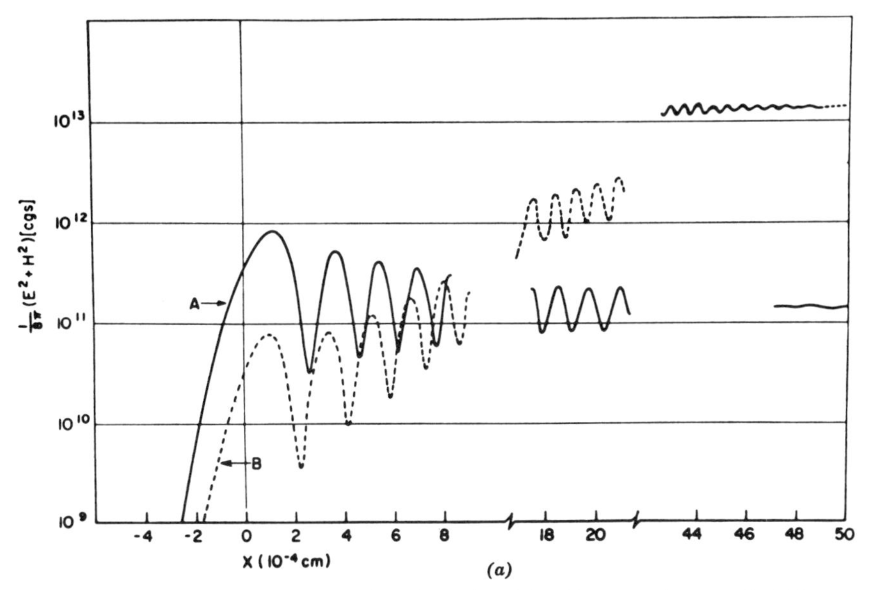

Figure 10.10a A laser beam is incident from the right side on a plasma of initial temperature of 100 eV and linear density increasing from zero at $x=50$ um to the cutoff density at $x=0$ and then increasing more rapidly. The exact stationary (time-dependent) solution without retardation of the Maxwell equations with a nonlinear refractive index n, based on a collision frequency [Eq. (6.59)], results in an oscillation due to the standing wave and dielectric swelling of the amplitude (curve A). At a later time, $(2\times10^{-13}$ sec), the laser intensity in 2×10^{16} W/cm^2 (curve B), where the relative swelling remains, but the intensity at $x=0$ is attenuated by dynamic absorption [1].

well, and the swelling factor $|n|^{-1}=6$ (taken from Fig. 10.10*a*) agrees with the value calculated from the actual refractive index.

For later times, curve A moves in nearly parallel fashion to higher values. However, its upward shift is reduced near the cutoff density ($x=0$). This decrease becomes so marked at curve B, that the intensity in the thin plasma ($x\simeq50$ μm) decreases by a factor of 10 up to $x\simeq20$ and to a thousandth part of the initial intensity at $x\simeq0$. The reason is very simple. The standing wave pushes the plasma toward its nodes with an ion velocity as high as 10^7 cm/sec even at $x\simeq48$ μm. The gasdynamic velocities reached at that time are 10^3 cm/sec or less. The result is shown in Fig. 10.10*b*.

At $x=46$ to 50 μm the initial linear density (upper part of Fig. 10.10*b*) acquires a ripple. The slight total increase is due to net motion driven by the nonlinear force. The ion energy increases, oscillating up to 200 eV (from the initial value of 100 eV), out of phase with the ripple. There is a curious effect seen in the electrons, in which the temperature increase is only a fifth of that

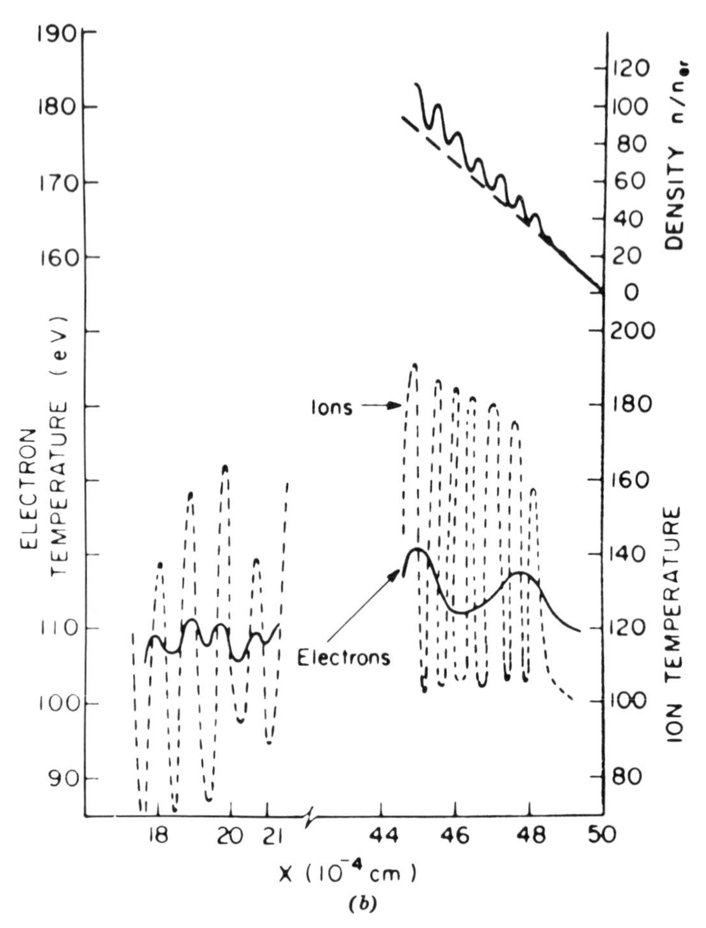

Figure 10.10b The initial density (dashed line) and the density along curve *B* of Fig. 10.10*a*, where a ripple is created by the nonlinear force, pushing plasma towards the nodes of the standing wave. The electron and ion temperatures are increased following the ripple by dynamic compression at conditions identical to curve *B* [1].

of the ions, as it should be for a collisionless shock; however, the periodicity is clearly different from that of the ions. At $x \simeq 20$ μm the periodicity of the temperature is again less than the ion temperature.

The density ripple explains the strong decrease of laser intensity with depth at later time. It causes strong reflection of light and a transfer of the optical energy into the ions by collisionless ripple shocks. Because it is not connected with decay of photons into microscopic acoustic modes, this process of dynamic ion absorption is called a "collisionless ion heating by nonlinear-force-induced macroscopic dynamic ion decay (MDID)." The depression of the laser intensity in interior regions becomes marked only after the density acquires pronounced maxima and minima (ripple).

An obvious idea is to start with a modified density profile, such that a change of the laser intensity I near cutoff results in a change of the actual nonlinear refractive index n, by virtue of the energy of the electron oscillation $\varepsilon_{osc}(I)$ determining the electron temperature (Eq. (6.57)). In Fig. 10.10*a* at 1.3 μm the absolute value of the refractive index is determined only by its real part (given by the actual density), while the imaginary part (determined by the temperature) is too small. Therefore, a computation is made for a density profile which is the same in the overdense region and which, in the underdense region varies linearly first from $x=0$ to 10 μm from cutoff to $1-10^{-6}$ times cutoff, then from there to $x=20$ μm to 0.99 times cutoff, and last linearly to zero at $x=50$ μm. Figure 10.11*a* shows the $(\mathbf{E}^2+\mathbf{H}^2)/8\pi$ values for an initial time (0.5×10^{-14} sec) at a laser intensity of 10^{15} W/cm^2 and at sub-

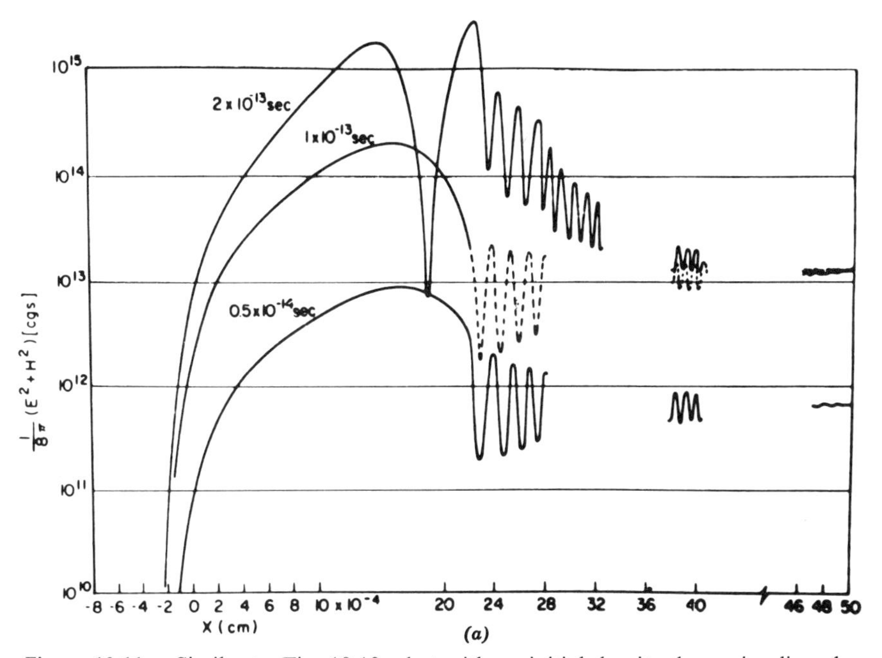

Figure 10.11a Similar to Fig. 10.10*a*, but with an initial density decreasing linearly from n_{ec} at $x=0$ to $n_{ec}\times(1-10^{-6})$ at $x=10$ μm, and to $n_{ec}\times(1-10^{-2})$ at $x=20$ μm and then to $n_e=0$ at $x=50$ μm. The $(\mathrm{E}^2+\mathbf{H}^2)$ values show an increase by a factor of 31.2 at time 5×10^{-15} sec. At 10^{-13} sec, the again-constant laser intensity of 2×10^{16} W/cm^2 is reached, with much less attenuation due to dynamic absorption at 2×10^{-13} sec. A swelling of $S\lesssim400$ occurs [1].

sequent times of 10^{-13} sec and 2×10^{-13} sec. The swelling factor at the beginning is

$$S=|n|^{-1}=31.2 \tag{10.14}$$

This is caused by the dielectric properties of this density profile. At 0.1 psec, the little depression of the curve at $x=20\ \mu$m indicates some tendency to decrease due to density rippling. The maxima and minima of $\mathbf{E}^2+\mathbf{H}^2$, however, change locally very quickly, and at 0.2 psec, a swelling of $S=|n|^{-1}=400$ is reached. The dynamic change of the density near $x=20\ \mu$m is shown in Fig. 10.11*b*. A process resembling tunnelling of the electromagnetic field through the overdense plasma near $x=20\ \mu$m at 0.2 psec is very strong, by virtue of the depth of the overdense plasma ($\sim1\ \mu$m) being small compared with the actual effective wavelength of more than 20 μm.

The resulting velocity profile at 0.2 psec is shown in Fig. 10.11*c*. The compression of plasma between $x=2\ \mu$m and $x=14\ \mu$m with velocities of 4×10^7 cm/sec does not change much within subsequent time intervals, while the velocities at $x=50\ \mu$m, for example, are still changing. The increased ion

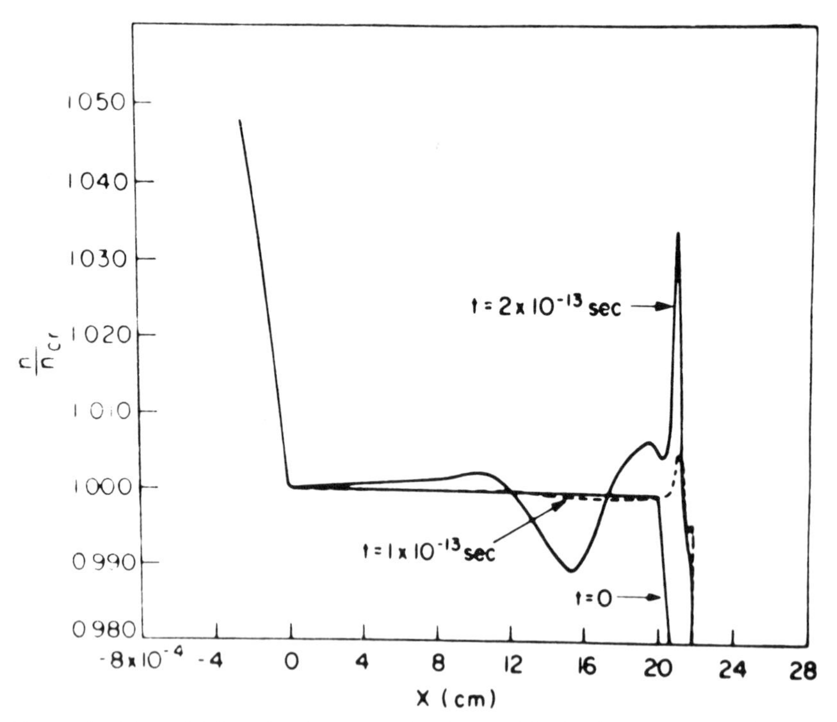

Figure 10.11b Initial and subsequent density profiles indicate motion of plasma toward the interior for x less than 14 μ [74].

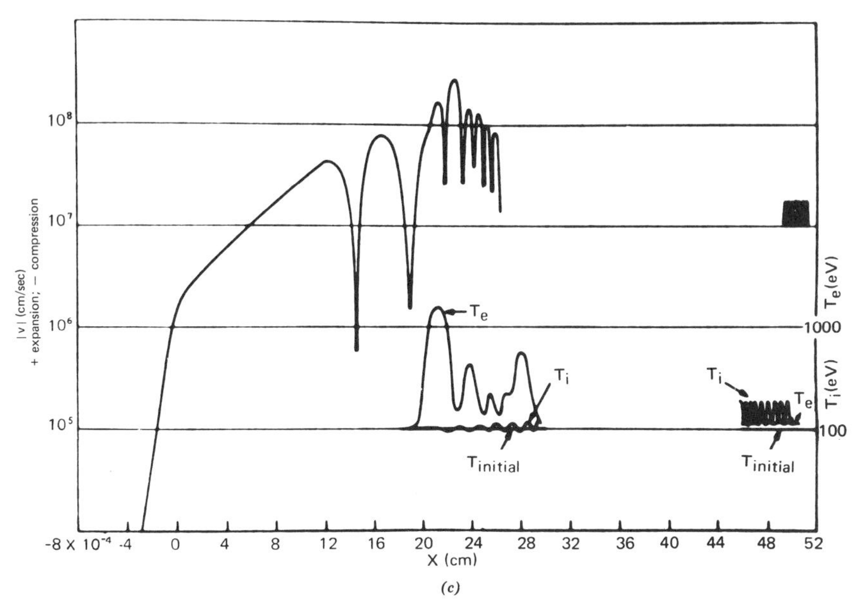

Figure 10.11c Dynamic heating of ions (at $x = 48$ μm) and of electrons ($x = 20$ μm) occurring at time $t = 2 \times 10^{-13}$ sec. the resulting plasma velocity v shows a compressing motion of the whole plasma between -4 and 14 μm with speeds up to 5×10^7 cm/sec, followed by expansion and alternating compression and expansion due to the standing-wave field [1].

temperature from collisionless processes at $x = 50$ μm is again much larger than the electron temperature, as could be expected from macroscopic dynamic ion absorption. A curious effect occurs in the interval from 20 to 28 μm, where the electrons are much hotter than the ions and display an irregular oscillation. It can be assumed that a longitudinal acceleration of the electrons occurs, due to the nonlinear force when the electrons have less interaction with the ions, comparable to hot electrons in solids at high electric fields. This phenomenon of dynamic electron absorption may be called "collisionless electron heating by nonlinear-force-induced macroscopic dynamic electron decay (MDED)."

The effects observed are highly sensitive to changes in the intensity I and the density profile. For an optimum case with $I = 4 \times 10^{16}$ W/cm^2 and the identical time and density profiles, the incident laser energy per cm^2 is 4.1 kJ during up to 0.2 psec. The compression front between $x = -2$ μm and $x = 14$ μm absorbs 0.96 kJ in kinetic energy, the ablating plasma take 0.68 kJ in kinetic energy of net motion and 2.2 kJ for dynamic heating. The remainder of 0.61 kJ goes into reflection and collisional heating of electrons.

The dynamic description of plasma by this numerical model for neodymium-glass laser intensities exceeding 10^{16} W/cm^2 indicates, that

1. A strong dielectric increase of the laser field and the effective wavelength occurs, described by a swelling factor $S=|n|^{-1}$ of up to 400 (n is a parameter resulting from rigorous solution of the Maxwell equations and is identical with the complex refractive index in the WKB approximation).
2. A compression of plasma driven by the nonlinear force occurs at a thickness of 15 μm with velocities exceeding 6×10^7 cm/sec. The mechanical energy in the compression front contains 43% of the incident laser radiation.
3. A rippling of the density occurs in the whole plasma at densities below cutoff with nonlinear-force generated velocities of 10^7 cm/sec or more. This may explain generation of laser produced fusion neutrons from peripheral plasma coronas. The ripple decreases laser light penetration of a plasma down to the cutoff density.
4. Rippling can be suppressed by selective profiling of the plasma density and the laser intensity with time.
5. A preferential transfer of laser energy to ions at a density gradient by collisionless shock occurs in the ripples (MDID).
6. A preferential transfer of laser energy to electrons can occur at nearly constant densities near cutoff (MDED).

10.4 Experimental Examples

The experimental study of the nonlinear force aims to find a direct proof. It is not very helpful, if indirect examples are seen, as, for example, Fig. 1.4. A nonlinear behavior of the keV ions in the fast group of plasma is evident, but a generation of the necessary intensities above the nonlinear force threshold, Section 9, is reached only indirectly including self-focusing or otherwise.

The direct action of the nonlinear force was shown first by the generation of cavitons by microwaves. Directed fast ions and a swelling of the electric field (by a factor of 700!) were measured. Wong and Stenzel [227] and Kim, Wong, and Stenzel [228] succeeded in these brilliant measurements, Fig. 10.12. The irradiation (40 kW at 1 GHz) into a 4 m long plasma with almost linearly increasing density resulted in fast ions moving against the irradiated field. Their energy was higher than the ion temperature. The latter was in the order of eV, because the interaction time was too short for a reasonable heating of the ions by collisions.

The first observation of the action of the nonlinear force in a laser irradi-

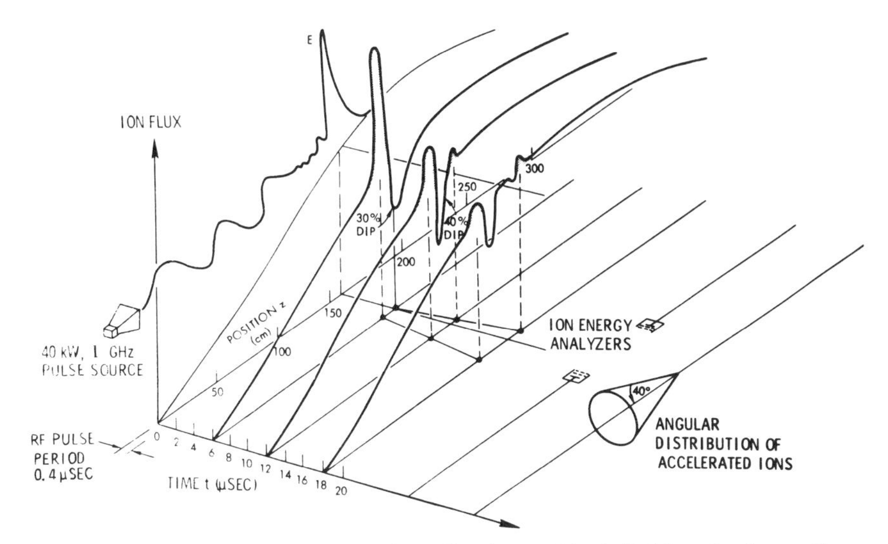

Figure 10.12 Space-time representation of ion bursts (shaded) driven by the nonlinear force. Density cavities are created as a result of ion expulsion. After Wong and Stenzel [227].

ated plasma was that by Marhic [229], Fig. 10.13. A consequent evaluation, based on the above derived theory, was in full agreement with the measurements.

A side-on measurement of the caviton produced by the nonlinear force (similar to the result of Wong et al. [227]) was seen first by Zakharenkov et al.

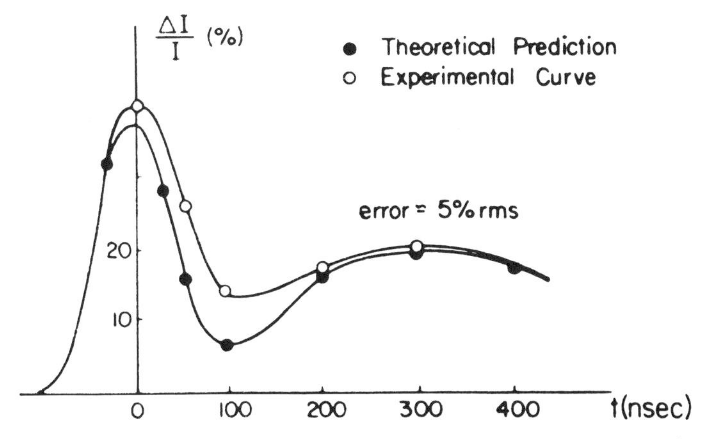

Figure 10.13 Relative change in light intensity at the focus indicating the action of the nonlinear force. $p=0.4$ Torr, $B=5$ kG, $J=100$ A [229]. After Marhic [156].

[230], Fig. 10.14, with neodymium glass lasers and by Fedosejevs et al. [231], Fig. 10.15, with CO_2 lasers. Similar results including the radial caviton, Fig. 2.10, were seen by Azechi et al. [232].

The stability of these cavitons were studied theoretically [233] and experimentally [234]. The nonlinear force in connection with the generation of fast ions was discussed by several authors [235 to 240] for cases where self-focusing could not have occurred, and where the ion energies were in the order of the values expected from the nonlinear force theory.

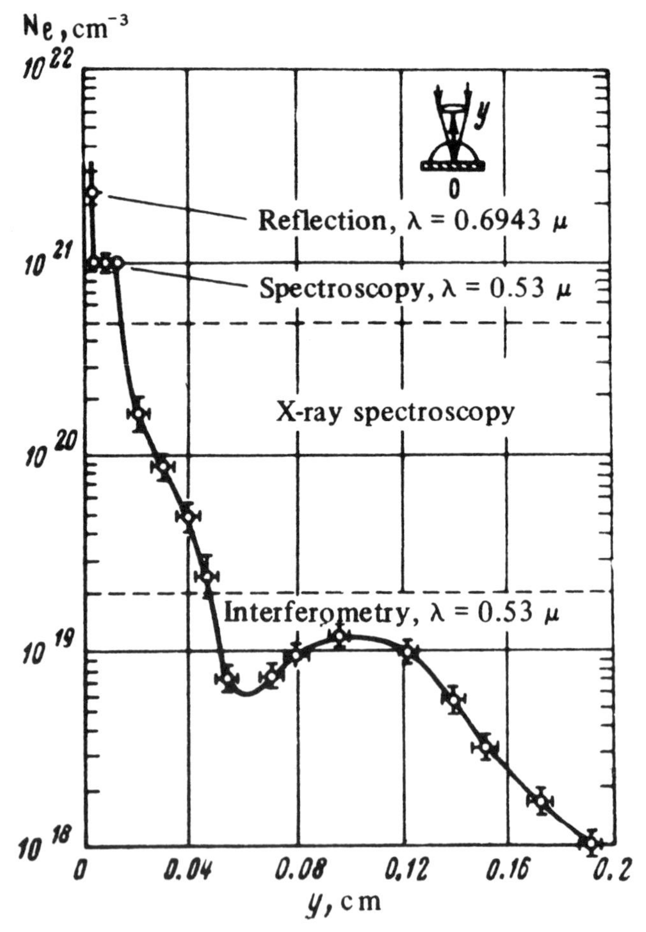

Figure 10.14 Electron density profile for laser plasma after 2 nsec, obtained for given flare by different methods. Target A, flux density 3×10^{14} W/cm^2. After Zakharenkov et al. [230].

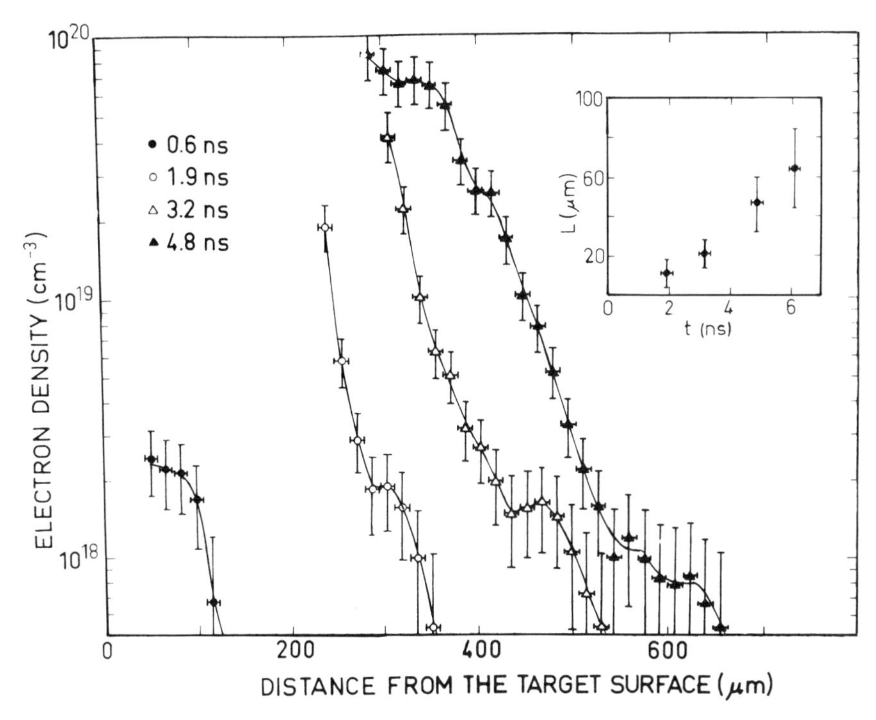

Figure 10.15 Plot of axial electron density versus distance from the original target surface at various times. The insert shows the scale length $L = n_e(dn_e/dx)^{-1}$ of the density plateau at various times relative to the leading edge of the CO_2-laser pulse. After Fedosejevs et al. [231].

10.5 Acceleration of Thick Blocks

This subsection reports on an extension of the fully gasdynamic calculations for plane electromagnetic waves at perpendicular incidence on stratified plasmas, including nonlinear forces as well as a nonlinear complex refractive index. The examples described in Figs. 10.10 and 10.11 were for short pulses of subpicosecond duration, which seems to be unrealistic for high intensities at present, although pulses of 170 femtoseconds (0.17 psec) have been generated [17, 241] and laser pulse powers of gigawatts have been achieved.

The following calculations are based on similar assumptions as used in that for Figs. 10.10 and 10.11; some numerical refinements were introduced and longer interaction times have been reached, before the computation became unstable. A parameter for this was the appearance of negative densities. The consistency of the computations was given by the conservation of energy, by the behavior of nonlinear momentum and energy transfer in agreement with global calculations, and last but not least by a numerical exact reproduction of a solitonlike behavior. The intensities of the laser pulses in all the cases were $\sin^2(mt)$-like, increasing to the maximum at 1 psec and where m is a constant. In all figures I corresponds to this constant intensity [242 to 244].

In order to avoid reflection, standing waves and the subsequent Brillouin-type density rippling and high-reflectivity—at least for the earlier stages—an initial density profile was chosen, which was expected to be very close to a case of very low reflection, namely the Rayleigh case, as discussed in Section 7. There a collisionless plasma was considered only. The fact that collisions are included can cause a superdense behavior near $x=0$ if the plasma temperature is 100 eV or less and the intensity is below the threshold for the predominance of the nonlinear force. The exponential decay of the intensity from 30 μm and below is seen in Fig. 10.16.

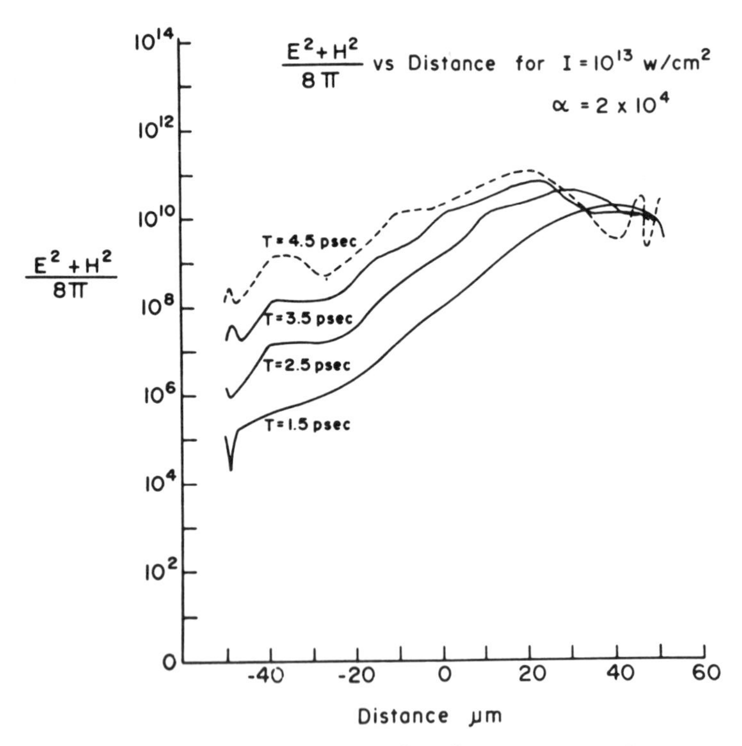

Figure 10.16 Time dependent solutions of $(\mathbf{E}^2+\mathbf{H}^2)/8\pi$ for an initial temperature of 100 eV and an initial profile as in Fig. 10.17 [243].

The fully dynamic calculation includes thermokinetic and nonlinear force motion. It causes an increasing transparency and a subsequent higher swelling in time. As shown in Fig. 10.6, at 4.5 psec the swelling is given by twice the ratio of E^2 at 50 μm at the earlier times, to that of the broad maximum at 4.5 psec. Its value is $S=82.2$. Note that $(\mathbf{E}^2+\mathbf{H}^2)/8\pi$ is always given in cgs units.

For the same initial density profile as in Fig. 10.17 and an initial temperature of 100 eV, Fig. 10.18*a* shows the energy densities at 1.5 psec for various intensities, and Fig. 10.18*b* shows the corresponding profiles of the plasma velocities. One recognizes the generation of fast moving thick blocks of plasma with velocities beyond 10^8 cm/sec, while the ion and electron temperatures have been increased to less than 500 eV. Figure 10.18*b* contains numerical extension to intensities of 10^{19} W/cm^2. This is not valid, as this is in the range of relativistic effects for the optical constants that are not yet included into the code.

The following rough estimation confirms that velocities above 10^8 cm/sec for intensities of 10^{18} W/cm^2 are achieved. Figure 10.18*a* leads to the

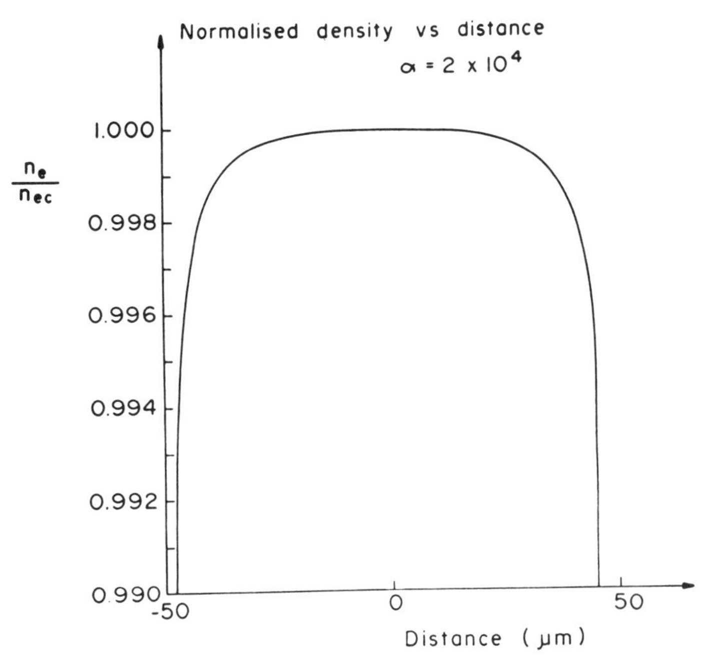

Figure 10.17 An inhomogeneous bi-Rayleigh density profile with $\alpha=2\times10^4$ cm^{-1} for neodymium glass lasers or CO_2 lasers, corresponding to Eq. (7.39). In all cases, initial temperatures are assumed to be uniform throughout the plasma [243].

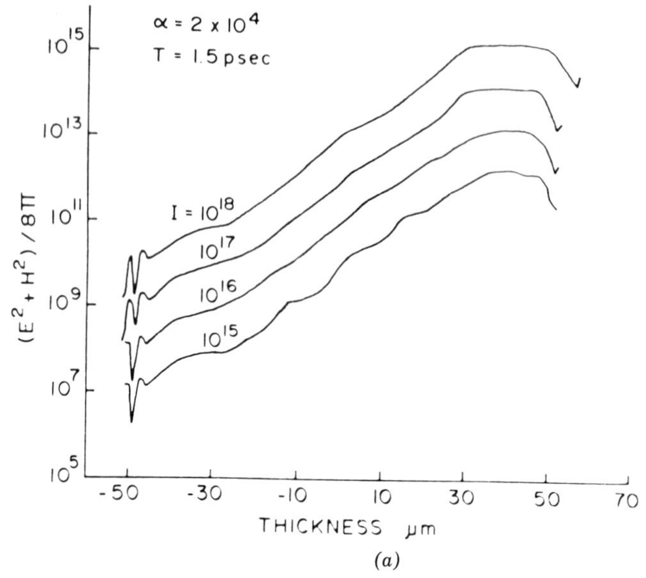

(*a*)

Figure 10.18a For an initial profile of Fig. 10.17 and an initial temperature of 100 eV, single broad maxima of the energy density $(\mathbf{E}^2+\mathbf{H}^2)/8\ \pi$ are generated at 1.5 psec, which vary directly with intensity of the incident laser light [244].

acceleration a from

$$\mathbf{f}_{\mathrm{NL}}=m_i n_i \mathbf{a}=\mathbf{i}_x \frac{\partial}{\partial x}(\mathbf{E}^2+\mathbf{H}^2)8\pi \tag{10.15}$$

$$\mathbf{a}=\mathbf{i}_x \frac{1}{m_i n_i}\frac{\Delta[(\mathbf{E}^2+\mathbf{H}^2)/8\pi]}{\Delta x} \tag{10.16}$$

For $m_i=2.5\times 1.67\times 10^{-24}$ g (DT-plasma) and the Nd glass cutoff density n_i with $\Delta[\mathbf{E}^2+\mathbf{H}^2]/8\pi=10^{15}$ cgs along 5 μm, $|\mathbf{a}|=4.79\times 10^{20}$ cm/sec². The velocity achieved during a time $t=1.5$ psec at constantly approximated acceleration is 7.1×10^8 cm/sec. Considering the temporally increasing acceleration, the velocities in Fig. 10.18*b* are then understood.

The total kinetic energy transfered to the plasma between the thicknesses x_1 and x_2 is given by

$$E=\frac{1}{x_2-x_1}\int_{x_1}^{x_2}\frac{m_i n_i(x)}{2}\,v_i^2(x)\,dx \tag{10.17}$$

It is evaluated numerically for the case of $\alpha=3000$ cm^{-1} $x_1=-0.05$ mm, and $x_2=+0.05$ mm. The result shows a superlinear increase (Fig. 10.19)

$$E_{\mathrm{kin}}=I^m; \qquad m=1.8 \tag{10.18}$$

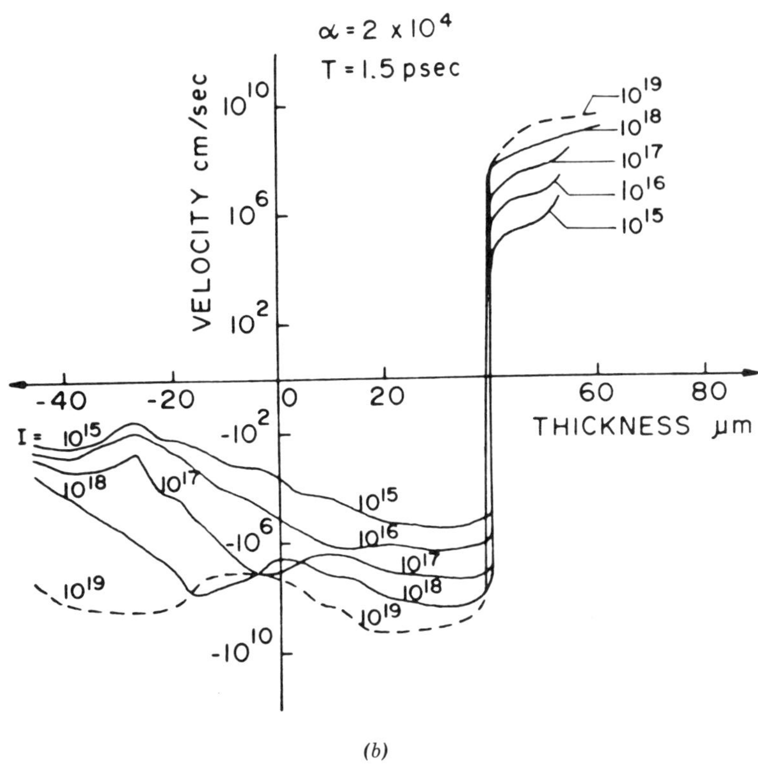

(*b*)

Figure 10.18b The velocity of plasma corresponds to the electromagnetic energy density of Fig. 10.18*a*. These blocks of plasma move toward the interior of the plasma due to the nonlinear force [224].

according to the nonlinear macroscopic absorption process of energy transfer by the nonlinear force.

For a longer interaction, the initially smooth profiles for the electromagnetic energy density and plasma density are split up (see Figs. 10.20*a* to *c*). For a Nd glass laser, the same behavior is observed in the case of a CO_2 laser (see Figs. 10.21*a* to *c*). It is remarkable that the maxima of the electromagnetic energy densities $(\mathbf{E}^2+\mathbf{H}^2)/8\pi$ are—as it must be—different by about 100 times (for CO_2 10^{16} and Nd 10^{18} W/cm^2), but the resulting maximum velocities of the blocks, Fig. 10.21*b*, are nearly the same in both cases. This confirms that the squares of the maximum plasma velocities, due to the nonlinear force acceleration are proportional to $I\lambda^2$.

$$v^2 \propto I\lambda^2 \tag{10.19}$$

This result had been confirmed experimentally for numerous quantities:

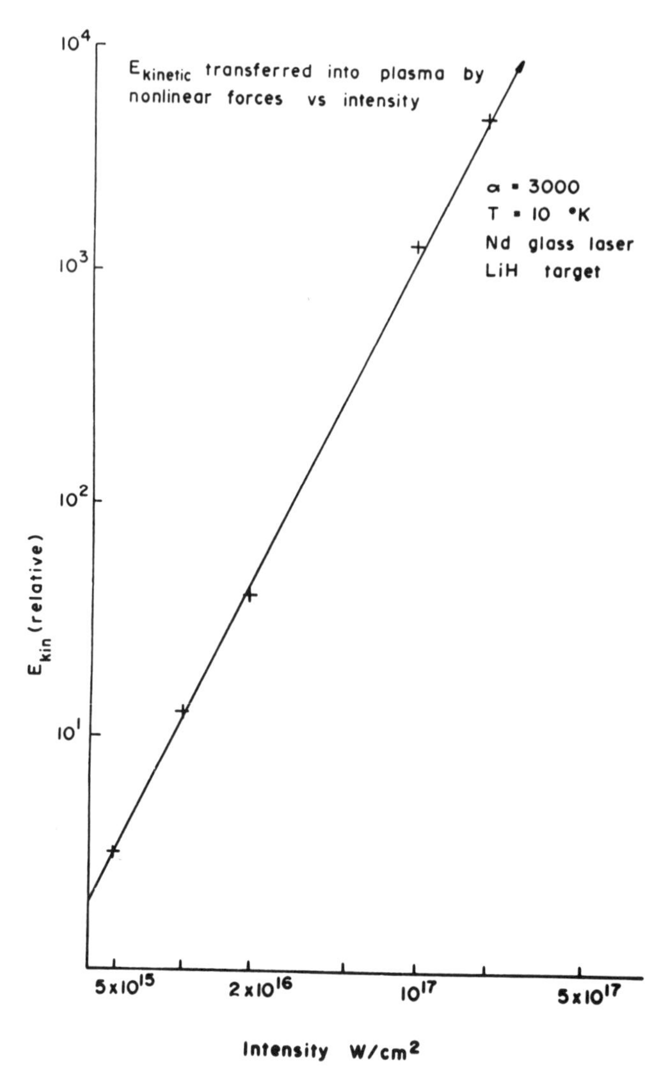

Figure 10.19 Kinetic energy transferred to the plasma depends on laser intensity and justifying the nonlinear nature of the interaction [243].

for the fast ion energy of laser produced plasmas, for the similar nonlinear "temperature," and for similar nonlinear quantities [245]. The explanation is very easy: these quantities, as well as the nonlinear force [see Eq. (9.23)] are proportional to the (nonrelativistic) oscillation energy [Eq. (6.56)].

$$\varepsilon_{osc} \sim I\lambda^2 \tag{10.20}$$

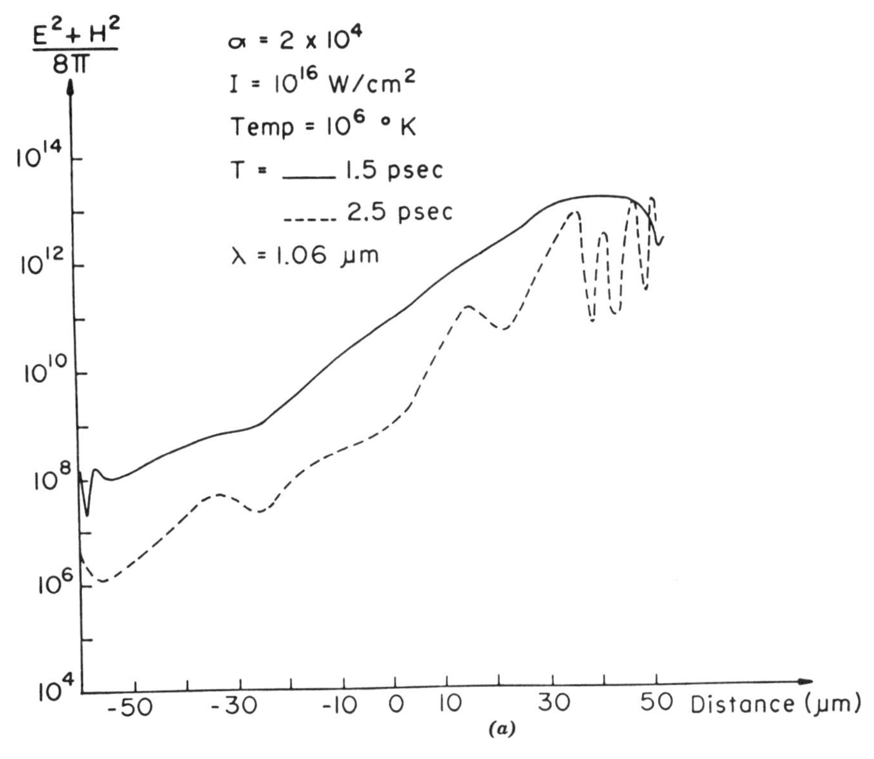

Figure 10.20a Electromagnetic energy density $(\mathbf{E}^2+\mathbf{H}^2)/8\pi$ for same initial conditions of a bi-Rayleigh profile $\alpha=2\times10^4$ cm^{-1} as in Fig. 10.18 but for time $t=1.5$ and 2.5 psec, at Nd glass laser intensity 10^{16} W/cm^2 [243].

This explains the results [245], which were gained by very expensive experiments.

10.6 Solitons

The previously described change of the smooth behavior into the oscillating one at later times (Fig. 10.20*a* to *c*) suggests the possibility of developing solitons at the plasma dynamics. The absorption process seems to follow a solitonlike behavior according to the Korteweg–de Vries equation with all the well-known consequences of solitons. This is shown by an evaluation of the computer results of the type in Fig. 10.20*a* to *c*.

There are various processes that have a solitonlike behavior and fulfill a Korteweg–de Vries (KdV) equation. If a system is within this regime, it

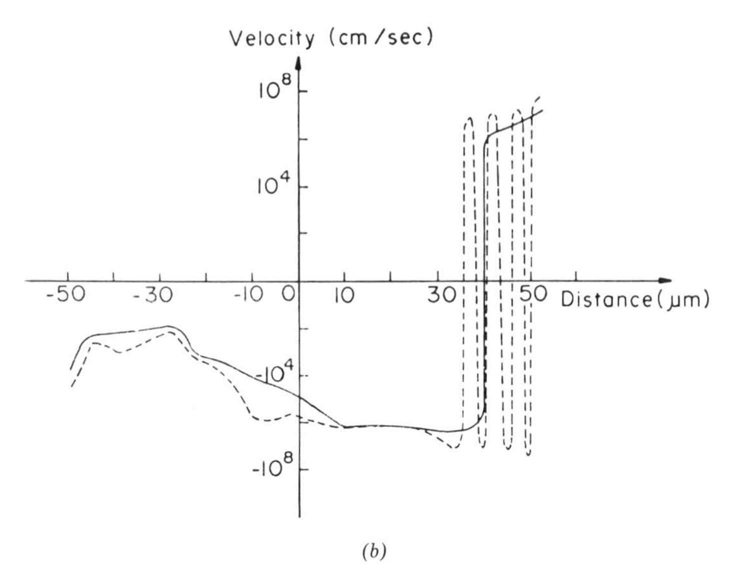

(*b*)

Figure 10.20b The velocity profiles for the cases of Fig. 10.20*a* [243].

behaves with some stable properties; but to arrive at these conditions or to deviate, dissipative processes (energy absorption or transfer) are necessary. In this case, the nonlinear force causes this dissipation.

One well-known example of solitons in plasma are acoustic waves. The KdV equation for ion acoustic waves in a plasma describes solitons in full agreement with experiments [246].

$$\frac{\partial}{\partial t} n_i + \frac{\partial}{\partial t} x \frac{\partial}{\partial x} n_i = -\mu' \frac{\partial^3}{\partial x^3} n_i \tag{10.21}$$

where n_i is the ion density. A dispersion function μ' is used, or a term representing the nonlinear force must be added. In the case considered here, numerical results of nonlinear laser plasma interaction of the type prescribed before are evaluated. The plasma velocity $\mathbf{v}$ is considered, following a KdV equation

$$\frac{\partial}{\partial t} \mathbf{v} + \mathbf{v} \frac{\partial}{\partial x} \mathbf{v} = -\mu' \frac{\partial^3}{\partial x^3} \mathbf{v} \tag{10.22}$$

These are the "velocity-type" solitons and not ion acoustic solitons as in Eq. (10.21). The left-hand side of Eq. (10.22) can be identified with the acceleration, given by the force density $\mathbf{f}$.

$$\frac{1}{m_i n_i} \mathbf{f} = \frac{\partial}{\partial t} \mathbf{v} + \mathbf{v} \frac{\partial}{\partial x} \mathbf{v} \tag{10.23}$$

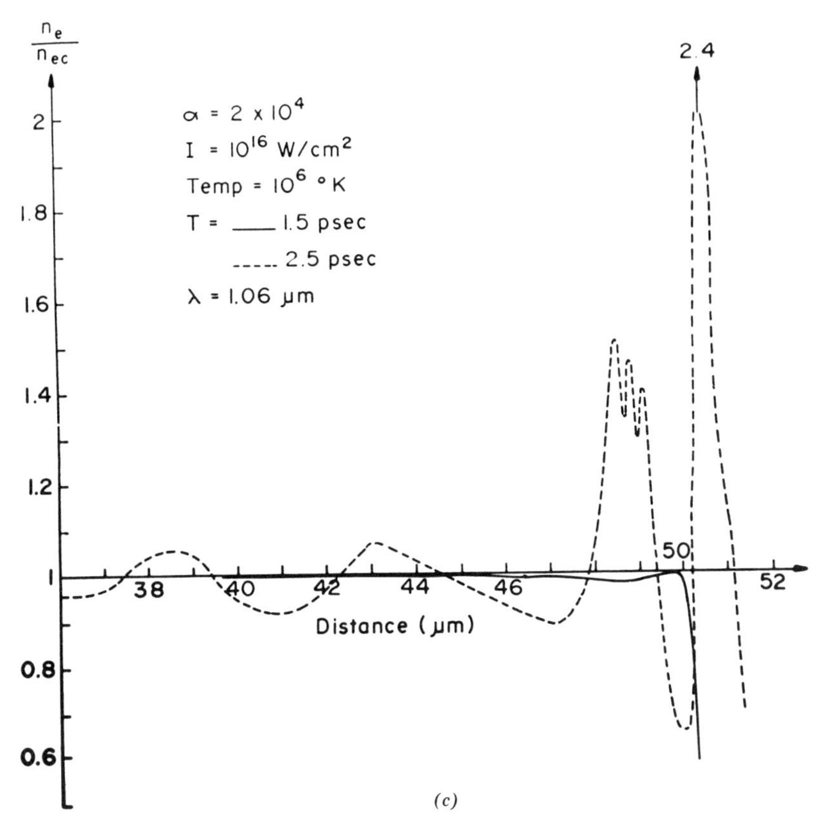

Figure 10.20c The density rippling due to the oscillations of the velocities corresponding to Figs. 10.20*a* and 10.20*b* [243].

For an evaluation of numerical results of the type in Fig. 10.20*a* to *c*, **f** has to be compared in Eq. (10.23)

$$\mathbf{f} = -\nabla p + \nabla(\mathbf{E}^2 + \mathbf{H}^2)/8\pi \tag{10.24}$$

with the third spatial derivation of **v**, Eq. (10.22). It will turn out that an agreement is possible only if the gasdynamic pressure p is neglected and only if the nonlinear force with the strengths **E** and **H** are included. The energy deposition by the radiation, based on the nonlinear intensity dependence of the optical constants, has to be considered [242, 243]. This results in examples such as in Figs. 10.20*a* to *c*.

The numerical evaluation of these results shows a soliton process of Eq. (10.22). Examining whether a similarity or a relation exists in the sense of

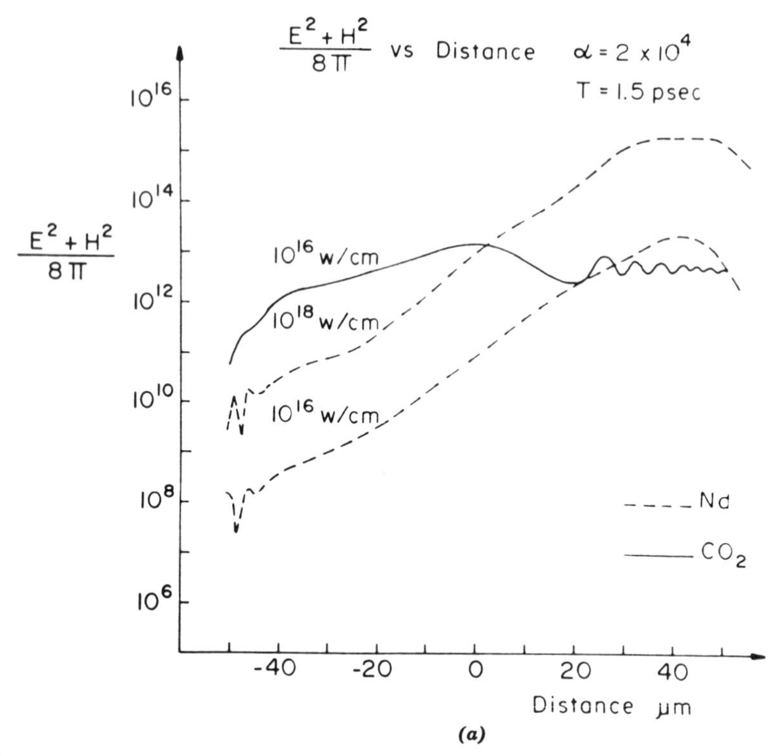

Figure 10.21a Electromagnetic energy density at 1.5 psec for various intensities for Nd glass and CO_2 lasers [243].

a Korteweg–de Vries equation, from Eqs. (10.22) to (10.24), we should have

$$-\mu' \frac{\partial^3}{\partial x^3} v = \frac{1}{8\pi m_i n_i} \frac{\partial}{\partial x} (\mathbf{E}^2 + \mathbf{H}^2) = \frac{|\mathbf{f}_{\mathrm{NL}}|}{m_i n_i} \tag{10.25}$$

The thermokinetic force has to be dropped as shown from the following.

For the evaluation, several cases of the type in Fig. 10.20 are used, where there results a very transparent decay of the $(\mathbf{E}^2 + \mathbf{H}^2)/8\pi$ field or the density into ripples, or the velocities into oscillations after an earlier very smooth behavior [247]. The evaluation of $\partial^3 \mathbf{v}/\partial x^3$ and of the right-hand side of Eq. (10.25) by numerical differentiation is shown in Fig. 10.22. There are polelike maxima of which every second is coincident.

After this result, several dispersion functions μ were tested and a complete satisfaction of Eq. (20.25) was found.

$$\mu' = \frac{\partial/\partial x(1 - n_e/n_{ec})}{1 - n_e/n_{ec}} \tag{10.26}$$

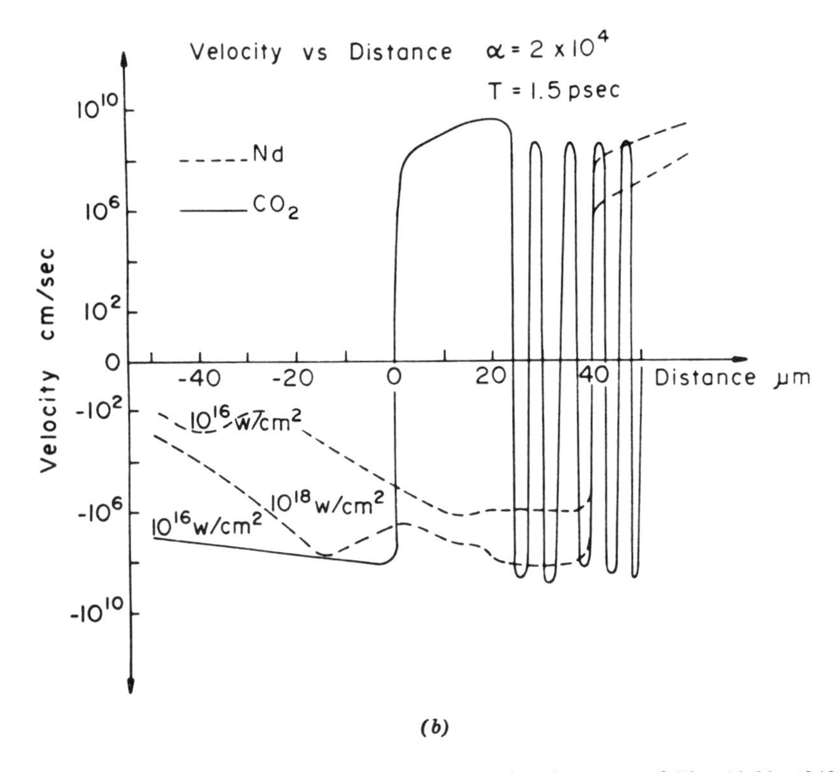

Figure 10.21b The velocity profiles at 1.5 psec for the cases of Fig. 10.20*a* [243].

where n_e is the electron density, and n_{ec} is the cutoff density. This can be seen from Fig. 10.21 when comparing the part of n_e/n_{ec} for the time of the plots of the quantities in the upper diagram. Going from A to B by multiplying $(\partial/\partial x)(\mathbf{E}^2+\mathbf{H}^2)/8\pi$ (dashed line) by $-1/\mu$, keeps the sign but drops the lines to zero at B to be coincident with the $\partial^3 v/\partial x^3$ values. Due to the pole of $-1/\mu$ in Eq. (10.26), one arrives at the same pole for the dashed curve multiplied by $-1/\mu$ as the pole of $\partial^3 v/\partial x^3$ shows at C. The same procedure can be followed from C to E.

Within the numerical accuracy, a complete identity of

$$\frac{\partial}{\partial t}v+v\frac{\partial}{\partial x}v=\frac{1}{8\pi n_i m_i}\frac{\partial}{\partial x}(\mathbf{E}^2+\mathbf{H}^2) \tag{10.27}$$

with a KdV equation (10.22) is found [limited to positive brackets, Eq. (10.26)], otherwise the more general expression [Eq. (10.26)] has to be used:

$$\frac{\partial}{\partial t}v+v\frac{\partial}{\partial x}v=\frac{\partial \ln \varepsilon'}{\partial x}\frac{\partial^3}{\partial x^3}v \tag{10.28}$$

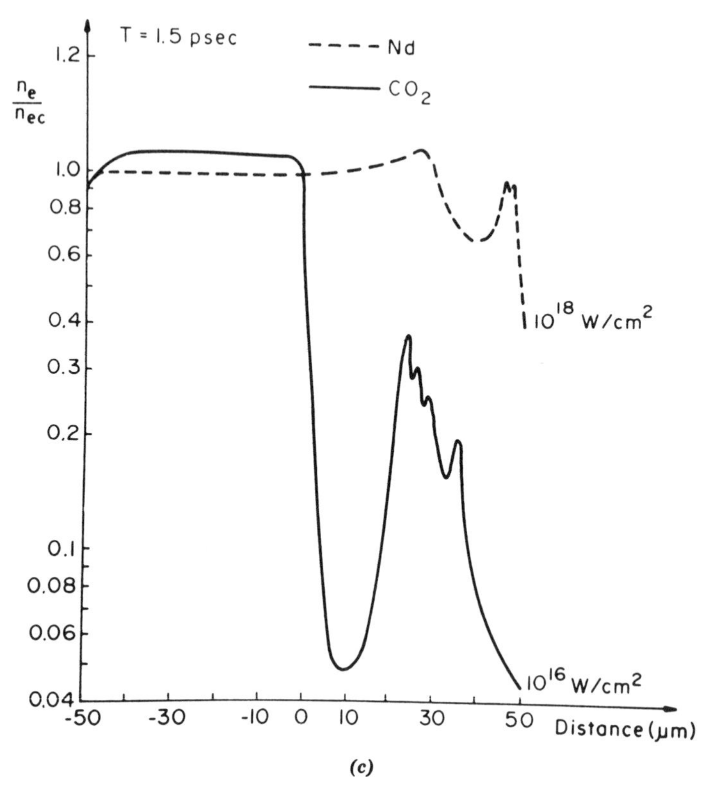

(c)

Figure 10.21c The density profiles for Ng glass lasers at 10^{18} W/cm^2 and for CO_2 lasers at 10^{16} W/cm^2 at 1.5 psec. Both density formations show the existence of a caviton [243].

The real part of the dielectric constant ε'

$$\varepsilon = \varepsilon' + i\varepsilon''; \qquad \varepsilon' = 1 - \omega_p^2/\omega^2 = 1 - n_e/n_{ec} \tag{10.29}$$

(ω_p = plasma frequency, ω = laser frequency) has to be used. The numerical accuracy is very low in the region of the function poles, the coincidence of which, however, is very sharp. The accuracy near the points B and D is very high and confirms that no imaginary part has to be used from ε.

Any correlation of Eqs. (10.27) and (10.28) for the special time is a proof of the numerical stability of the calculations described in the preceding subsection. Furthermore, as expected, the gasdynamic pressure p for the motion in Eq. (10.24) is negligible compared with the nonlinear force. The only strangeness of the result is that the dispersion factor is not the dielectric constant but a logarithmic derivative of the real part of the dielectric constant

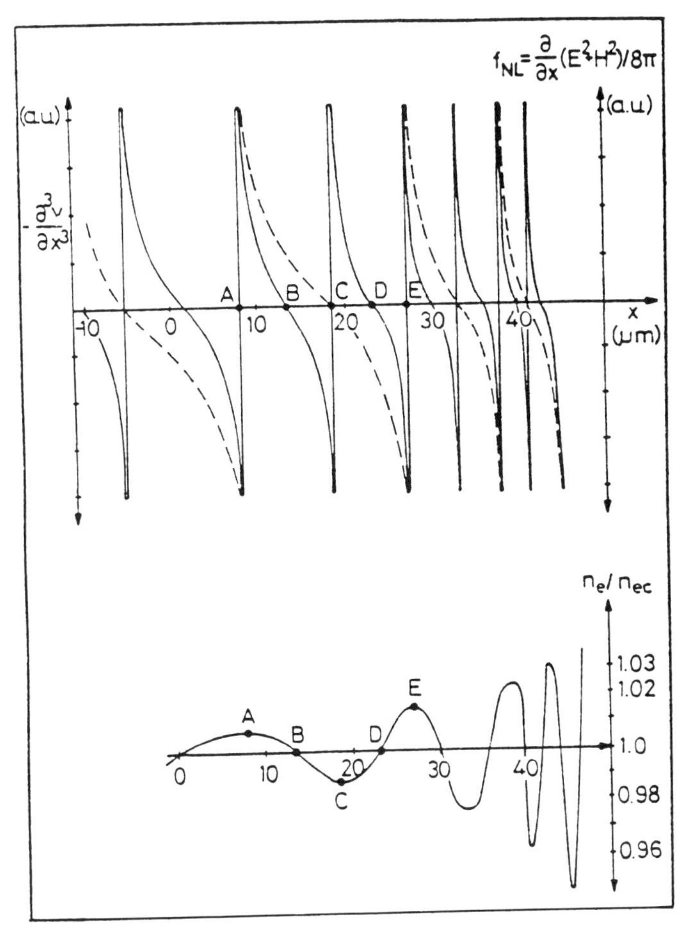

Figure 10.22 $\partial^3 v/\partial x^3$ (——) and f_{NL} (- - -) are evaluated from numerical examples of nonlinear dynamic calculations of laser plasma interaction [247] and compared with the generated density ripple. Introducing the special dispersion function, Eq. (10.26), a behavior according to the Korteweg de Vries equation is established [266].

only. However, the similarity to the expression $\partial \ln \varepsilon/\partial x$ in the theory of the resonance absorption [248] is not surprising. It has an importance similar to the attempts to explain the density rippling [242, 249] as a structure resonance [250, 251].

It has to be noted that at earlier times of the interaction, correlation to a KdV equation is not possible. The change from this case to the soliton case is due to nonlinear-force produced macroscopic and nonthermalizing absorption [252].

In order to gain further numerical facts about the solitonlike process, Lalousis [253] calculated cases with the same parameters as in Fig. 10.20*a* but with varying parameter α of the bi-Rayleigh initial density profile. For relatively low α, for example, $\alpha = 10^{-3}\ \text{cm}^{-1}$, the cutoff density is not reached up to 3 psec along the whole plasma of about 100 μm thickness. Attention is then due to usual and nonlinear absorption only, and a wave field with a high degree of reflection (standing wave) is created. It is remarkable that at 2 psec, the profiles of the nonlinear force and of the derivatives of the velocity are following roughly the Korteweg–de Vries equation (see Fig. 10.23) but the lower range from 40 down to -50 μm follows the Banjamin–Ono equation [254]

$$\frac{\partial}{\partial t} v + v \frac{\partial}{\partial x} v = -H \frac{\partial^2}{\partial x^2} v = \frac{\partial}{\partial x} \frac{\mathbf{E}^2 + \mathbf{H}^2}{8\pi} \tag{10.30}$$

where the Hilbert transform H has to be unity. For an analysis of these processes, more numerical examples are necessary for understanding the change into the solitonlike behavior by the dissipation processes.

The result of this subsection permits the following conclusion. It is shown numerically and experimentally that the nonlinear force can transfer optical energy into thick fast-moving blocks of plasma without heating. This highly efficient transfer of optical energy into kinetic energy of compressed plasma is the basis of a high-efficiency concept of laser fusion [255]. It avoids the difficulties of the interaction (absorption) and transport processes for laser

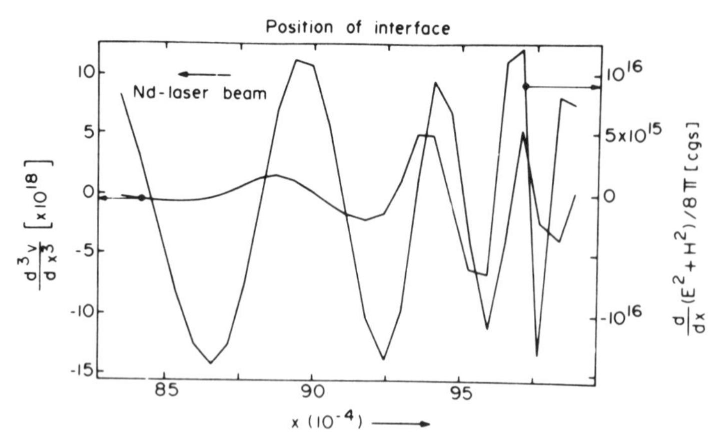

Figure 10.23 Output for same initial conditions as in Fig. 10.20 for $I = 10^{16}\ \text{W/cm}^2$ but for $\alpha = 10^3\ \text{cm}^{-3}$. The peripheric part (43 to 49 μm) follows nearly the Korteweg–deVries equation [253].

fusion [255], but it needs better laser technology (shorter and more precise pulses). The result of the soliton decay, however, provides much faster thermalization of laser radiation to ions in the corona than by Coulomb collisions (see Eq. (10.8). This gives one hope of extending the gasdynamic ablation-compression scheme of Nuckolls [219] to higher laser intensities.

ELEVEN

Striated Motion and Resonance Absorption

Sections 9 and 10 discuss the nonlinear forces, the photon momentum, the created ion energies, and several fully dynamic, numerical treatments of the plasma expansion following laser interaction, including the nonlinear forces and optical properties. Plane waves from a laser are assumed to be perpendicularly incident onto a stratified (inhomogeneous) plasma. This section discusses the case of plane waves obliquely incident onto a stratified plasma.

There are two questions of interest. While the net motion of the plasma corona under laser irradiation is found to be directed toward the decreasing density (parallel to the negative density gradient) and no net motion can happen in the plane of the plasma surface because of conservation of momentum, the standing wave field in the plasma corona can produce a straited motion. Parts of the plasma can move toward the one direction and other parts to the contrary direction with vanishing net motion. The general formulas for these forces from the general equation of motion are valid for all kind of wave fields, even for the evanescent wave field beyond the turning point of the wave in the corona. But the evaluation of the wave fields is restricted to the WKB approximation only for an angle of wave propagation of less than 50° [256].

The other question is the evaluation of the classical and linear solution of the wave field for the densities below the turning point of the wave. It has been found by Denisov [248], see also Ginzburg [257], that for *p*-polarization the component of the electric field perpendicular to the plane of the plasma has an extraordinarily high maximum at the cutoff density [258]. This is far below the turning point at oblique incidence, inside the evanescent region of the wave field. The reason is the resonance with the electrostatic waves. This process has been called resonance absorption [257]. Thus the process of aborption has to be calculated separately.

11.1 Striated Motion

Using the x-direction along the depth of the stratified plasma and using linear polarized radiation, the case of the p-polarization of an obliquely incident wave is then given by field components E_x, E_y, and H_z only. From the equation of motion in the formulation of Eqs. (8.81) or (8.82); the y-component of the nonlinear force is then [256]

$$f_{\mathrm{NL}py} = \frac{1}{8\pi}\frac{\partial}{\partial x}[E_x E_y - (1-n^2)E_x E_y] + \frac{1}{16\pi}\frac{\partial}{\partial y}[-E_x^2 + (2n^2+1)E_y^2 - H_z^2) \tag{11.1}$$

For the case of the s-polarization (**E** perpendicular to the plane of incidence) the wave field has the components E_z, H_x, and H_y only. The resulting component of the nonlinear force in the y-direction is then

$$f_{\mathrm{NL}sy} = \frac{1}{8\pi}\frac{\partial}{\partial x}H_x H_y + \frac{1}{16\pi}\frac{\partial}{\partial y}(-E_z^2 - H_x^2 + H_y^2) \tag{11.2}$$

There will be no time-averaged forces in the z-direction. The x-component of the general forces for oblique incidence are discussed in Section 8.

The solution for **E** and **H** for stationary solutions with a frequency according to Eq. (6.19) is given from Eq. (6.30)

$$\Delta\mathbf{E} + \frac{\omega^2}{c^2}n^2\mathbf{E} - \nabla\frac{2}{n}\mathbf{E}\cdot\nabla n = 0 \tag{11.3}$$

The magnetic field can be derived from the Maxwellian equation

$$\dot{\mathbf{H}} = c\nabla\times\mathbf{E} \tag{11.4}$$

It is evident, for the case of s-polarization, that the last term in Eq. (11.3) vanishes, as there is a z-component of **E** only and the refractive index n is dependent on the x-coordinate only. In this case the solutions of the WKB case, Eqs. (7.28) and (7.30), can be used in Eq. (11.3). A vanishing time-averaged y-component of the nonlinear force follows, according to Eq. (11.2).

Any force can result for the p-polarization only $\mathbf{f}_{\mathrm{NL}} = 0$. For this case, the last term of Eq. (11.3) has to be taken into account. This term couples the transversal components of the electromagnetic wave with the longitudinal components generated in the plasma when turning the direction of the obliquely incident wave. As this coupling is not necessary for the discussion of the net force component in the x-direction (to lower plasma density), this part is neglected in Sections 7 and 8. Now this coupling is included, and the WKB approximation of the wave field in the corona of the plasma is derived for p-polarization. The geometry is described in Fig. 11.1, which shows the resulting nodes of the standing wave and the finally resulting forces for the striated motion.

From Eq. (11.3) for p-polarization—the index p will be neglected within this section—the following equations are found

$$\left(\frac{\partial^2}{\partial x^2}+\frac{\partial^2}{\partial y^2}+\frac{\omega^2}{c^2}n^2(x)\right)E_x+\frac{\partial}{\partial x}\frac{2E_x}{n}\frac{\partial n}{\partial x}=0 \tag{11.5}$$

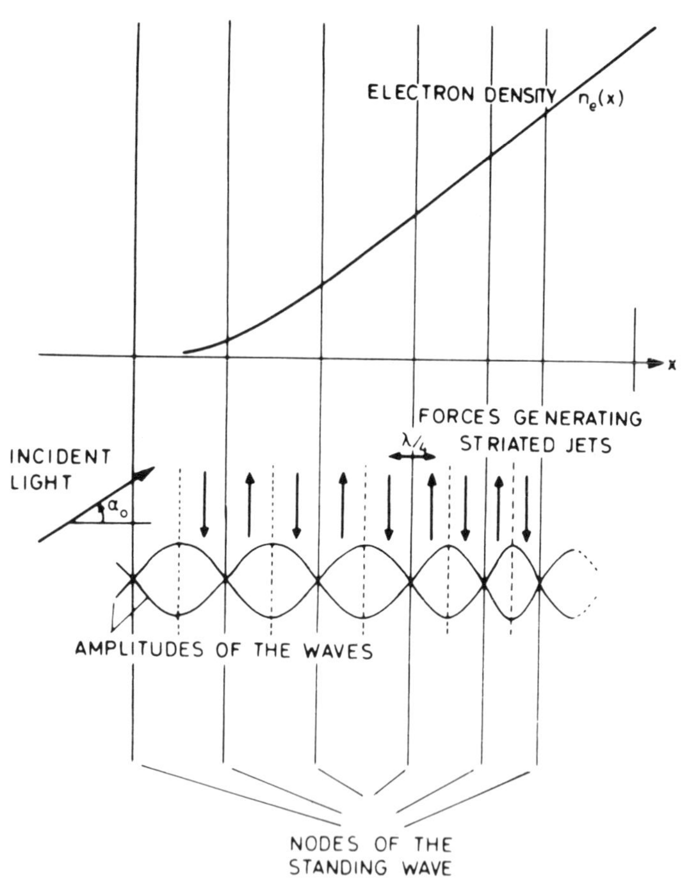

Figure 11.1 A plane p-polarized wave incident at an angle α_0 on a stratified plasma with an electron density $n_e(x)$ together with the corresponding reflected wave generates spatially averaged forces and a net motion along the fixed nodes of the standing wave (striated jets) [256].

$$\left(\frac{\partial^2}{\partial x^2}+\frac{\partial^2}{\partial y^2}+\frac{\omega^2}{c^2}n(x)\right)E_y+\frac{\partial}{\partial y}\frac{2E_x}{n}\frac{\partial n}{\partial x}=0 \tag{11.6}$$

These equations can be separated by the ansatz

$$E_x=e_{xx}(x)e_{xy}(y); \qquad E_y=e_{yx}(x)e_{yy}(y) \tag{11.7}$$

The appropriate boundary conditions for plane waves incident at an angle α_0 in vacuum (Fig. 11.1) leads to Snell's law

$$\sin\alpha_0=n(x)\sin\alpha(x) \tag{11.8}$$

From Eq. (11.5) is obtained

$$\frac{\partial^2}{\partial y^2}e_{xy}+\frac{\omega^2}{c^2}(\sin^2\alpha_0)e_{xy}=0 \tag{11.9}$$

$$\frac{\partial^2}{\partial x^2}e_{xx}+\frac{\omega^2}{c^2}n^2(x)\cos^2\alpha(x)e_{xx}+2\frac{\partial}{\partial x}\frac{d\ln n}{dx}e_{xx}=0 \tag{11.10}$$

For the separation of Eq. (11.6), the relation

$$e_{xx}=e_{yx}\frac{\sin\alpha(x)}{\cos\alpha(x)} \tag{11.11}$$

has to be, which is used in agreement with the case of homogeneous plasma, and which leads to

$$\frac{\partial^2}{\partial x^2}e_{yx}+\frac{\omega^2}{c^2}n^2[\cos^2\alpha(x)]e_{yx}=0 \tag{11.12}$$

$$\frac{\partial^2}{\partial y^2}e_{yy}+\frac{\omega^2}{c^2}(\sin^2\alpha_0)e_{yy}+2\tan\alpha(x)\frac{d\ln n}{dx}\frac{\partial}{\partial y}e_{xy}=0 \tag{11.13}$$

A necessary condition for this separation is

$$\tan\alpha(x)\frac{\partial\ln n(x)}{\partial x}=\text{const} \tag{11.14}$$

which is immediately given by Snell's law, Eq. (11.8).

Using the integration constants $C_n(n=1, 2, \ldots)$, the exact solution of Eq. (11.9) is

$$e_{xy}=C_1\cos[y(\omega/c)\sin\alpha_0]+C_2\sin[y(\omega/c)\sin\alpha_0] \tag{11.15}$$

The solution of Eq. (11.10) is given for the assumptions of the WKB approximation and by neglecting the last term of Eq. (11.10), which is justified at this point because it does not produce forces in the y-direction, but it must not be neglected at the other points of this derivation.

$$e_{xx}=[n(x)\cos\alpha(x)]^{-1/2}[C_3\cos(G_x)+C_4\cos(G_x)] \tag{11.16}$$

where

$$G_x = -\frac{\omega}{c}\int^x n(\xi)\cos\alpha(\xi)\,d\xi \tag{11.17}$$

The WKB solution of Eq. (11.12) is

$$e_{yx} = [n(x)\cos\alpha(x)]^{-1/2}[C_5\cos(G_x) + C_6\sin(G_x)] \tag{11.18}$$

In Eq. (11.13) the solution (11.15) has to be substituted. The exact solution is obtained [202, 256]:

$$e_{yy} = C_7\sin[y(\omega/c)\sin\alpha_0] + C_8\cos[y(\omega/c)\sin\alpha_0] + 2\tan\alpha(x)\frac{c\ln n(x)}{dx}(\omega/c)(\sin\alpha_0)(y-y_0) \times\left[C_1\cos\left(y\frac{\omega}{c}\sin\alpha_0\right) - C_2\sin\left(y\frac{\omega}{c}\sin\alpha_0\right)\right] \tag{11.19}$$

The constant y_0 is coupled with the integration constants.

We try to fit the solution in the inhomogeneous plasma with the propagating wave in the vacuum at an angle of incidence α_0 (Fig. 11.1) and an amplitude E_v and a magnetic vector

$$\mathbf{H} = \mathbf{i}_3 E_v\cos[G_{yt} - (\omega/c)y\sin\alpha_0] \tag{11.20}$$

where

$$G_{yt} = -(\omega/c)y\sin\alpha_0 + \omega t \tag{11.21}$$

One finds the conditions for the integration constants with a propagating wave at an angle of incidence α_0 in Eqs. (11.13) to (11.16). With Eq. (11.5) is obtained

$$E_x = \frac{E_v\sin\alpha_0\cos^{1/2}\alpha_0}{(n\cos)^{1/2}}(\cos G_x\cos G_{yt} - \sin G_x\sin G_{yt}) \tag{11.22}$$

$$E_y = \frac{E_v\cos\alpha_0\cos^{1/2}\alpha_0}{(n\cos)^{1/2}} \left[\cos(G_x+G_{yt}) + 2\tan(x)\frac{d\ln n}{dx}\frac{\omega}{c}(y-y_0)(\sin\alpha_0)(\cos G_x\cos G_{yt} + \sin G_x\cos G_{Nt})\right] \tag{11.23}$$

The difference in the solutions of E_x and E_y from the case in vacuum is the spatial variation of the amplitude depending on x according to the WKB condition, and further the term with the factor $d\ln n/dx$, which determines the longitudinal component of the electric vector with respect to the actual direction of propagation of the wave. This expression, called the "longitudinal term," increases linearly on y, where y_0 is constant. Although an infinite plane wave is assumed, the real conditions are determined by a wave bundle of finite width. To describe exactly this complex problem with a wave bundle in the optical case (in contrast to the wave mechanical case [257]), the plane wave must be cut laterally to form a bundle; the side of lower y can be identified by y_0, so the longitudinal component increases monotonically from one side. To facilitate the discussion one can restrict this treatment to the condition where the longitudinal term is much smaller than the remaining term in the bracket of Eq. (11.23): that is,

$$\tan(x)\frac{d\ln n}{dx}\frac{\omega}{c}(y-y_0)\sin\alpha_0=p_0\leqslant 0.30 \tag{11.24}$$

With this restriction, the magnetic field in the plasma can be approximated by

$$\mathbf{H}=\mathbf{i}_3\frac{E_v(n\cos\alpha_0)^{1/2}}{\cos^{1/2}\alpha}(G_x+G_{yt})-\mathbf{i}_3\frac{c}{\omega}\frac{E_v\cos^{1/2}\alpha_0(0.5-\sin^2\alpha)}{n^{3/2}\cos^{3/2}\alpha}\frac{\partial n}{\partial x}\sin(G_x+G_{yt}) \tag{11.25}$$

With this result for a plane wave propagating at an angle of incidence α_0 (Fig. 11.1) we can derive the field for a standing wave by adding a wave with an angle of propagation resulting from total reflection of the first wave. The complications due to the Goos–Haenchen effect are excluded from the discussion here and may not be of significant influence [257].

The reflected wave results in expressions similar to those given in Eqs. (11.23), and (11.25), where the sign of G_x and of $\mathbf{H}$ have to be changed. Adding the solutions for the incident and reflected wave, the standing wave for the polarization of $\mathbf{E}$ parallel to the plane of incidence (p-polarization) results in

$$\begin{aligned}\mathbf{E}=&\frac{2E_v\cos^{1/2}\alpha_0}{(n\cos\alpha)^{1/2}}\cos\left(-\frac{\omega}{c}\int_0^x n(\xi)\cos\alpha(\xi)\,d\xi\right)\\&\times\cos\left(-\frac{\omega}{c}y\sin\alpha_0+\omega t\right)\Bigg[\mathbf{i}_7\sin\alpha_0+\mathbf{i}_2\cos\alpha_0\\&\times\left(1+2\tan\alpha(x)\frac{d\ln n}{dx}\frac{\omega}{c}(y-y_0)(\sin\alpha_0)\right)\Bigg]\end{aligned} \tag{11.26}$$

$$\begin{aligned}\mathbf{H}=\mathbf{i}_3&\frac{2E_v(n\cos\alpha_0)^{1/2}}{\cos^{1/2}\alpha}\sin\left(-\frac{\omega}{c}\int^x n(\xi)\cos\alpha(\xi)\,d\xi\right)\\&\times\cos\left(-\frac{\omega}{c}y\sin\alpha_0 t+\omega t\right)\left(1-\frac{dn}{dx}\frac{\omega}{c}\frac{E_v\cos^{1/2}\alpha_0(0.5-\sin^2\alpha)}{n^{3/2}\cos^{3/2}\alpha}\right)\end{aligned} \tag{11.27}$$

The evaluation of the nonlinear force in the y-direction follows Eq. (11.1), where terms averaged by the y-coordinate are considered. This averaged value of the second term on the right-hand side of Eq. (11.1) vanishes, because the y-dependence of $\mathbf{H}^2$, E_x^2, and E_y^2 is given by terms containing only $\sin^2[y(\omega/c)\sin\alpha_0]$, $\cos^2[y(\omega/c)\sin\alpha_0]$, or $(y-y_0)\sin[y(2\omega/c)\sin\alpha_0]$ as factors. The derivation by y results in spatially averaged vanishing values. This vanishing does not hold if condition (11.24) is not fulfilled. From the first term on the right-hand side of Eq. (11.1) and Eqs. (11.22) and (11.23) is found, after time averaging

$$\bar{f}_{\mathrm{NL}py} = \frac{1}{8\pi}\frac{\partial}{\partial x}\frac{E_v^2\omega\sin^3\alpha_0\cos^2\alpha_0}{2cn^2\cos^2}\sin\left(2\frac{\omega}{c}\int^x n(\xi)\cos\alpha(\xi)\right)\frac{\partial\ln n}{\partial x}(y-y_0) \tag{11.28}$$

Differentiation and neglect of second and higher order terms in $\partial n/\partial x$ leads to

$$\bar{f}_{\mathrm{NL}sy} = \frac{1}{8\pi}\frac{E_v^2\sin^3\alpha_0\cos^2\alpha_0\omega^2}{n\cos\alpha}\frac{\omega^2}{c^2}(y-y_0)\frac{\partial\ln n}{\partial x} \times\cos\left(2\frac{\omega}{c}\int^x n(\xi)\cos(\xi)\,d\xi\right) \tag{11.29}$$

At a constant x-value, the force increases along the y-coordinate and is proportional to the gradient of the refractive index n. An increase in the force occurs especially at densities close to the cutoff density n_{ec}. The validity of this condition can be assumed within the depth given by a large number of vacuum wavelengths $\lambda_0 = 2\pi c/\omega$ as a special calculation demonstrates [223]. Also at very small dn/dx, the force can be large owing to sufficiently high y. The nonlinear force component $f_{\mathrm{NL}py}$ oscillates in time and is comparable to the field of a standing wave (Fig. 11.1). The force is zero at the nodes and antinodes of the standing wave, which have a distance of a quarter of the actual wavelength,

$$\lambda^* = \lambda_0\cos(x)/n \tag{11.30}$$

Between the nodes, the force is directed toward the positive or negative y-direction for $y>0$ and $y_0<0$. The maximum (or minimum) value of the force is, using $p_0 = 0.3$ from Eq. (11.24) in cgs units,

$$f_{\mathrm{NL}py} = f_0 = 0.075(E_v^2/\lambda)|n|\sin\alpha_0\cos^2\alpha_0 \tag{11.31}$$

The resulting force density can reach remarkably high values. The restriction to $\alpha<40^\circ$ was only necessary [138] in decoupling the waves of linear polarization. In the following only linear p-polarization is considered. For a reasonable example of an intensity $I_v = 10^{14}$ W/cm^2 of neodymium laser irradiation, with $\alpha = 25^\circ$ and a value of the refractive index $|n| = 0.9$ is found

$$f_0 = 2.25\times 10^{14}\ \mathrm{dyn/cm^3} \tag{11.32}$$

The thermokinetic force of a deuterium plasma with a temperature

$T_e = 100$ eV and a density $n_e = 1.9 \times 10^{20}$ cm^{-3}, dropping linearly to vacuum within 10^{-2} cm, results in

$$f_{\text{th}} = 3.2 \times 10^{12} \text{ dyn/cm}^3 \tag{11.33}$$

The essential term in the nonlinear force in the y-direction of a propagating wave in an inhomogeneous plasma is the nonlinear term. Without it, the spatially averaged force vanishes. The case of p-polarization is determined by Eqs. (11.3) and (11.4), where the last term in Eq. (11.3) vanishes. Therefore, no longitudinal component is created for the conditions used here in the case of s-polarization. The y-component of the nonlinear force density, Eq. (11.2), results, as mentioned, in a vanishing spatially averaged value similar to that for p-polarization without the longitudinal term.

Several conditions have to be discussed in connection with striated motion in the plasma due to the force given by Eq. (11.1). Besides the questions of how the standing wave is created and why the gasdynamic motion of the plasma through the standing wave pattern is slow enough to build up a striated motion, the *mean free path* of the particles with respect to the thickness of the layers must be considered, and, furthermore, the limitation between a *laminar motion of the layers and a turbulent motion* must be determined.

The thickness of one layer is one-quarter of the vacuum wavelength multiplied by $n^{-1}(x) \cos \alpha(x)$. Because we consider conditions of $S = 1/n < 10$ and $\alpha_0 < 45°$, we can begin with neodymium glass laser radiation from a thickness $d > 2.3 \times 10^{-5}$ cm. The mean free path for ions in fully ionized deuterium is

$$\bar{l} = v_i / \nu_{ii} \tag{11.34}$$

In a partially ionized plasma, the mean free path is less because of collisions with neutrals. For our purposes, l of Eq. (11.34) is an upper bound, where v is the thermal velocity of the ions $v_i = (2kT_i/m_i)^{1/2}$, while T_i is the ion temperature. The ion collison frequency $\nu_{ii} = \nu_{ee}(m_e/m_i)^{1/2}$ (with the electron mass m_e), which is used from Spitzer's derivation, Eq. (3.35), results in

$$\nu_{ii} = 9 \times 10^{-9} n_i \ln \Lambda / T_i^{3/2} \tag{11.35}$$

where n_i is in cm^3 and T_i is in eV. The Coulomb logarithm $\ln \Lambda$ is simply set equal to 10. To get a mean free path l smaller than d, the following condition is obtained

$$n_i \geqslant n_{ic} = 4.4 \times 10^{17} T_i^2 \tag{11.36}$$

Figure (11.2) contains a plot of the critical ion density n_{ic} as function of temperature T_i. Densities lower than n_c are excluded in the following considerations.

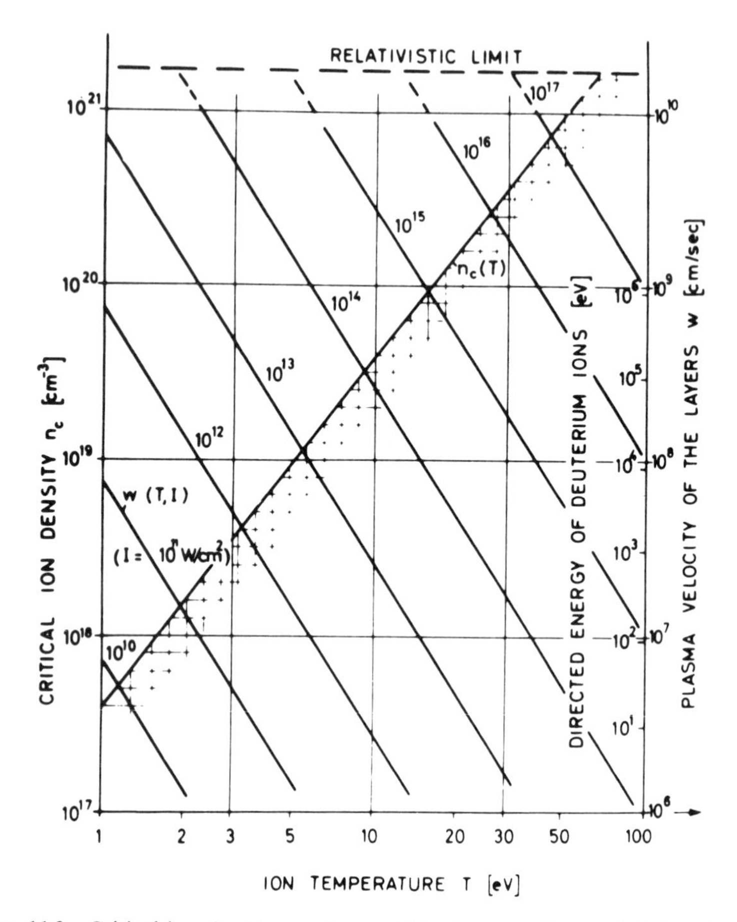

Figure 11.2 Critical ion density n_{ic} above which the mean free path is less than the thickness d of one layer. Plasma velocities w and ion energies of laminar motion for two subsequent layers depending on temperature $T=T_i$ and neodymium glass laser intenities I [256].

The striated motion acted on by the nonlinear forces (11.29) under laminar conditions is calculated. At a distance d,

$$d=|n|(\lambda_0/4)\cos\alpha_0 \tag{11.37}$$

the stationary velocity $\bar{w}$ due to a force F at a viscosity η is

$$\bar{w}=2Fd/\eta \tag{11.38}$$

The factor 2 takes into account that at both sides of d, forces F act in opposite directions. F is the force per cm^2 of the layer of laminar motion, and it is given from the force density f by

$$F = fd \tag{11.39}$$

Using the viscosity η of the liquid of the ions with an ion temperature T_i

$$\eta = \frac{n_i K T_i}{\nu_{ii}} \tag{11.40}$$

From Eqs. (11.37) to (11.40) and Eq. (11.25) is found

$$w = \frac{\lambda_0}{107|\eta|} E_v^2 \sin \alpha_0 \cos^4 \alpha_0 \frac{\nu_{ii}}{nkT} \tag{11.41}$$

E_v is in V/cm, T_i is in eV, and the wavelength λ_0 is in cm. The velocity of the laminar motion does not depend on the density, as is well known from the kinetic theory of gases. Figure 11.2 contains the plots of the difference of the velocity w of deuterium between one maximum and one minimum of the forces at the maximum value $u_0 = 25.5°$, for a reasonable $|n| = 0.25$ and $\ln \Lambda = 10$ for various intensities of neodymium glass laser radiation. The corresponding ion energies are also given by the ordinate on the right-hand side of the diagram.

The striated motion of the plasma is calculated for laminar conditions. This is limited by the critical Reynolds number Re_{kr}, above which the motion is turbulent. In the case of turbulence, the final velocities of the layers will be much higher under stationary conditions than in the laminar case. The turbulent state is of very complex nature, and the straited motion will then be disturbed by Helmholtz–Kelvin instabilities. Anyhow, the larger velocities of the layers and the stronger, subsequent thermalization has to be taken into account for the processes in the corona with maximum effects around 25° incidence for p-polarization.

The following evaluation, for the restricted case of laminar motion, should give some knowledge of what processes of straited motion can occur in the corona. The necessary restriction to low plasma temperatures is only a limitation for this case. Under real conditions, the striated motion will then happen at higher temperatures with higher velocities in the regime of turbulent motion. The critical Reynolds number has a value

$$\mathrm{Re}_{kr} = \frac{\bar{w} d n_i m_i}{\eta} \tag{11.42}$$

between 10^3 and 10^6. The higher limit may be more realistic because of the fact that the motion of the layers is not disturbed by surfaces or corners of solid boundaries. Using Eqs. (11.37) to (11.41) for deuterium is found

$$\mathrm{Re}_{kr} = 2.5 \times 10^{-3} \frac{\lambda_0^2}{kT^5} (\ln \Lambda)^2 E_v^2 n_k m_k$$

$$= 8.7 \times 10^{-30} \frac{E_v^2 n_i}{T^5} \tag{11.43}$$

E_v is in V/cm, T in eV, and the ion density n_i is in cm^3. Using a reasonable Reynolds number of 3×10^5 (about 100 times its minimum value), one arrives at maximum ion densities n_i in Fig. 11.3 at maximum laser intensities at given temperatures where laminar conditions are fulfilled.

From the results of Figs. (11.2.) and (11.3), the conditions for the case are selected, where an ion velocity in the various layers of the laminar striated motion representing 1 keV ion energies is reached. Figure 11.4 then gives a synopsis; the necessary laser intensity I_1 is a well-defined function of the plasma temperature. The density n_i of the ions first must be larger than the

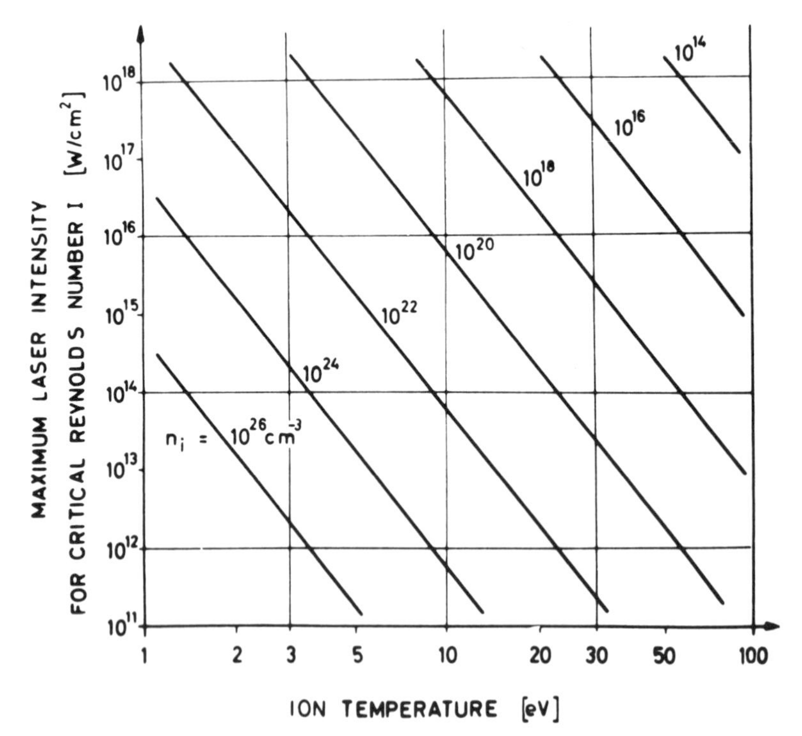

Figure 11.3 Maximum neodymium glass laser intensities at miximum ion densities n_i for laminar motion of the striated jets [256].

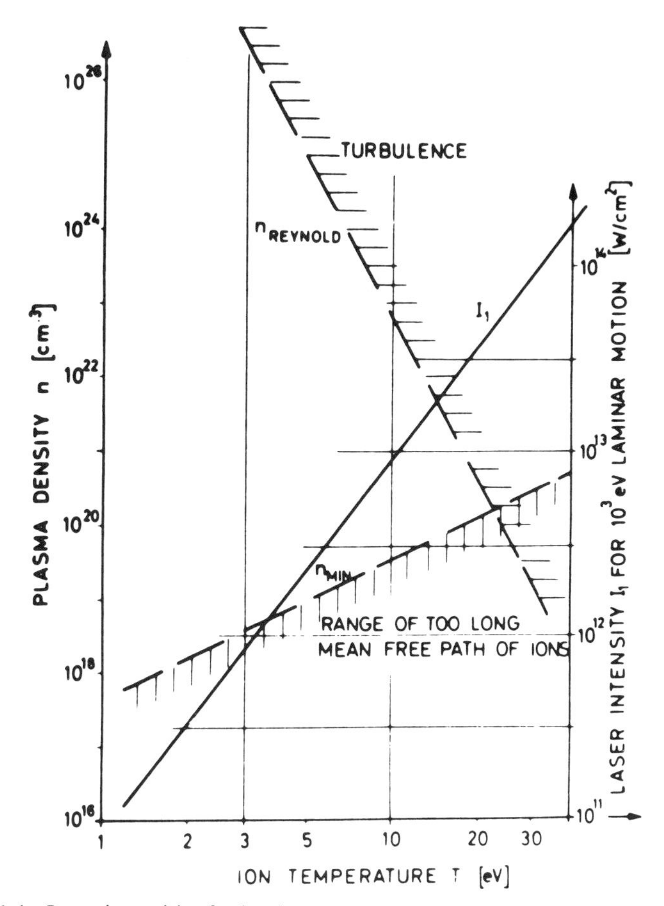

Figure 11.4 Laser intensities for laminar motion of 1 keV ion energy depending on the plasma temperature. Limitation of the ion densities by the mean free path and by Reynolds number [256].

minimum value n_{min} defined by the mean free path $l \leqslant d$. A further limitation of the density is given by n_{Re}, which distinguishes between the cases of laminar and turbulent motion. Reasonable conditions of keV energy of the deuterium ions for obtaining laminar motion can be reached for densities up to 100 times lower than that of the cutoff density of 10^{21} cm^{-3} (neodymium glass laser) at reasonable laser intensities between 10^{12} and 10^{14} W/cm^2 and for temperatures below 20 eV. This is not unrealistic for the conditions of a laser produced plasma. Otherwise, turbulent faster motions occur.

Now is calculated the time within which the striated motion is built up. The force density f accelerated the plasma, if viscosity is neglected, up to the velocity w.

$$\bar{w}=at_a; \qquad a=\frac{f}{n_i m_i} \tag{11.44}$$

where a is the acceleration. The acceleration time t_a is given by

$$t_a=\frac{wn_i m_i}{f}=1.11\times10^{-29}\frac{n_i}{T^{5/2}|n|^3}\text{ sec} \tag{11.45}$$

This acceleration time does not depend on the laser intensity I, because each of w and f are linear with I according to Eqs. (11.30) and (11.41). In the numerical factor of Eq. (11.41), the angle of incidence is $\alpha_0=25°$. The reasonable temperature $T=20$ eV results in

$$t_a=6.2\times10^{-33}n_i\text{ sec} \tag{11.46}$$

It shows a buildup time for the striated motion of less than a picosecond for densities $n_i<1.6\times10^{20}$ cm^{-3}. At these low temperatures and high densities, the Coulomb collision time is high enough to conserve the conditions of the equation of motion, Eq. (8.77), for the space charge neutrality of the plasma. The time t_a for establishing the striated motion is shorter than the time the plasma takes to penetrate the length between the thickness of one layer by its thermokinetic expansion, if the nodes of the stationary wave are assumed to simply rest in space. The thermokinetic expansion results in plasma velocities not seriously exceeding 10^7 cm/sec at the laser intensities in question. To penetrate the length of a quarter of the wavelength, a time of less than 2 psec is then necessary. Therefore, if the propagating wave process has built up a striated motion, it can persist for a time until being destroyed by thermokinetic processes, if some more favorable stable state is not created to suppress the thermokinetic properties. Rethermalization of the striated motion with keV ions should then de detectable by the production of fusion neutrons from the plasma irradiated, even at such low intensities as 10^{13} W/cm^2, if oblique incidence and p-polarization occur. In this connection, it should be remarked that fusion neutrons have been produced even at intensities of 100 times less [258].

A further evaluation of the net nonlinear absoportion by the striated motion will now be done. The radiation energy which is gained under stationary conditions, by the layers in the laminar striated motion due to its viscosity can be calculated. A layer of a thickness d and a cross section of 1 cm^2 is considered. The consumed power density of the radiation is then equal to

$$I_c=F\bar{w}=\frac{2f^2d^3v_{ii}}{n_i kT_i} \tag{11.47}$$

For the optimum angle of incidence of 25.4° [with the maximum forces from Eq. (11.29)] is found

$$I_c = 7.2 \times 10^{-11} \frac{\lambda_0 I^2 \ln \Lambda}{T_i^{5/2} |n|^5} \tag{11.48}$$

The wavelength λ_0 is in cm, I is in W/cm^2, and T is in Ev. Using a Coulomb logarithm of 10, $T = 10$ eV, and an ion temperature of 10 eV,

$$I_c = 2.2 \times 10^{-17} I^2 \frac{I^2}{|n|^5} \; [\text{W/cm}^2] \tag{11.48a}$$

From this, the amount of laser power absorbed within a layer of d can be calculated from I_c/I. The absorption within a depth of one vacuum wavelength is about six times this value (absorbed energy)/(cm^3 sec along one wavelength) $= 1.3 \times 10^{-15}$ I/n^5. This is a reasonable value, for example, 1.3% at $I = 10^{13}$ W/cm^2 and 13% at $I = 10^{14}$ W/cm^2. The *nonlinear nature of the absorption* is seen in Eq. (11.48) by the factor I^2. A comparison with the exponential laws of linear absorption is not readily possible, but the reasonable absorption within the depths of a few wavelengths at the usual intensities confirms the possibility of reality for this process. This process would also confirm a very fast thermalization of the laser energy for oblique incidence and p-polarization.

We here summarize the results of this subsection. For p-polarization of obliquely incident laser radiation, a striated motion is generated in the standing wave field of the plasma corona. This process is established within times of picoseconds and creates relative ion velocities, in a laminar motion, which correspond to ion energies of more than 1 keV for neodymium glass laser intensities of 10^{13} W/cm^2.

The mechanism is a typical nonlinear absorption process. For conditions of turbulent motion, the ion energies may be larger, but the details of this process will include the generation of Helmholtz–Kelvin instabilities very soon, so that a detailed analysis will depend on the individual conditions of intensities, angles of incidence, and so on, for which numerical studies are necessary.

11.2 Resonance Absorption

This subsection is devoted to the process of the so-called resonance absorption, which was treated first by Denisov [248], and which was studied subsequently by numerous authors. The process happens when a plane p-polarized electromagnetic wave is obliquely incident on a stratified plasma. It should be noted from the beginning that resonance aborption does not occur at perpendicular incidence, and it does not occur for s-polarization. It must be considered in two steps. First there is the linear solution of the Maxwell equations for the p-polarization, which leads to a strong, resonancelike increase of the longitudinal component of the **E**-field at the cutoff density. This resonance field is a simple consequence of the temporally harmonic solution of the wave equations (6.26) for $\mathbf{H}=\mathbf{i}_z\ H_z$ and Eq. (6.35) for $\mathbf{E}=\mathbf{i}_x E_x+\mathbf{i}_y E_y$ corresponding to p-polarisation (x is the direction of the gradient of the stratified plasma).

Denisov [248] calculated the **H**-vector for his treatment and discussed the interesting longitudinal component E_x subsequently, using complex coordinates for plasmas with absorption. A more straightforward way is to treat **E** from the beginning. The difficulty is the coupling of E_x and E_y, see Eq. (11.6). However, if E_x alone is discussed, Eq. (11.5) permits an independent solution. This approach was used by White and Chen [259]. Under the assumption of a stationary solution with a harmonic ansatz (6.19), after separation (11.7) of Eq. (11.6), the x-dependent factor of the E_x-component follows Eq. (11.10).

$$\frac{\partial^2}{\partial x^2} E_{xx}+\frac{\omega^2}{c^2} n^2(x)\cos^2 \alpha(x)E_{xx}+2\frac{\partial \ln n}{\partial x}\frac{\partial E_{xx}}{\partial x}+2\frac{\partial^2 \ln n}{\partial x^2} E_{xx}=0 \qquad (11.49)$$

The solution of the y-dependent factor, E_{xy}, is given in Eq. (11.15). In contrast to the case of the plasma corona [where the last two terms in Eq. (11.49) can be neglected], the discussion of a depth

$$x>x_t; \quad 1-n^2(x_t)\sin^2 \alpha_0=0 \qquad (11.50)$$

has to include the last terms of Eq. (11.49). x_t is the turning point, up to which the wave field is the superposition of a standing and a propagating wave. For $x>x_t$, below the turning point x_t, the evanescent wave field is dominant. x_t is uniquely defined by Snell's law, if n is monotonically increasing with x.

While White and Chen [259] immediately gave the transformation of Eq. (11.49) into such without a linear term, it is instructive to follow the general steps of this "reduction of a linear differential equation" of second order for $y(x)$

$$y''+g(x)y'+h(x)y=0$$

It is transformed by [260]

$$u(x)=y(x)\exp\frac{1}{2}\int -g(x)\,dx$$

into

$$u''(x)+Iu=0$$

where

$$I=h-\tfrac{1}{4}g^2-\tfrac{1}{2}g'$$

This is immediately evident by resubstitution. In the case of Eq. (11.49) the functions

$$g=2\frac{\partial \ln n}{\partial x}$$

and

$$h=(n^2-\sin^2\alpha_0)\frac{\omega^2}{c^2}+2\frac{\partial^2 \ln n}{\partial x^2}$$

are used. The resulting function $u(x)$

$$u=E_{xx}\exp\frac{1}{2}\int 2\frac{\partial \ln n}{\partial x}\,dx=E_{xx}n$$

turns out to be of a very special formula. After calculations of the function $I=\tilde{K}^2$, the reduction of Eq. (11.49) arrives at

$$\frac{\partial^2}{\partial x^2}(E_{xx}n)+\tilde{K}^2(E_{xx}n)=0 \tag{11.51}$$

The wave vector $\tilde{K}$ defines an "effective dielectric constant" $\tilde{\varepsilon}_{\text{eff}}=\mathbf{N}^2$

$$\tilde{K}^2=\frac{\omega^2}{c^2}\tilde{\varepsilon}_{\text{eff}}=\frac{\omega^2}{c^2}\left[n^2-\sin^2\alpha_0+\frac{c^2}{\omega^2}\frac{\partial^2}{\partial x^2}\ln n-\frac{2}{\omega^2}\frac{c^2}{n^2}\left(\frac{\partial n}{\partial x}\right)^2\right] \tag{11.52}$$

This is formally identical with the expression of White and Chen [259], using $n^2=\varepsilon$

$$\tilde{\varepsilon}_{\text{eff}}=\tilde{\varepsilon}-\sin^2\alpha_0+\left(\frac{1}{2}\frac{\varepsilon''}{\tilde{\varepsilon}}-\frac{3}{4}\frac{\varepsilon'^2}{\tilde{\varepsilon}^2}\right)\frac{c^2}{\omega^2} \tag{11.53}$$

however, taking a dielectric constant for a plasma with collisions. This generalization causes a basically different behavior.

The fact that $|n|$ is going to very low values at $n_e=n_{ec}$ causes a very high maximum of E_{xx} at this resonance density. This has been pointed out first by Denisov [248] and is shown by a special example with a linear decrease of the plasma density in Fig. 11.5 [261]. It has been noted that the numerical solution of the wave equation for Fig. 11.5 was a new way to fit the complex amplitude of the reflected wave: a variation of phase and modulus has been done for arriving at a minimum "reflected wave" in the evanescent region. This is a special solution of the Osterberg problem (see Section 7 [152]). The result of ε_{eff} for a collisionless plasma, $\nu=0$, was discussed by White and Chen [259]. Near x_0, the value of ε_{eff} passes a negative pole, see Fig. 11.6.

The discussion for $\nu\neq 0$ will be done now for $\tilde{\varepsilon}_{\text{eff}}=\varepsilon_{\text{eff}}(n=|n|)$ where the complex refractive index is substituted by its absolute value. This procedure is not too strange, as the nonlinear force for perpendicular incidence is

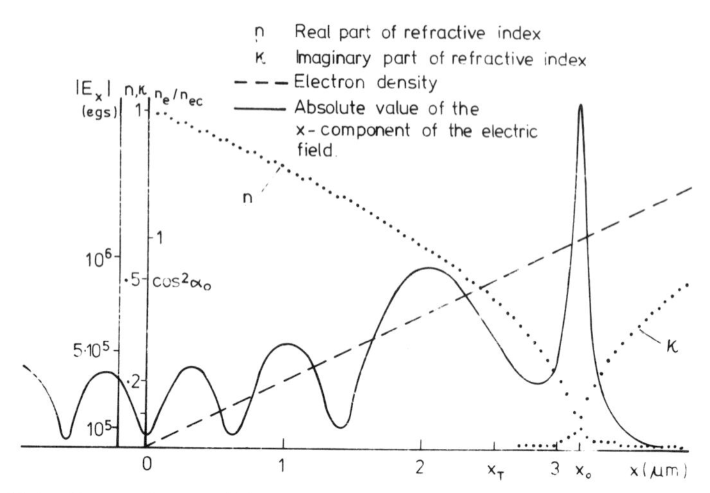

Figure 11.5 Exact solution of E_{xx} [Eq. (11.51)], for Nd glass laser radiation for a linear density profile, $T_e = 100$ eV and an angle of incidence $\alpha_0 = 26°$. The swelled maximum is reached before the turning depth x_t. The evanescent field for $x > x_t$ shows the resonance maximum at $x = x_0$, where $n_e = n_{ec}$. The figure shows an exact solution with respect to reflectivity determined by only penetrating waves at $x \gg x_0$, derived by Ladrach [261].

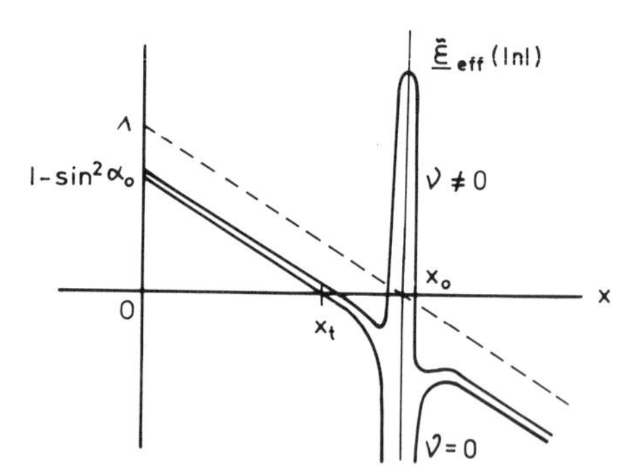

Figure 11.6 Schematic slope of $\tilde{\varepsilon}_{eff}$ near the turning point x_t and the cutoff density x_0 for $\nu = 0$, following White and Chen [259]. With collisions $\gamma \neq 0$, a positive $\tilde{\varepsilon}_{eff}$ ($|n|$) results from the discussion in this section, Hora, [4].

achieved correctly by using the absolute values of n, see [138, Eq. (25a)]. It has the advantage, and is justified that we shall be able to derive the Denisov length from our generalized equation (11.53) for *any* density profile of a plasma with collisions. The agreement of our generally considered profiles with those of a linear decay of the electron density will be shown after the general derivation.

Figure 11.6 shows $\tilde{\varepsilon}_{eff}$ schematically compared with the case of $\nu = 0$. At the critical density $x=x_0$, $\tilde{\varepsilon}_{\text{eff}}$ jumps from negative infinity to a maximum which can have high positive values, even larger than unity, reminiscent of a dielectric medium with dipoles, corresponding to the electrostatic oscillation.

Following Eq. (11.52), a positive $\tilde{K}^2$ is possible only for the evanescent region if

$$\frac{1}{|n|}\frac{\partial^2}{\partial x^2}|n| > \frac{2}{|n|^2}\left(\frac{\partial |n|}{\partial x}\right)^2 - \frac{\omega^2}{c^2}\left[|n|^2(x) - \sin^2 \alpha_0\right] \tag{11.54}$$

As mentioned, for complex numbers n the absolute values were used. The wave field $E_{xx}n$ will then have a penetrating wave in a narrow zone near x_0, as in a wave guide [262].

This is possible only if there is some absorption ($\nu \neq 0$) or any kind of damping. As Landau damping is unavoidable in principle, this condition is fulfilled. In Fig. 11.7, $|n|$ is given for the collisionless case and for that with collisions ($\nu \neq 0$). The sharp minimum near x_0 causes a very steep positive slope of the first derivative and therefore a very small but sharp and high positive peak of the second derivative. As

$$\frac{\partial |n|}{\partial x} \sim \frac{\partial n_e}{\partial x}$$

a small damping and a strongly steepened profile of the electron density near x_0 causes the wave guide effect there [262].

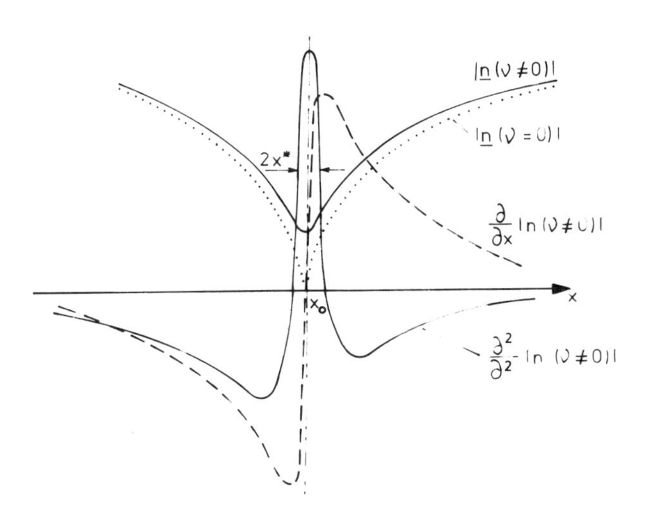

Figure 11.7 $|n|$ and its spatial derivatives near $x=x_0$ without collisions (dotted line) and with collisions (other lines). Obviously, the first and second derivatives for $\nu=0$ have poles at $x=x_0$ [4].

A further property of Eqs. (11.51) and (11.52) is the proportionality of $E_{xx}n$ and the square root of the electromagnetic energy density in the plasma. To preserve the constant energy flow into the y-direction (as is well known from the evanescent wave and the Goos–Haenchen effect at total reflection [263]), either the transport velocity of the energy has to decrease in this wave guide, or the value of E_{xx} has to be very large. This would immediately confirm the conclusion of Denisov [248] that E_{xx} has a maximum near x_0 and a very steep slope to small values for larger or smaller values of x.

Following Fig. 11.7, the thickness $x^* = L/2$ is calculated. x^* is half the range, where the second-order term in Eq. (11.52) can be positive starting from the relation of Eq. (6.48)

$$|n|\bigg|_{\min} = \left(\frac{\nu}{\omega}\right)^{1/2} \qquad (n_e = n_{ec}) \tag{11.55}$$

for $\nu \neq 0$, and from the fact that $|n(\nu=0;\, x = x_0 + x^*)| \approx |n||_{\min}$ is found

$$\frac{1 - n_e(x + x^*)}{n_{ec}} = \frac{\nu}{\omega} \tag{11.56}$$

and by expanding $n_e(x + x^*) = n_e(x_0) + (\partial n_e/\partial x)|_{x_0} \cdot x^*$

$$x^* = \frac{\nu}{\omega} \frac{1}{(\partial \ln n_e/\partial x)_{x=x_0}} \tag{11.57}$$

In order to find the maximum value of the second derivative $\partial^2 |n|/\partial x^2$, the values of the differences for the limiting case of small x^* is used

$$\frac{\partial^2}{\partial x^2} |n(\nu \neq 0)| = \frac{1}{x^*}\left[\frac{\partial}{\partial x} |n(\nu \neq 0)|_{x = x^* + x_0} - \frac{\partial}{\partial x} |n(\nu \neq 0)|\bigg|_{x = x_0}\right] \tag{11.58}$$

The second term in the bracket on the right-hand side has to be zero by definition (Fig. 11.7). Approximating the other term with sufficient accuracy by its value without collisions ($\nu = 0$), from Eq. (11.49) is obtained

$$\frac{\partial}{\partial x} |n|_{x = x_0 + x^*} = \frac{\partial}{\partial n_e} |n| \frac{\partial n_e}{\partial x} = \frac{\partial \ln n_e}{\partial x} \frac{1}{2(\nu/\omega)^{1/2}} \tag{11.59}$$

The discriminating value for the second-order term in Eq. (11.52) is then

$$\frac{1}{|n|} \frac{\partial^2}{\partial x^2} |n| = \frac{1}{2} \frac{\omega^2}{\nu^2} \left(\frac{\partial \ln n_e}{\partial x}\right)^2 \tag{11.60}$$

which is always positive. Using this result, Eq. (11.52) is rewritten for $x = x_0$, where $|n| = \nu/\omega \ll 1$, and $\partial |n| \partial x = 0$

$$\tilde{K}_{\max} = \left[\frac{\omega^2}{\nu^2} \frac{1}{|n|^2} \left(\frac{\partial \ln n_e}{\partial x}\right)^2 - \frac{\omega^2}{c^2} \sin^2 \alpha_0\right]^{1/2}$$

or using the width x^* [Eq. (11.57)]

$$\tilde{K}_{\max} = \left(\frac{1}{x^{*2}} - 4 \frac{\pi^2}{\lambda^2} \sin^2 \alpha_0 \right)^{1/2}$$

(where $\omega^2/c^2 = 2\pi/\lambda$) and finally

$$\tilde{K}_{\max} = \frac{2\pi}{\lambda} \left(\frac{\lambda^2}{4\pi^2 x^{*2}} - \sin^2 \alpha_0 \right)^{1/2} \equiv \frac{\omega}{c} \mathbf{N} \tag{11.61}$$

Positive radicals in Eq. (11.61) are realized if

$$\frac{\lambda}{2\pi x^*} > |\sin \alpha_0| \tag{11.62}$$

As soon as x^* is sufficiently small (one wavelength or less, depending on θ_0), the wave equation (11.51) results in a wave guide type layer of $2x^*$ thickness around x_0 with propagating waves. The effective thickness of the wave guide is extended by the Goos–Haenchen effect [263] in a similar way to that discussed extensively for dielectric wave guides [262]. Thus, the solution discussed was $E_{xx}n$ and not E_{xx} alone. Finally, it is very important to remark that the length $2x^*$ is identical with the length L, which Eliezer and Schuss [264] have derived for the density profile

$$n_e = n_{ec} \exp \left(\frac{x - x_0}{L} \right)$$

if ν/ω is interpreted as the same coefficient as in the analysis at the critical density near x_0. This result was derived from the stochastic interaction of the nonlinear force (Section 8) for the conditions of resonance absorption, tacitly including the result of the exact value of the nonlinear force (Section 8) for these conditions.

Eliezer and Schuss evaluated the special cases for L (and therefore for x^*) for 10^{15} W/cm^2 intensity for CO_2 lasers and neodymium glass lasers to 1.5 μm and to 1.5 nm, respectively. Neglecting $\sin^2 \alpha_0$ would result in "effective refractive indices" $N = 112$ for both lasers. The case of a refractive index of 112, or similar values near the conditions assumed, leads to a very new concept for the discussion of resonance absorption. The very high energy density accumulated in the wave guide drastically influences the conditions of the nonlinear forces. The detailed study, under these extreme refractive indices, needs clear understanding of the Abraham–Minkowski problem with absorption (Section 9) at the interface to the plasma.

We have derived the Denisov length $L' = 2x^*$ in Eq. (11.62) without specifying the slope of the electron density. Our derivation was based only on a schematic slope and the calculations were done on the distances of maxima and turning points only. It would not need a further specification, because proof by comparison with the Denisov length is absolute. Only for further illustration, we show here the results Läderach and Balmer [265] achieved continuing their work by elaborating the real and imaginary part of $\tilde{\varepsilon}_{\text{eff}}$, Eq. (11.53), Fig. 11.8, for a plasma with linearly increasing electron density with collisions. The behavior of our quantity $\tilde{\varepsilon}_{\text{eff}}$ $(n=|n|)$ is similar to $\text{Re}(\varepsilon_{\text{eff}})$. The maximum can exceed 1, as was shown for another case in [265, Fig. 1*a*]. For illustrating the schematic (but for our derivation of L sufficient) curves of Fig. 11.7 with a special case of linear density, Fig. 11.9 is shown to confirm full agreement.

After the phenomenon of the resonance field with a strong maximum of

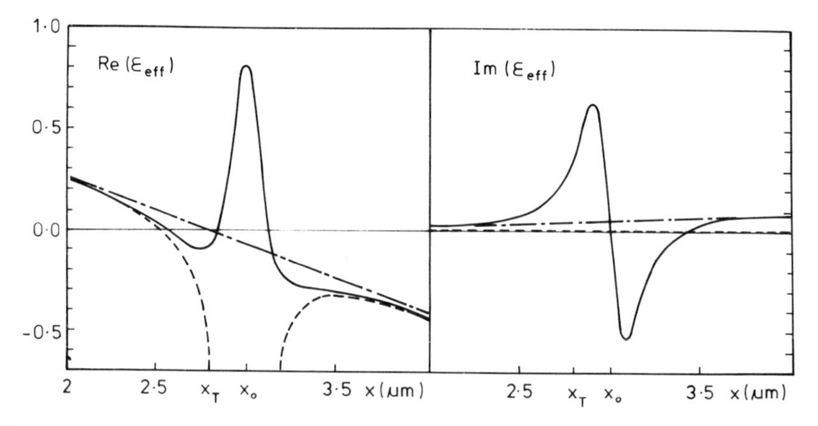

Figure 11.8 Numerical result of ε_{eff}, (Eq. 11.53) for a plasma with collisions and with a linear density profile. This is an extension of the work of Lädrach and Balmer [265] which was kindly provided for this discussion.

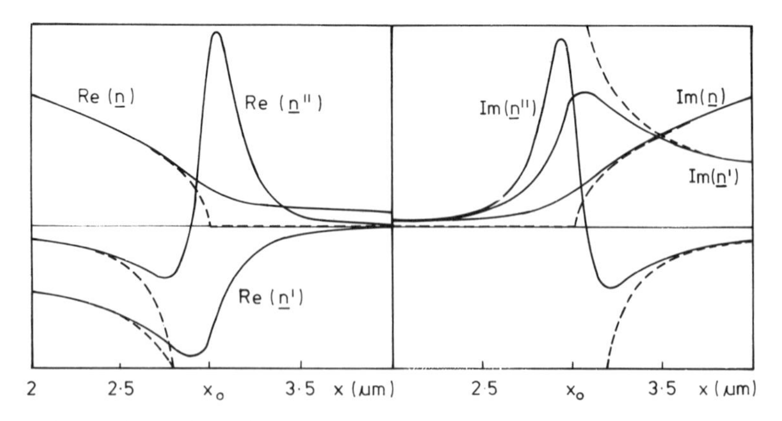

Figure 11.9 Numerical result of n for the same case of Fig. 11.8 [265].

the (longitudinal) E_x-component near x_0 has been shown, it should be mentioned in what sense this phenomenon is a "resonance *absorption*." The resonance field can contribute to an absorption due to the coupling with plasma waves at $x = x_0$. Several models have been discussed to describe this absorption. A very transparent calculation was the evaluation of the collisional absorption in the region between x_t and x_0 [267]. The absorption rate calculated by this way is shown in Fig. 11.10.

Another process is the single-particle simulation of the penetration of an electron through the E_x-resonance maximum [268]. It turns out that an electron moving at a certain initial velocity v_{in} into the maximum is accelerated. Maki [269] has an analytical solution showing that the superfast electrons of a Maxwellian distribution can gain energy. Ten keV electrons can gain energies up to 200 keV, directed perpendicular to the resonance layer. This is in agreement with single-particle simulations [268]. The same process was described by a wave braking mechanism [270] or by a soliton process [271].

Another description is the acceleration of an initially resting or slowly moving electron (anyhow without the need of an initial velocity v_{in}) near the E_x-resonance maximum. The resulting quivering motion leads to a drift and acceleration of the electrons similar to the quivering process [272] resulting in the nonlinear force for an inhomogeneous high-density plasma.

The high electromagnetic energy density $E_x^2 + E_y^2 + H_z^2$ and its strong gradient obviously act as a potential generating a nonlinear force. For a more careful analysis, the immediate application of an equation of motion of the type of Eq. (8.82) must be taken into account, as the conditions of space charge neutrality and restriction by the Debye length may not be fulfilled. Similar difficulties and the possibility of a simple $\mathbf{E}^2 + \mathbf{H}^2$ description arise from the theory, when the nonlinear forces in a plasma lateral to a laser beam are considered [273]. To be on the safe side, the quivering model is

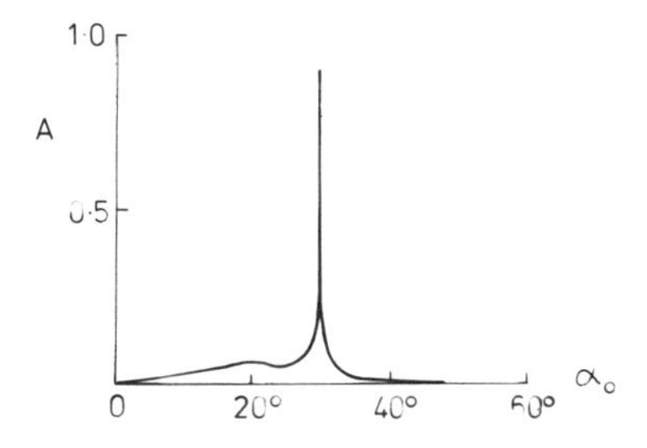

Figure 11.10 Coulomb collisional absorption in the evanescent region between the turning point and the critical density following Maki and Niu [267], at x_0 for varying angle of incidence.

followed for the description of the dynamics due to nonlinear forces in the region of the resonance absorption.

Physically, the E_x does not produce an "electrostatic field" immediately, as E_x oscillates at high frequency. Only the fact that the quivering electron is moving into areas of a smaller electric field and back into a larger field results in a net drift motion. This is an essentially different process from the drift due to the quivering motion of the electron in a plane wave, which is perpendicularly incident onto a stratified inhomogeneous plasma. Then the phase between **E** and **H** is the reason for the drift [138]. These facts show that the nonlinear forces and typical nonlinear properties of laser-plasma physics are not just simple extensions of the theory of force densities, however sophisiticated the refinements in Eq. (8.3) by additional nonlinear terms may have been. The nonlinear forces are a more general phenomenon, and the model of quivering motion is one of the guiding tools for analysis only.

The electric field maximum in the area of the resonance absorption x_0 has a very large value of E_x with a very steep decay for larger and smaller x. A drifting motion of quivering electrons is there not so much considered, but is determined by any phase difference between E_x and H_z, which together cause an eight-like motion. Gradients of time-averaged H_z values along y are relatively small and therefore neglected here. The net drift-motion of electrons is determined by the fact that the electric field E_x at a point $x=x_1$ near x_0 strongly depends on x by [266]

$$E_x(x)=E_x(x_1)[1+c_1(x-x_1)+c_2(x-x_1)^2+\cdots]\cos\omega t \tag{11.63}$$

The equation of motion for a single electron is then

$$\ddot{x}=\frac{e}{m}E_x(x_1)[1+c_1(x-x_1)+c_2(x-x_1)^2+\cdots]\cos\omega t \tag{11.64}$$

The first approximation ($c_i=0$, $i=1, 2, 3\cdots$) is

$$x^{(1)}=-\frac{e}{m}\frac{E_x(x_1)}{\omega^2}\cos\omega t \tag{11.65}$$

which determines the second approximation

$$\ddot{x}^{(2)}=\frac{e}{m}E_x(x_1)\left[1-c_1\frac{eE_x(x_1)}{m\omega^2}\cos\omega t\right]\cos\omega t \tag{11.66}$$

resulting, by integration, in

$$\dot{x}^{(2)}=\frac{eE_x(x_1)}{m\omega}\sin\omega t-c_1\left[\frac{eE_x(x_1)}{m\omega}\right]^2\left(\frac{t}{2}+\frac{1}{4\omega}\sin 2\omega t\right) \tag{11.67}$$

The time-averaged (nonquivering) part of the next integration is taken only, giving

$$\overline{x^{(2)}} = -c_1 \left(\frac{eE_0}{2m\omega}\right)^2 t^2 \tag{11.68}$$

E_0 is the maximum of the resonance field E_x at $x = x_0$. Any symmetric part given by c_2, c_4, does not contribute to the drift.

Now the special case of a linear decrease of E_x with x from x_0 (to both sides) is considered.

$$E_x = \begin{cases} E_x(x_0)(1 - 2|x - x_0|/d_0) & \text{if } |x - x_0| \leqslant d_0/2 \\ 0 & \text{if } |x - x_0| \geqslant d_0/2 \end{cases} \tag{11.69}$$

At the end of the steep profile of E_x, at x^* with $|x^* - x_0| = d/2$.

The kinetic energy of the drifting electron is then given by

$$\frac{m}{2}\dot{x}^2 = \frac{1}{2}\frac{e^2 E_x^2(x_0)}{m\omega^2} = \frac{E_x^2(x_0)}{8\pi n_{ec}} \tag{11.70}$$

This is equal to the maximum oscillation energy of the quivering motion of the electrons due to the field E_x at $x = x_0$. Though a very restrictive specialization for the E_x-profile, Eq. (11.69) is assumed. Any other general profile, however, results in the same relation (11.70), due to the energy conservation law for an electron passing around a closed circle, accelerating from x_0 to translational motion and back to x_0 using two different $E_x(x)$-profiles. The similar result for general profile for ions was derived for the completely different case of a net nonlinear force with perpendicular incidence, Eqs. (9.21 and (9.26)). Note that in Eq. (11.70), the kinetic energy is larger by a factor of 2.

In this sense of electrons drifting due to the quivering motion at the localized maximum of the high-frequency E_x-field near x_0, one can see the expansion of the plasma at the resonance absorption as if there were an explosion due to a "quasi-electrostatic" potential. One then has an "absorption," which is the same kind of absorption as in a macroscopic nonlinear motion without collisions, as when the net acceleration of the stratified inhomogeneous plasma by perpendicularly incident plane waves resulted numerically in a dynamic absorption (Fig. 10.19).

If the Debye length is less than half the thickness x^* of the resonance field, the accelerated electrons are coupled electrostatically to the ions, and the plasma is accelerated. The ion energy of translation is then

$$\varepsilon_i^{\text{transl}} = Z\varepsilon_{\text{osc}} \tag{11.71}$$

From Eq. (11.70), the total oscillation energy, in contrast to the half in Eq. (9.26) for the deconfining nonlinear force acceleration, is transferred into ion energy per ion charge. In Eq. (11.71), ε_{osc} is the maximum oscillation energy of the electrons due to the E_x-field.

The ion energies generated by the striated motion of the plasma corona at p-polarization for 10^{14} W/cm^2 neodymium glass laser radiation can reach 20 keV, especially if turbulent motion is generated. The process of resonance absorption, based on the quivering (nonlinear force) motion in the E_x-maximum, can easily result in similar energies. One has to take into account that the maximum of E_x can be 10 to 100 times the E_v-value, corresponding to a "swelling" of 100 to 10,000. The range of ε_{osc} in (11.71) will be near 20 keV at 10^{14} W/cm^2 neodymium glass laser intensities.

It should be mentioned that a further competitive anomaly was discovered by Goldman and Nicholson [274]. Whenever a caviton is generated during the nonlinear force interaction, this density minimum self-focuses electrostatic (Langmuir) waves in a stimulated way. The amplitudes of the electrostatic waves in the caviton can then reach high values and cause collapses. If the density in the center of the caviton is very low or even zero, as in the case of nonlinear-force-driven self-focusing, the number of electrons involved is small. For laser-produced plasmas, the effect might be not very strong.

TWELVE

Laser Beams in Plasma

The discussion of laser plasma interaction was based on plane electromagnetic waves in Sections 9 to 11, exclusively. Some minor exemptions were parametric instabilities, where any deviations from plane wave or stratified geometries were marginal fluctuations only. This section discusses the behavior of laser beams in plasmas. This is the domain of self-focusing of the laser beams in plasmas of high or moderate density dominated again by the nonlinear force or by relativistic effects. At low densities, the nonlinear forces will expel the plasma out of the beam which results in interesting measurements. The finite diameter of the laser causes the spontaneous generation of magnetic fields, the interaction of which (including Alfvén waves) will then be considered.

While ideal plane (or spherical) wave fronts might be necessary for the application of lasers for nuclear fusion (see next section) and any self-focusing would have to be avoided, many other applications, such as treatment of material makes use of self-focusing. Especially at very high intensities the generation of MeV ions or the expected electron-position pair production and similar effects are of interest. A discussion of these questions is included at the end of this section.

There is an essential and basic problem we must mention at the beginning. This is the difficulty of an exact description of the laser beam. We shall see that several contradictions will arise if a solution for the complex nonlinear problems is not based on the fully exact description.

There is a long history for the derivation of the exact light beam. Debye [275] tried to describe an optical wave bundle by superposition of plane waves of various directions with a certain spectrum of intensities. To arrive at a very transparent description of an exact beam, the Schrödinger equation for electron waves permits more possibilities [276] than the optical case, as used in the description of the Goos–Haenchen effect for matter waves [263]. This better situation for matter waves is due to a basic difference of second order between optical and matter waves [277]. If a special approximation is used for optical wave beams with a Gaussian radial intensity profile as in the following, one must be aware from the beginning, that this is only an approximation of the exact Maxwellian case and the possibility of deviations from the basic nature with exact beams must be watched carefully.

Historically, the first quantitative theory of self-focusing of laser radiation, resulting in a threshold for the laser power was successful for dielectric materials (nonionized solids, liquids, and gases) [278]. The essential mechanism was the nonlinearity of the dielectric constant. Another mechanism of self-focusing in organic tissues with a much lower threshold was observed and can be related to the breakdown field strength [279]. For self-focusing in plasmas, there was no similar nonlinearity of the dielectric constant which could be used. The first successful way of deriving the self-focusing threshold was an application of nonlinear forces [205]. To arrive at a first qualitative threshold for the laser power it has to be mentioned that the first ideas on self-focusing of lasers in plasma was published by Askaryan [205]. He considered the energy momentum flux density of the laser beam $(\mathbf{E}^2+\mathbf{H}^2)/8\pi$ by which the whole plasma has been expelled and where the pressure is then balanced by the plasma pressure profile acting against the center of the laser beam. The balance should be given in this way in the form of an equation

$$\frac{\mathbf{E}^2+\mathbf{H}^2}{8\pi}=n_e KT_e\left(1+\frac{1}{Z}\right) \tag{12.1}$$

Askaryan was able to compare the necessary optical intensities to balance or compensate the gasdynamic pressure. The correct justification of this formula will be derived in Section 12.3.

12.1 Nonlinear Force (Ponderomotive) Self-Focusing

In order to derive the threshold for self-focusing of a laser beam in a plasma, three physical mechanisms (Fig. 12.1) have to be combined. Assuming that the laser beam has a Gaussian intensity profile along the y-axis while propagating in x-direction, the generated nonlinear force $\mathbf{f}_{\mathrm{NL}}$ in the y-direction has to be compensated by the thermokinetic force $\mathbf{f}_{\mathrm{th}}$ [206]

$$f_{\mathrm{th}}=f_{\mathrm{NL}}; \quad \nabla\cdot\left(\mathbf{T}-\frac{n^2-1}{4}\mathbf{EE}\right)=\nabla n_e KT(1+Z) \tag{12.2}$$

where use is made of Eq. (8.82). The second physical mechanism is the total

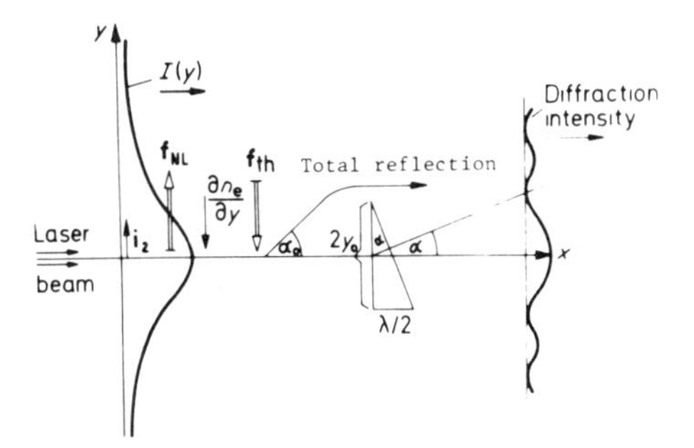

Figure 12.1 Scheme of a laser beam in plasma of a lateral intensity decrease $I(y)$, producing nonlinear forces $\mathbf{f}_{NL}$ in the plasma, rarifying the axial regions until being compensated by the thermokinetic force $\mathbf{f}_{th}$ of the gasdynamic pressure. The density gradiant causes a total reflection of partial beams. The diffraction condition for permitting partial beams of an angle of propagation less than total reflection for achieving the first diffraction minimum is the final condition for deriving the self-focusing threshold [206].

reflection of the laser beam components starting under an angle α_0 from the center of the beam and being bent into a parallel direction to the axis due to the density gradient of the plasma. The third condition is the diffraction requirement that the main part (e.g., as defined by the first diffraction minimum) of the beam has to have an angle of propagation α, which is less than the angle of total reflection. These three conditions are sufficient to calculate the threshold.

A Gaussian density profile including the refractive index n is described by the formula.

$$\overline{E_y^2} = \frac{E_v^2}{2|n|} \exp\left(-\frac{y^2}{y_0}\right); \qquad \overline{H_z^2} = |n|^2 \overline{E_y^2} \tag{12.3}$$

y_0 can be interpreted as the radius of the laser beam. This is only an approximation of the exact Maxwellian formulation. In analogy to the suggestions of Askaryan [205] and Schlüter [280], the nonlinear force in the direction of y is

$$\mathbf{f}_{\mathrm{NL}} = -\frac{1}{8\pi} \nabla(\overline{E_y^2} + \overline{H_z^2}) \tag{12.4}$$

It is important to note that this formulation is approximately valid only for conditions of strong swelling, that is, for refractive index differing from unity, which is the case for plasma exceeding the cutoff density in the region around the laser beam. The knowledge of difficulties for low-density plasma will be discussed in Section 12.3. The calculations [206] are repeated by several authors, on the basis of different assumptions, with always the same result achieved.

Using Eq. (12.3) in Eq. (12.4), the maximum nonlinear force in the y-direction is

$$\overline{\mathbf{f}_{\mathrm{NL}}}=\mathbf{i}_y \frac{1+n^2}{16\pi n}\frac{E_y^2}{y_0}\sqrt{2}\exp\left(-\frac{1}{2}\right) \tag{12.5}$$

If this has to be compensated by a thermokinetic force under the assumption of a spatially constant plasma temperature (the general treatment for varying temperature was studied extensively by Sodha and co-workers [281]. The formulation is found

$$\mathbf{f}_{\mathrm{th}}=-\mathbf{i}_y kT_{\mathrm{th}}\left(1+\frac{1}{Z}\right)\frac{dn_e}{dy} \tag{12.6}$$

Equating this force and the nonlinear force of Eq. (12.5) provides an expression for the electron density gradient of the plasma at the laser beam.

$$\frac{\partial n_e}{\partial y}=\frac{\sqrt{2/\exp(1)}}{16\pi KT_{\mathrm{th}}}(1+|n|^2)\frac{E_v^2}{y_o|n|(1+1/Z)} \tag{12.7}$$

The second physical condition of total reflection is given by the refractive index in the center of the beam, n and its value at y_0, for which with Eq. (12.7) is formulated [206]

$$\sin\left(\frac{\pi}{2}-\alpha_0\right)=\frac{|n|}{|n_{y_0}|} \tag{12.8}$$

Using the following Taylor expansion, for the case of a negligibly small collision frequency

$$n_{y_0}=n+\frac{\partial n}{\partial y}y_0; \qquad n^2=1-\omega_p^2/\omega^2 \tag{12.9}$$

gives from Eq. (12.8)

$$\sin\alpha_0=\left(\frac{2}{n}\frac{\partial n}{\partial n_e}\frac{\partial n_e}{\partial y}y_0\right)^{1/2} \tag{12.10}$$

If—as a third physical condition—a particular wave with an angle α for the first minimum of diffraction has to be reflected totally, the condition is found

$$\sin\alpha=\frac{\pi c}{2\omega y_0}\leqslant\sin\alpha_0 \tag{12.11}$$

Expressing the right-hand side by Eq. (12.10) and using there Eq. (12.7) and the relation for the electrical laser field amplitude $E_{v0}-c_1P^{1/2}/y_0$ (where P is the averaged laser power and c_1 is a constant of 1.63×10^{-5} cgs) one arrives at [206]

$$P \geqslant \frac{(\pi c)^2 n^3 m_e}{e^2[2/\exp(+1)]^{1/2} c_1^2 (1+n^2)} \tag{12.12}$$

It is remarkable that this threshold for self-focusing is a laser power and not an intensity. This is surprising, but not strange, as the threshold for the self-focusing of a laser beam in a dielectric nonionized medium is also a power and not an intensity [278], although both processes are basically different.

For an evaluation of Eq. (12.12), one can use the value of n given by Eq. (6.32), valid for temperatures above 10 eV. Expressing the plasma temperature T in eV and the laser power P in watts, one arrives at

$$2P \geqslant \begin{cases} 1.46 \times 10^6\ T^{-5/4} & \text{for } \omega_p \lesssim \omega \\ 1.15 \times 10^4\ T & \text{for } \omega_p \ll \omega \end{cases} \tag{12.13}$$

In the derivation given, only a laser slab has been used and no cylindrical beam. The use of a cylindrical beam results in a modification of Eq. (12.11) by the use of the Rayleigh diffraction factor 1.22 for the diffraction condition. Instead of Eq. (12.11) we then have for the beam the diffraction condition

$$\sin \alpha = \frac{1.22\pi c}{2\omega r_0} \leqslant \sin \alpha_0 \tag{12.14}$$

All the preceding derivations with the coordinates y and y_0 can be substituted by the radial coordinate r and the beam radius r_0 of a cylindrical coordinate system. The power threshold for the beam instead of that of the light slab is then

$$P \geqslant \frac{(1.22\pi c)^2 n^3 m_e}{e^2[2/\exp(1)]^{1/2} c_1^2 (1+n^2)} \tag{12.15}$$

and in numbers with P in watts

$$P \geqslant \begin{cases} 1 \times 10^6 T^{-5/4} & \text{for } \omega_p \lesssim \omega \\ 8 \times 10^3\ T & \text{for } \omega_p \ll \omega \end{cases} \tag{12.15a}$$

This was the first quantitative theory of the nonlinear force self-focusing or, as it was initially called [206], the ponderomotive self-focusing. This result was rederived by several authors [282] and fully reproduced. The same thresholds are also achieved by Chen's nonlinear force treatment of the filamentation instability; see Section 9.5 [193].

A further surprising result [206] is the fact that the power threshold for self-focusing in plasma is very low, in the range of megawatts or less. This is in agreement with measurements first published by Korobkin and Alcock [283] and other authors [282, 284]. The measurements of Richardson et al.

[283] especially demonstrated in detail that the beam center shows a depletion of plasma. Another success of the theory is the agreement of the measured beam diameters [282] of a few microns for a laser power of 3 MW. If one can assume that the stationary conditions for self-focusing are reached when all plasma is moved out of the center of the laser beam, the electromagnetic energy density is then equal to $n_e(1+1/Z)KT$. This is the case for densities close to the cutoff density, where the laser intensity is equal to the threshold intensity I^*, as given in Fig. (9.2) for neodymium glass laser radiation. It is evident that the beam has then to shrink down to such a diameter to reach the necessary 10^{14} W/cm^2 from a laser power of 3 MW. The resulting beam diameter is then a few micrometers, in full agreement with the measurements.

12.2 Relativistic Self-Focusing

Another type of self-focusing happens [16] if the relativistic effects are considered. The relativistic change of the electron mass, due to oscillation energies close to or above $m_e c^2$, causes a modification of the optical constants, as shown in Eqs. (6.77) to (6.81). For the optical constants, Eqs. (6.33) and (6.34), or for the absolute value of the refractive index, Eq. (6.47), have to be used. With this relativistic intensity dependence, the effective wavelength of propagating laser radiation in a plasma is then given by

$$\lambda = \frac{\lambda_0}{|n(I)|} \tag{12.16}$$

where λ_0 is the vacuum wavelength. In Fig. 12.2, a Gaussian-like intensity profile of a laser beam moving through a homogeneous plasma is considered. The relativistic refractive index results in the condition

$$|n(I_{\max})| > |n(I_{\max}/2)| \tag{12.17}$$

showing, that the effective wavelength (12.16) is shorter for the higher laser intensity in the center of the beam than at the lower intensity of the half maximum intensity value. As shown in Fig. 12.2, an initially plane wave front is then bent into a concave front, which tends to shrink down to a diffraction limited beam diameter of about one wavelength. From the geometry of Fig. 12.2 this shrinking can be approximated by an arc resulting in a self-focusing length l_{SF}

$$l_{SF} = \left[d_0 \left(\rho_0 + \frac{d_0^2}{4} \right) \right]^{1/2} \tag{12.18}$$

d_0 is the initial beam diameter, and the radius of the arc with ρ_0 is given by

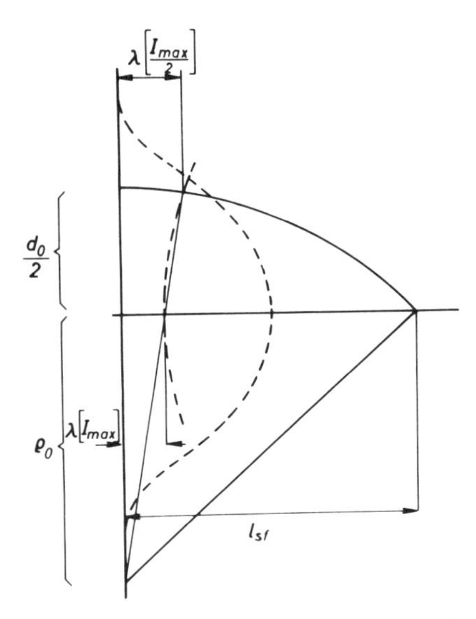

Figure 12.2 Evaluation of the relativistic self focusing length from the initial beam diameter d_0 and from the effective wavelengths. The relativistic effects cause a shorter wavelength at the maximum laser intensity I_{max} than at the half maximum intensity [16].

the effective wavelengths of the various intensities. From the geometry of Fig. 12.2, the following relation is derived:

$$\frac{|n(I_{max}/2)|^{-1}}{(d_0/2+\rho_0)}=\frac{|n(I_{max})|^{-1}}{\rho_0} \tag{12.19}$$

In combination with Eq. (12.18), this results in the ratio of the self-focusing length related to the beam diameter [16]

$$\frac{l_{SF}}{d_0}=0.5\left(\frac{|n(I_{max})|+|n(I_{max}/2)|}{|n(I_{max})|-|n(I_{max}/2)|}\right)^{1/2} \tag{12.20}$$

Using the exact absolute value of the refractive index n, as given by Eq. (6.47), with the intensity dependent relativistic values of the plasma frequency and the collision frequency [Eqs. (6.77) to (6.81)], a numerical evaluation of Eq. (12.20) is given in Fig. 12.3 for neodymium glass laser radiation for plasma densities of 10, 1, and 0.1% of the nonrelativistical cutoff density value. It is remarkable that the self-focusing length is as low as seven times the beam diameter for 10% of the cutoff density if the laser intensity is 3×10^{18} W/cm^2. This intensity is the relativistic threshold corresponding to an electron oscillation energy of m_ec^2. It is further interesting to note that the process of the relativistic self-focusing also occurs for laser intensities that are much less than the relativistic threshold, even 1000 times less. This

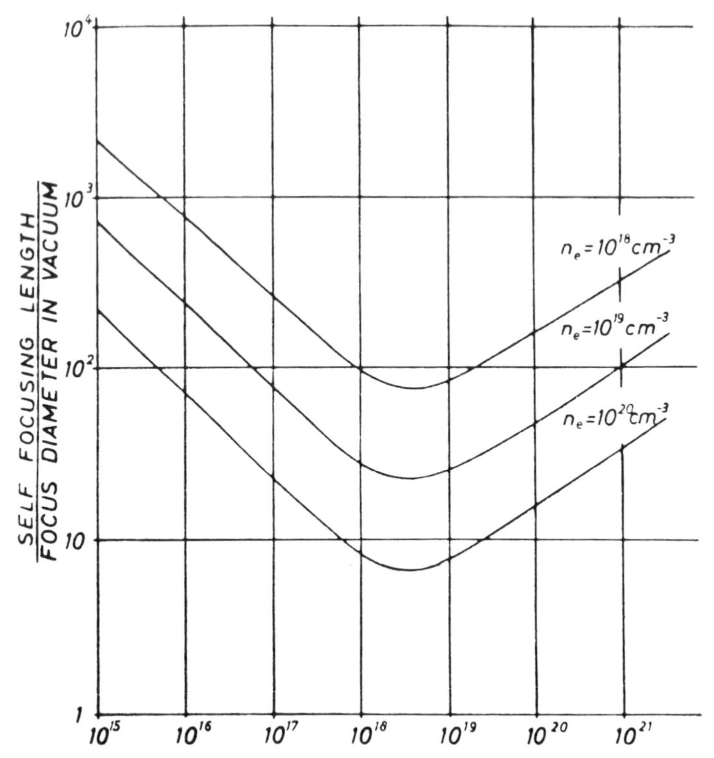

Figure 12.3 Calculated self-focusing lengths divided by the laser beam diameter for neodymium glass laser radiation for various plasma densities depending on the laser intensity [16].

phenomenon of the occurrence of relativistic effects at intensities much lower than the relativistic threshold was not new, as could be seen from the work of Tsindsatse et al. [285] for relativistic instabilities in plasmas.

The relativistic self-focusing has its maximum effect at the relativistic threshold. Its effect is lower for higher intensities. This can be easily understood from the fact that, at these higher intensities, there is an intensity dependent increase of the cutoff density, Eq. (6.78), so to speak, the plasma becomes transparent for propagating waves at densities, where the non-relativistic conditions would require evanescent waves.

The extension of the calculation of Eq. (12.20) to higher densities, shown in Fig. 12.3, is possible numerically. Simultaneously, the dependence on the plasma temperature and on the degree of ionizations is included. It is very surprising that self-focusing lengths of the same value as the initial beam diameter result, see Fig. 12.4 [286]. For lower intensities, a numerical cutoff

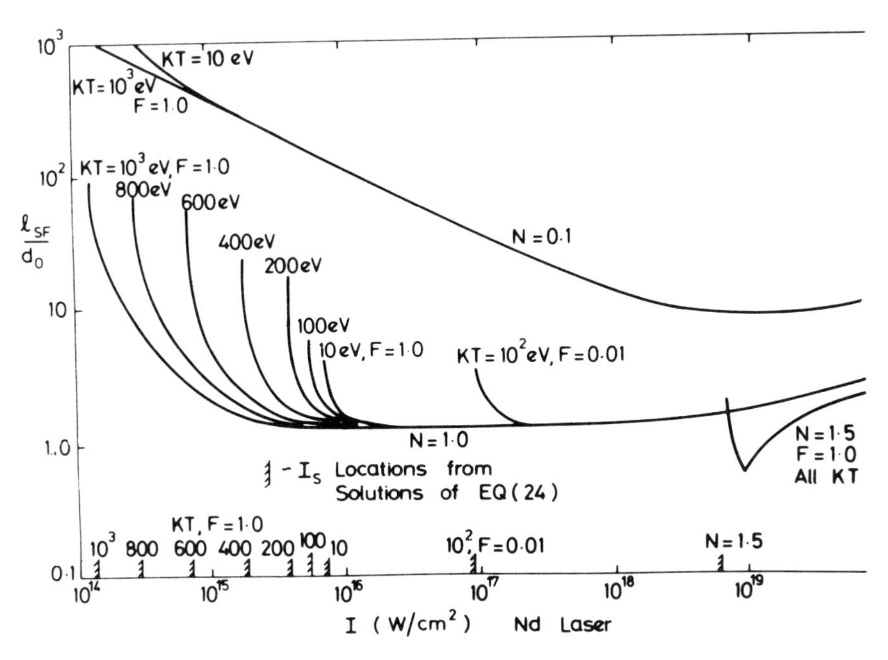

Figure 12.4 Ratio of the self-focusing length l_{SF} over the initial laser beam diameter d_0 for laser intensties near the relativistic threshold of 3×10^{18} W/cm^2 for neodymium glass laser radiation for varying plasma temperatures. The plasma density is equal to the nonrelativistic cutoff value ($N = n_e/n_{ec} = 1$) and 10% of this value ($N = 0.1$), respectively. The factor F is given by an effective collision frequency $\nu_{eff} = \nu/F^{2/3}$ to understand an eventual increase by anomalous effects [286].

of the plots is observed, where nevertheless the action of relativistic self-focusing is still working for intensities of less than 1% of the relativistic threshold. This is a remarkable result. While the well-known difficulties of designing an optical lens systems for focusing a laser beam in vacuum limit the minimum beam diameters to about 10 wavelengths, the plasma at the cutoff density realizes the very fast shrinking of a laser beam down to one wavelength diameter automatically by relativistic plasma effects.

It is worth noting that a later theory of relativistic self-focusing based on a completely different model which, however, is restricted to intensities below the relativistic threshold [287], results in nearly the same self-focusing length, as in the calculation based on the generally valid Eq. (2.20) [16], see Fig. 12.5.

It must be noted that both models mentioned, for nonlinear force self-focusing and for relativistic self-focusing, do not describe the complete process. The first model describes the threshold condition for a stationary

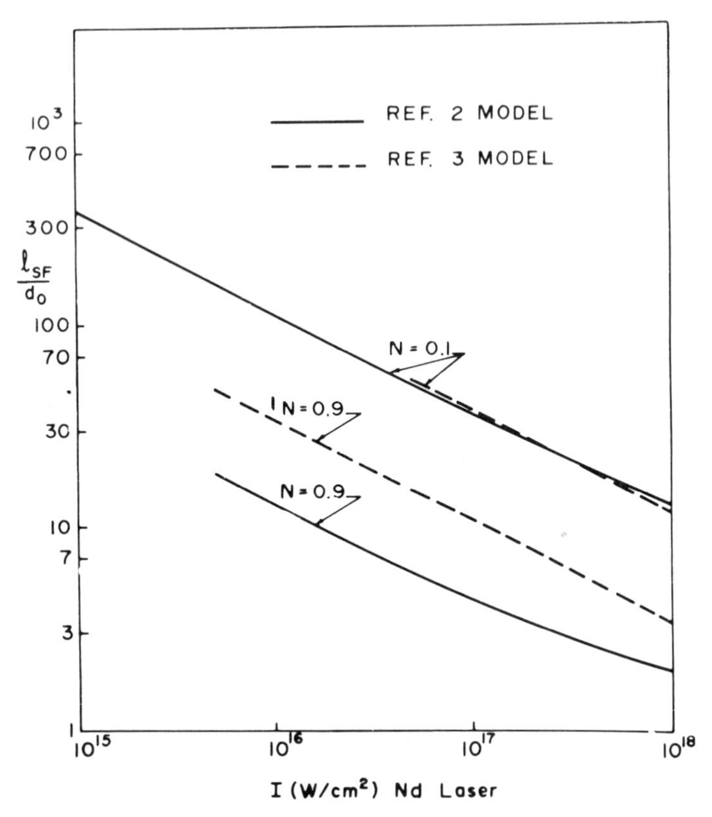

Figure 12.5 Comparison [288] of the relativistic self-focusing length l_{SF} per initial beam diameter d_0 depending on the initial neodymium glass laser intensity I following the general model (2) [16] and the subrelativistic model of Spatschek (3) [287].

case after sufficiently long interaction. The transient mechanism, the generation of the self-focusing tunnel and the resulting self-focusing length is not covered. The second model describes the relativistic self-focusing process in a homogeneous plasma, where the nonlinear forces disturb the homogeneity very quickly. The combination of all these mechanisms can be studied numerically. A very instructive example was derived by Siegrist [289], see Fig. 12.6. A very general numerical study including the nonlinear force, the relativistic self-focusing, and the transient behavior was performed by Kane [290], of which Figs. 12.7 and 12.8 are examples. The results describe the stationary solution of the axial intensity of a beam depending on the propagation length Z of a neodymium glass laser wavelength in a plasma, where the initial density and temperature are given and where the initial beam diameter of a Gaussian intensity profile is 30 μm [206].

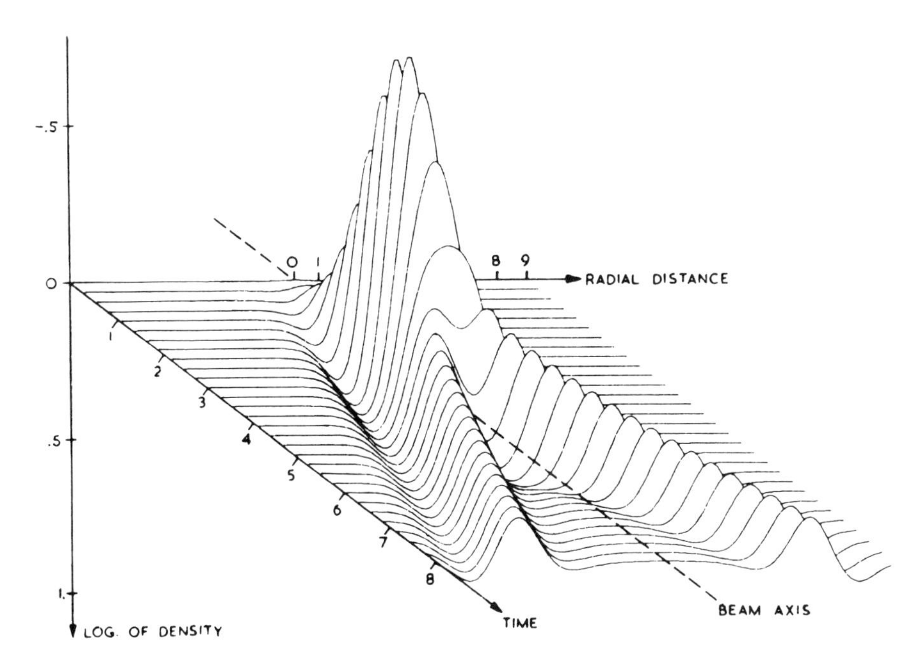

Figure 12.6 Density profile in a self-focusing channel in a plasma calculated by Siegrist [289].

There is nearly no effect for MW laser powers. The self-focusing can be seen at the next higher case of 5 MW and reaches the saturation beam diameter at a faster rate, the higher the laser power is. The saturation laser intensity at the beam center is the same for all powers: nearly 10^{15} W/cm^2, with a slight increase due to thermal effects, as for the threshold of the nonlinear force interaction, see Fig. 9.2, in full agreement with the result of nonlinear-force self-focussing [206]. see Eq. (12.15a).

The complete numerical calculations of the relativistic and nonlinear force self-focusing was done by a very general numerical treatment [290]. A Nd glass laser beam of nearly Gaussian radial intensity profile with 10 μm diameter of 30 psec duration is incident on an initially homogeneous hydrogen plasma of 80% of the cutoff density. The relativistic effect causes an immediate shrinking of the beam, while the nonlinear forces expel plasma from the interior of the beam. The radial velocities exceed 10^7 cm/sec. Because of the small radial field gradients in the center of the beam, a residual plasma remains and causes a hollow beam after 9.2 psec operation, see Fig. 12.9. This central plasma disappears, for example, at 18.4 psec due to fast axial motion, see Fig. 12.10. Axial ion energies of 5 MeV are achieved at 9.2 psec [290]. An example with a 10^{13} W-5 psec neodymium glass laser

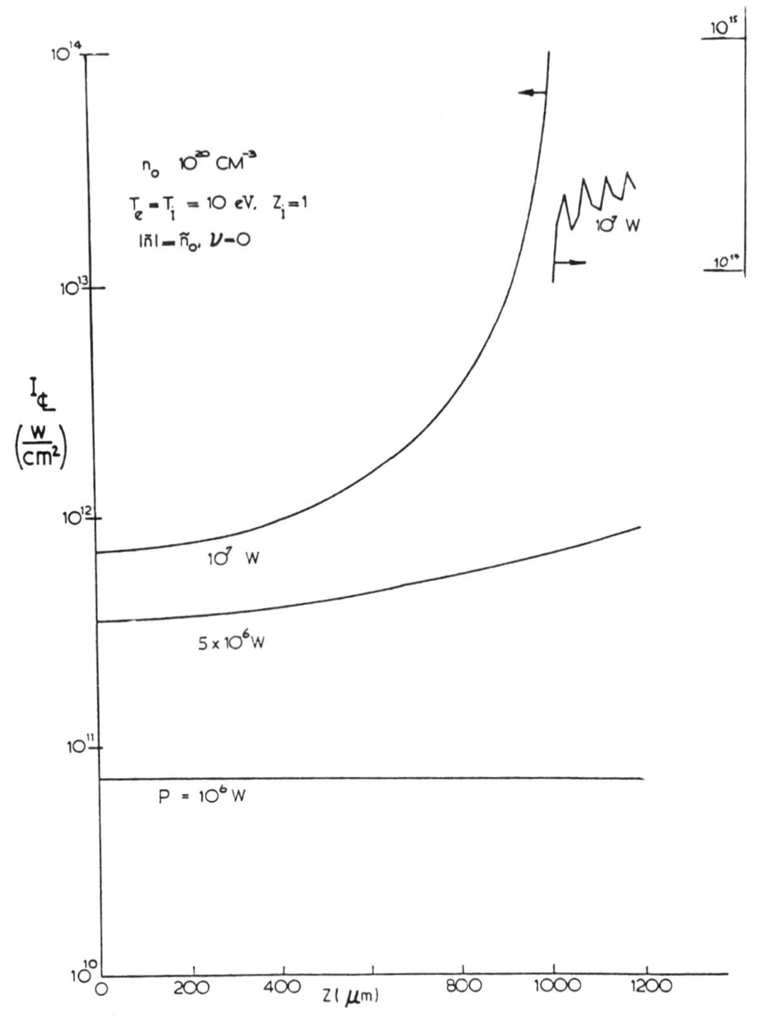

Figure 12.7 Stationary central beam intensity I_c for Nd glass laser beams of powers P in plasma of given density and temperature [290].

pulse of 20 μm diameter incident on a deuterium target of cutoff density showed the fast shrinking along an axial distance of 18 μm to a cross section of about one wavelength diameter. Ion energies exceeding 100 MeV were the result [291]. Applications in nuclear reactions and in the safe breeding of fission material by nuclear photoeffect are very promising.

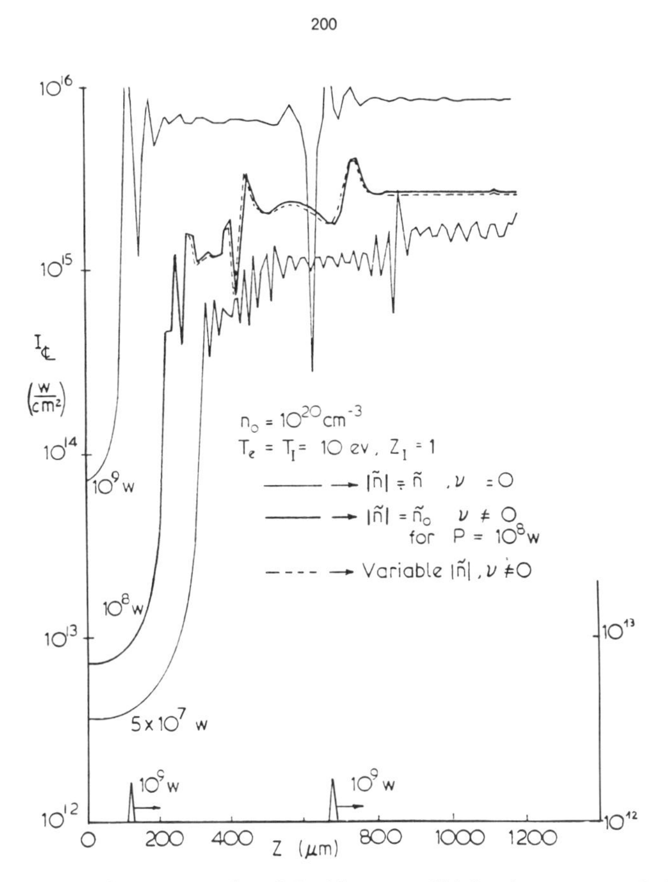

Figure 12.8 The same as Fig. 12.7 with cases of higher laser powers P [290].

12.3 Tenuous Plasmas, Exact Beams, and Free Electron Lasers

While laser beams in plasmas at or below the cutoff density will cause a dynamic reaction and result in self-focusing, a very low-density gas or the plasma (after ionizing of the gas atoms) will not modify the beam severely but will expel electrons by the nonlinear force. If the plasma dimension (diameter of laser beam) is less than the Debye length, the nonlinear force acts at the electrons only, leaving the ions behind untouched. The macroscopic plasma theory is then no longer valid, but the model of the quivering electron in the laser field and its drift results in formally the same nonlinear forces [274]. These results were used to demonstrate the action of the non-

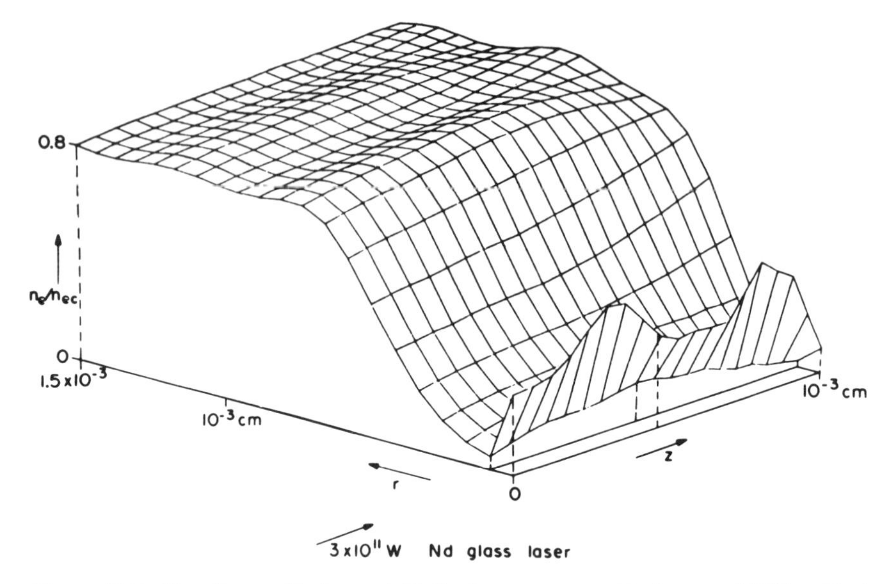

Figure 12.9 Density profile in an initially homogeneous H-plasma of 80% critical density at interaction with a beam of 10 μm half-radial width and 30 psec duration after 9.2 psec. A hollow beam has been created with a residual plasma in the beam center [291].

linear force immediately. Hollis [292] and Boreham [14, 293] focused neodymium glass laser beams into helium and other gases at about 10^{-4} Torr pressures. The laser beams had a focal intensity of about 10^{15} W/cm^2. The electrons emitted were measured along the direction of the laser light **E**-vector. Later, nearly no difference was found for other directions. It was found that the maximum electron energy

$$\varepsilon_e^{\text{transl}} = \varepsilon_{\text{osc}}^{\text{kin}} = \frac{\varepsilon_{\text{osc}}}{2} = \frac{I}{2cn_{ec}} \tag{12.21}$$

was equal to the average kinetic energy of the oscillation of the electrons according to Eq. (6.56), which was about 100 eV. This process can be explained in the same way as the energy gained by the drift of a quivering electron in a high-frequency field with a spatial gradient in field strength along the field direction, as was successful in the case of the resonance absorption for calculating the generated electron energy (Z times the ion energy), see Eq. (11.71) and the preceding derivation. For electrons in the low-density laser focus, we get the same results for electrons only if the Debye length is larger than the focus diameter.

The influence of the ions can be seen when the density produced by the

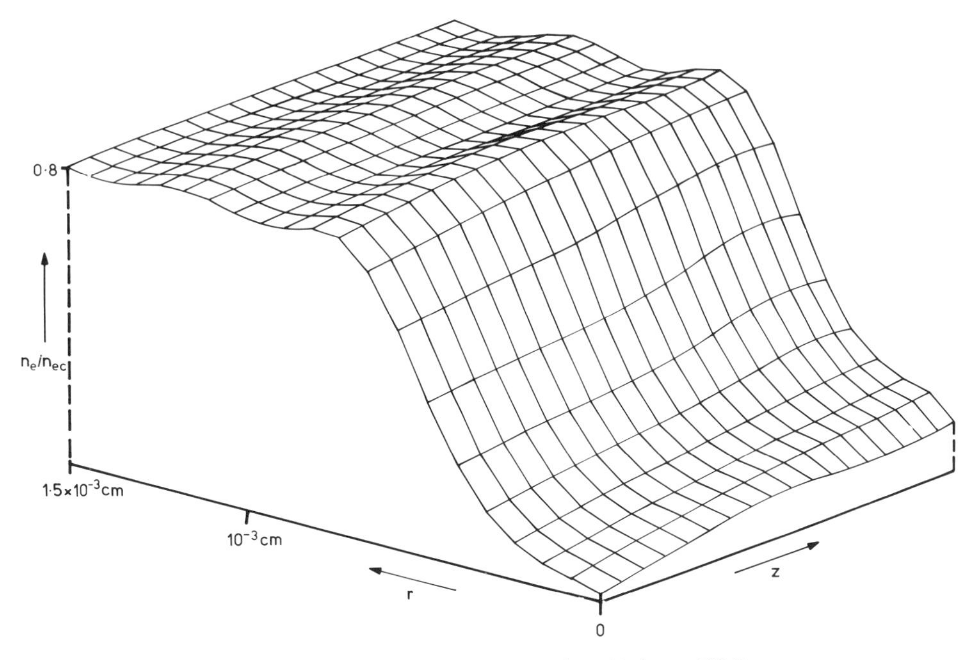

Figure 12.10 Same as Fig. 12.9 at time 18.4 psec [291].

gas pressure p is increased. Calculating a "Debye length" given by a temperature corresponding to the maximum electron energy of emission in terms of a pressure $p=p^*$, follows where the focus diameter d is equal to the Debye length, that

$$p^* = 1.37 \times 10^{-24} \frac{Z\bar{P}}{d^4} \tag{12.22}$$

where p is in torrs, $\bar{P}$ in watts, and d in centimeters. The laser intensity, given by a laser power $\bar{P}$, determines the maximum electron energy. For the conditions of the experiment by Boreham [293], the critical pressure p^* from Eq. (12.22) was 2.1×10^{-4} Torr. Figure 12.11 shows the result of the measurements. At these pressures, the emission of ions begins with the electrostatic attraction of the emitted electrons.

The experiment could be used to measure the delay of the ionization processes of the electrons bound to the helium atom at different energy levels. From this time delay it was possible for the first time [294] to measure the tunnel-type ionization process [295] according to the Keldysh theory [296] where the conditions of avalange ionization or the multiphoton ionization [297] are not given.

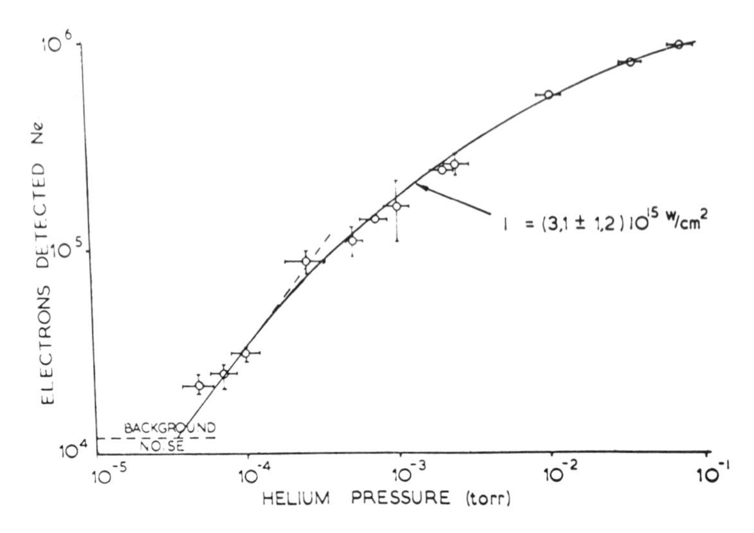

Figure 12.11 Number of electrons emitted from the focus of a laser beam in helium gas of varying pressure if the laser intensity is about 10^{15} W/cm^2. The linear increase on the number of electrons is saturated at about 2×10^{-4} Torr, due to the Debye length becoming equal to the focus diameter [14].

We come back to the question of whether the radial force of a laser beam interacting with plasma is correctly given by Eq. (12.4). There were no doubts for Askaryan [205] and all the following authors when using Eq. (12.1) for self-focusing. If one would have used the general expression of the Maxwellian stress tensor, the force in the y-direction (of the geometry of Section 12.1) is the gradient of $(E_y^2 - H_z^2)/8\pi$, if no x-component would be taken into account. It is necessary to emphasize that the field given by Eq. (12.3) is an approximation which does not fulfill the Maxwellian equations exactly.

We shall see now how a Maxwellian exact formulation of a laser beam (or slab) will result in a lateral nonlinear force given by Eq. (12.4). While plane electromagnetic waves are transversal only, as known from optics, the exact beam of finite diameter in vacuum needs a longitudinal component. Though it is well known that electromagnetic waves in media can have a longitudinal component, see Eq. (11.23) for p-polarization at obliquely incident plane waves, or in Dirac's quantum electrodynamics [298], we find the longitudinal component in vacuum as a simple result of the Maxwellian equations; only then will the lateral nonlinear force in the beam be found in agreement with Boreham's experiment [14].

To reproduce Eq. (12.21) according to the experiment, it was possible to use the drift of the quiver motion in the laser field along the E-direction even with the approximation (12.3)[273]; only for the perpendicular polarization, the inclusion of the longitudinal component of the field was necessary.

Using a slab of laser radiation with a decay in the x-direction (parallel to the polarization direction of **E**) we reproduce the earlier result [273]. The E_y field is then

$$E_y = E_0(1 + \alpha_1|y| + \alpha_2 y^2 + \cdots) \cos\left(\frac{\omega}{c}x - \omega t\right) \tag{12.23}$$

with y being the displacement in the y-direction if the center of the beam is at $y=0$. The acceleration of electrons therefore has the form

$$\frac{d^2y}{dt^2} = -\frac{e}{m}E_0(1 + \alpha_1 y + \alpha_2 y^2 + \cdots)\cos\left(\frac{\omega}{c}x - \omega t\right) \tag{12.24}$$

Neglecting second and higher orders, a solution by small perturbation results in

$$\frac{d^2y}{dt^2} = -\frac{e}{m}E_0\cos\left(\frac{\omega}{c}x - \omega t\right) + \alpha_1\frac{E_0^2 e^2}{m^2\omega^2}\cos^2\left(\frac{\omega}{c}x - \omega t\right) \tag{12.25}$$

Using $\alpha_1 = (1/E_0)(\partial E_y/\partial y)$ from a Taylor expansion at $x = x_0$ and averaging over a period we have [273]:

$$m\overline{\frac{d^2y}{dt^2}} = -\frac{e^2}{4m\pi\omega^2}\frac{\partial}{\partial y}\overline{E_y^2} = -\frac{\partial}{\partial y}\overline{(\mathbf{E}^2 + \mathbf{H}^2)}/8\pi \tag{12.26}$$

The generality of this solution for higher orders has been proved by reasons of energy conservation for the net acceleration of the electron to large values of x [273].

For the case of perpendicular polarization, we use a laser slab where the electric field polarization is again in y-direction, but the field decay is in the perpendicular direction z. Similarly we can write:

$$E_y = E_0(1 + \beta_1 z + \beta_2 z^{-2} + \cdots)\cos\left(\frac{\omega}{c}x - \omega t\right) \tag{12.27}$$

with z being the displacement from the beam center at $z=0$. The Maxwell equations arrive not only at the usual transversal component

$$H_z = E_0(1 + \beta_1 z + \beta_2 z^2 + \cdots)\cos\left(\frac{\omega}{c}x - \omega t\right) \tag{12.28}$$

but also in a longitudinal component. The longitudinal z-component of the magnetic field is given by

$$H_x = \frac{E_0}{\omega}(\beta_1 + \beta_2 z + \cdots)\sin\left(\frac{\omega}{c}x - \omega t\right) \tag{12.29}$$

which usually has been neglected. The acceleration of electrons, therefore, has the form

$$\mathbf{i}_y \frac{d^2y}{dt^2}+\mathbf{i}_z \frac{d^2z}{dt^2}=-\mathbf{i}_y \frac{e}{m} E_0(1+\beta_1 z+\beta_2 z^2+\cdots) \cos\left(\frac{\omega}{c}x-\omega t\right)$$
$$+\mathbf{i}_z \frac{e}{m}\frac{dz}{dt}\frac{E_0}{\omega}(\beta_1+2\beta_2 z+\cdots) \sin\left(\frac{\omega}{c}x-\omega t\right) \quad (12.30)$$

Considering again only the first-order effect and substituting the zero-order solution of (12.30) for y, z, and dz/dt, we obtain

$$\mathbf{i}_y \frac{d^2y}{dt^2}+\mathbf{i}_z \frac{d^2z}{dt^2}=-\mathbf{i}_y \frac{e}{m} E_0 \cos\left(\frac{\omega}{c}x-\omega t\right)$$
$$+\mathbf{i}_z \frac{e}{m}\frac{E_0\beta_1}{\omega}\left(-\frac{e}{m}E_0\right) \sin^2\left(\frac{\omega}{c}x-\omega t\right) \quad (12.31)$$

with $\beta_1=(1/E_0)(\partial E_y/\partial z)$ at $y=0$ and averaging over a time period results in a vanishing y-component but an acceleration in the z-direction

$$\overline{m\frac{d^2y}{dt^2}}=-\frac{e^2}{4m\omega^2}\frac{\partial}{\partial z}\overline{E_y^2}=-\frac{\partial}{\partial z}\frac{\overline{(\mathbf{E}^2+\mathbf{H}^2)}}{8\pi} \quad (12.32)$$

This confirms the correctness of the use of Eq. (12.4) for the polarization-independent lateral nonlinear forces at electrons or in a plasma due to a laser beam. It is remarkable that the very tiny longitudinal component changes the action from no to yes. This teaches how carefully one has to proceed with nonlinear phenomena. One might have assumed that non-linear extension should be less sensible against neglections. The inverse is the case: the basic assumptions have to be correct or nonapproximative to a greater degree than in linear theory.

The general formulation of an optical beam in vacuum which satisfies the Maxwellian equations exactly arrives always at a longitudinal component. The general formulation was given by Legendre functions [299].

We add here the result that the longitudinal component represents the property of the internal diffraction of the beam. Again a slab in vacuum is discussed where the symmetric terms are used only

$$E_y=E_0(1+\alpha_2 z^2+\alpha_4 z^4+\cdots) \cos\left(\frac{\omega}{c}x-\omega t\right) \quad (12.33)$$

where $\alpha_\nu=0$ for $\nu=4$. The Maxwellian equations then result in the usual component

$$H_z=E_0(1+\alpha_2 z^2+\cdots) \cos\left(\frac{\omega}{c}x-\omega t\right) \quad (12.34)$$

and in the longitudinal component

$$H_x = 2\frac{c}{\omega} E_0 \alpha_2 z \sin\left(\frac{\omega}{c} x - \omega t\right) \tag{12.35}$$

Obviously, E_y and H_x are out of phase, resulting in a zero lateral Poynting vector for this completely parallel beam $\bar{S}_z = 0$.

The definition of E_y in Eq. (12.33) is only for

$$z \leqslant z^* = \frac{1}{\alpha_2^{1/2}} \tag{12.36}$$

resulting in a cross section given in Fig. 12.12. The longitudinal component of H, Fig. 12.13 results in an angle α'' of a converging wave field. We find from Eqs. (12.35) and (12.36)

$$\tan \alpha' = \frac{H_x(z = z^*)}{H_z} = \frac{2c\alpha_2 z^*}{\omega} = \frac{\lambda}{\pi} \alpha_2^{1/2} \tag{12.37}$$

For the diffraction limitation the far-field first minimum of the slab results in the angle α_d of the deviation from the axis

$$\sin \alpha_d = \frac{\lambda}{2z^*} = \frac{\lambda}{2} \alpha_2^{1/2} \tag{12.38}$$

For small angle α' and α_d and neglecting a factor $\pi/2$, the converging wave front of the longitudinal field of H is the same as the diffraction limitation needs for convergence of the beam. It should be possible by phase shifts to

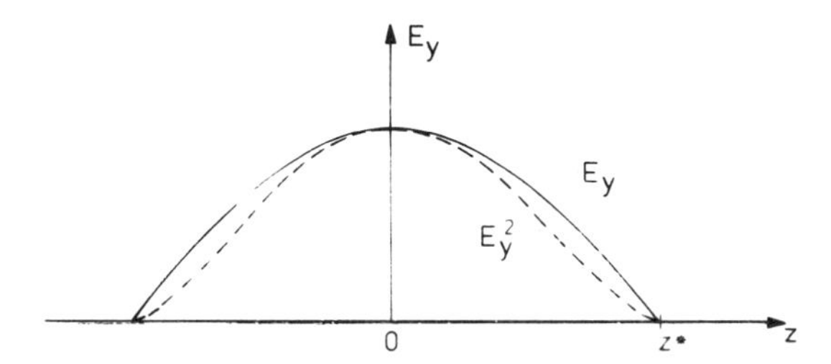

Figure 12.12 Lateral decay of E_y and E_y^2 of a slab beam according to Eq. (12.33).

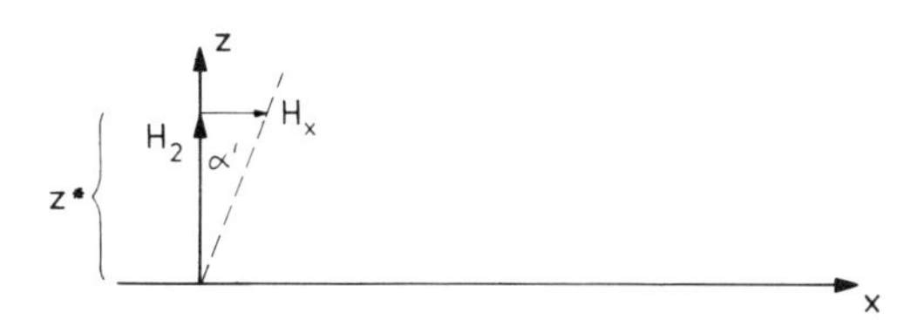

Figure 12.13 Longitudinal component H_x of the field for E_y of Fig. 12.12.

produce such absolutely parallel beams as described. This agrees with Einstein's needle radiation [189] including the finite width due to quantization.

The main result is that the radial nonlinear force due to a laser beam in a plasma is given by the gradient of $(\mathbf{E}^2+\mathbf{H}^2)/8\pi$ and all preceding calculations of self-focusing are unchanged. The confirmation of the polarization independency by Boreham's experiment is, however, to 30% accuracy only. An unexpected longitudinal component appeared in the exact description of laser beams in vacuum. We further experienced that only exact solutions lead to reasonable results at nonlinear effects.

This result is an example of how the nonlinear forces of laser radiation acting on electrons can be extended to low-density plasmas, where the macroscopic magnetohydrodynamic theory is no longer valid (the same model was successful for plane electromagnetic waves perpendicularly incident on dense plasmas [300]). It is also an example of the concept for a new type of free electron laser [301].

The presently known successful free electron laser is based on a synchrotron radiation process [302], where an electron beam has to move along the rippled magnetic field of a superconducting solenoid. The free-electron laser of the nonlinear force type, according to this new concept, can be explained by the experiment of Boreham [14]. The emission of electrons from the focus of the beam requires energy and results in an absorption of laser radiation of a nonlinear dynamic type. If—by inverting this process—an electron of energy equal to the maximum oscillation energy in the center of the laser beam is fired perpendicularly into the laser beam, the kinetic energy will be changed totally into oscillation energy. If the laser beam is switched off, when the electron is in the center of the laser beam, its oscillation energy will be transferred into optical energy of the laser beam. The different behavior of the electron in an axial direction compared to the radial direction of the laser beam follows from the general theory [138], especially from the results of Klima and Petrzilka [184].

In contrast to the synchrotron free-electron laser, the nonlinear force free-electron laser is an amplifier only of an otherwise produced laser beam. The interacting laser pulse has the same duration t_L as the crossing electron beam, which has to be incident in the direction of polarization. The focal radius is given by the energy spread ΔE of the electron beam, as $r=t_L(\Delta E/m)^{1/2}$. Using ΔE in cgs units, and the electron charge e in the same units as defines the electron beam density j (e is 1.602×10^{-19} Cb if j is in A/cm^2), the amplification factor is then

$$A=\frac{j}{n_{ec}\sqrt{\Delta E}}\,1.19\times10^5 \qquad (12.39)$$

The amplification increases with the square of the wavelength, as can be seen from the cutoff density in the denominator. This is similar to the synchrotron free-electron laser. The amplification is possible with presently available CO_2 lasers, if the amplification is repeated in a totally reflecting cavity a large number of times. The free-electron laser concept has the advantage of working at intensities, where any solid-state or molecular amplifier would break up or would be ionized.

This method could amplify 10^5 J/nsec CO_2 laser beam to 10^7 J by a successive pumping during a period of seconds or more [303].

12.4 Spontaneous Magnetic Fields—Alfvén Waves

The model of electron acceleration along the **E**-vector of the laser beam in low-density plasma could be used to explain self-generated magnetic fields, at least with respect to the presently known experiments. Self-generated dc magnetic fields in laser produced plasmas have been discovered by Stamper et al. [304]. The observation of megagauss fields [305] has been established and several differing models, partly on the basis of the nonlinear force, have been developed [306]. The recent measurements by Key et al. [307] and by Yamanaka et al. [308] show that the magnetic field in the megagauss range is directed perpendicularly to the plane of the irradiated target and at least a dipole field has been observed.

Using these facts, the following model seems to describe the phenomenon, if the observation of a dipole field is extended to a quadrupole field, Fig. 12.14, and if the Boreham experiment would confirm a degree of 30% of polarization dependence of the radial force in a beam (in which case the

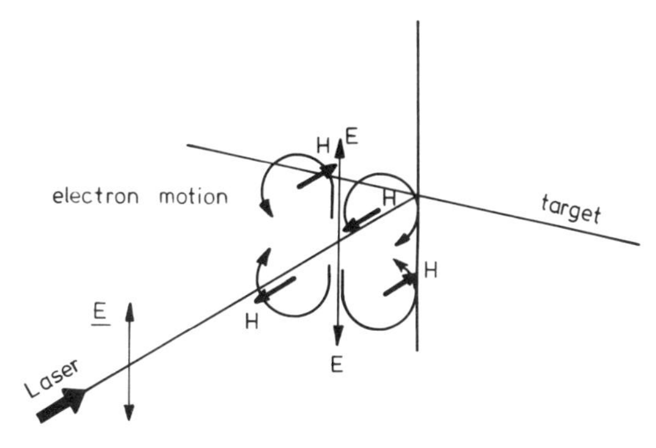

Figure 12.14 Laser radiation incident on a plane target producing a nonlinear force drift motion in the low-density corona along the *E*-field causing a quadrupole magnetic field **H** (double lined arrows.

general theory would have to be extended). The **E**-gradient will cause a quivering drift motion of the electrons according to Eq. (12.21), or according to the relation of Eq. (11.71).

The drift motion in the plasma corona of a density below the cutoff density results in a motion along the **E**-field, which causes a quadrupolelike circular motion of the electrons, as seen in Fig. 12.14. The magnitude of the magnetic field is given by converting the entire energy of oscillation (6.56) into magnetic field energy. This then gives

$$\frac{H^2}{8\pi}=N\frac{I}{c} \tag{12.40}$$

where $N=n_e/n_{ec}$. Using $N=0.1$ and an intensity $I=10^{22}$ cgs $=10^{15}$ W/cm^2, a magnetic field of 1.1 MG is reached in fair agreement with the experiments. It is further remarkable that the increase of the magnetic field follows a square root law on the laser intensity I, Eq. (12.40). The suggestion that a quadrupole field is produced, may not be too far away from the observations [307, 308], and special attention to this fact may lead to more precise experiments. The fact of a quadrupole field may explain why the experiment with the Argus laser [309] did not show a magnetic field, even if linear polarized laser radiation was used. If the direction of the Faraday rotation is very precisely parallel or perpendicular to the **E**-vector, no rotation is observed. Only an inaccuracy, a deviation from this direction results in the observation of the fields. A further confirmation of the theory can be given, if the use of circular polarized laser beams prevents the asymmetric nonlinear force motion described in Fig. 12.14.

We are aware that the model for explaining the spontaneous magnetic field in this subsection is very hypothetical. The fields are a fact, and we mention here for completeness the mechanism of compensating gasdynamic pressures by static magnetic fields on which extensive work of confinement of fusion plasmas (e.g., the tokamak) is based. The force density in a plasma is from Eqs. (8.4), (8.5), and (8.24)

$$\mathbf{f}=-\nabla p+\nabla\cdot\mathbf{T}+\frac{1}{4\pi}(n^2-1)\mathbf{EE}-\frac{\partial}{\partial t}\frac{\mathbf{E}\times\mathbf{H}}{4\pi c} \tag{12.41}$$

for equilibrium, $\mathbf{f}=0$ and $\partial/\partial t=0$. In a plasma with no HF fields, $\mathbf{E}=0$, and the magnetic field is assumed of the form $\mathbf{H}=(H_x(y), 0, 0)$. This results in Eq. (12.41)

$$\frac{\partial}{\partial y}\left(p-\frac{\mathbf{H}^2}{8\pi}\right)=0 \tag{12.42}$$

or by integration

$$n_e\left(1+\frac{1}{Z}\right)KT=\frac{\mathbf{H}^2}{8\pi} \tag{12.43}$$

where an integration constant has to be added. This compensation of a gasdynamic pressure by a static magnetic field can be important in laser produced plasmas where magnetic fields have been produced spontaneously.

We consider now the generation of the magnetohydrodynamic waves, or Alfvén waves, generated at motion of plasma with a velocity v_0 (y-direction) perpendicular to a static magnetic field H_0 (x-direction). The Lorentz force will then cause a current density j_z (in z-direction) which results in an acceleration of the motion by

$$n_i m_i \frac{\partial v_y}{\partial t} = \frac{1}{c} j_z H_0 \tag{12.44}$$

further differentiation by time gives

$$\frac{\partial j_z}{\partial t} = C \frac{n_i m_i}{H_0} \frac{\partial^2 v_y}{\partial t^2} \tag{12.45}$$

In the diffusion equation (6.8), Ohm's law, we neglect nonlinear terms and the fast oscillations of $j(\partial j/\partial t = 0)$ and collisions ($\nu = 0$)

$$\begin{aligned} \frac{m}{e^2 n_e}\left(\frac{\partial j}{\partial t} + \nu j\right) &= \mathbf{E} + \frac{1}{c}\,\mathbf{v}\times\mathbf{H} + \cdots \\ 0 &= E_z - \frac{1}{c} v_y H_0 \end{aligned} \tag{12.46}$$

The electric field E_z is that generated by the current j_z due to the motion of v across $\mathbf{H}_0$. Differentiating (12.46) twice by time results in

$$\frac{\partial^2}{\partial t^2} v_y = \frac{c}{H_0} \frac{\partial^2}{\partial t^2} E_z \tag{12.47}$$

Any fast motion of $\mathbf{E}$ will follow a wave equation (from the Maxwellian equations)

$$\nabla^2 \mathbf{E} = \frac{1}{c^2} \frac{\partial^2}{\partial t^2} \mathbf{E} + \frac{4\pi}{c^2} \frac{\partial j}{\partial t} \tag{12.48}$$

or from (12.45) and (12.47)

$$\nabla^2 \mathbf{E} = \frac{1}{c^2} \frac{\partial^2}{\partial t^2} \mathbf{E} \left(1 + 4\pi \frac{n_i m_i c^2}{H_0^2}\right) \tag{12.49}$$

This is a wave equation with a wave velocity

$$v_A = \frac{C}{(1 + 4\pi n_i m_i c^2 / H_0^2)^{1/2}} \tag{12.50}$$

called the Alfvén velocity. If $4\pi n_i m_i c^2 / H_0^2 \gg 1$, we find

$$v_A = \frac{H_0}{(4\pi n_i m_i)^{1/2}} = \frac{H_0}{\sqrt{4\pi\rho}} \tag{12.51}$$

The Alfvén waves are of importance in plasmas with static magnetic fields. However, there is a formal similarity with the high-frequency fields and the velocity an ion gains by the nonlinear force. Following Eq. (9.21) for high swelling $(1/|n| \gg 1)$, the energy gained by the ion after being accelerated along the inhomogeneous plasma surface is

$$\frac{m_i}{2} v_i^2 = \frac{\overline{\mathbf{E}^2}}{8\pi} \frac{Z}{n_{ec}} \tag{12.52}$$

using $n_e = Zn_i$ we arrive at

$$v_i = \frac{|\mathbf{E}|}{\sqrt{4\pi n_i m_i}} \tag{12.53}$$

Here, $\mathbf{E}$ and N_i have to be taken from an area close to the cutoff density. The result (12.53) is similar to the Alfvén velocity, if $\mathbf{E}$ is considered instead of $\mathbf{H}_0$. For densities below cutoff, $|\mathbf{E}| \approx |\mathbf{H}|$ of the laser field. The direction of v_i is that of v_0 in the initial derivation of the Alfvén wave. An interpretation of the connection of the nonlinear-force acceleration by the HF laser field with the Alfvén velocity is possible by considering the stepping through of the plasma along the HF wave maxima. The velocity v_i given by the nonlinear force to the ions is the electric analogy to the Alfvén velocity.

12.5 Conclusions for Medium Laser Intensities

Though the discussion of the self-focusing was described here in a short chapter only, based on the extensively described theory of nonlinear forces in the preceding sections, its influence on the destruction of materials at moderate laser intensities is of eminent importance. This is the reason why discussion of the confusing experiments mentioned in Section 1.4 about the earlier experiments of laser plasma interaction is so complex.

There are no ideal plane wave fronts incident on the targets, but focused laser beams having a certain number of "hot spots" due to insufficiencies of diffraction and birefringent properties in the laser amplifiers. Apart from the fact that in the most cases the power threshold for self-focusing is reached, the hot spots are an additional mechanism for producing more than one self-focusing channel.

Intensities of 10^{15} W/cm^2 are very easily reached in the plasma filaments, and the nonlinear force acceleration produces the keV ions, the nonlinear recoil [69], and the electron emission current densities 1000 times higher than permitted by the space charge limitation laws [71]. The mechanism of overcoming the space charge limitation can be seen very easily in the fact that the electron acceleration by the nonlinear force in the plasma filament works only on the plasma electrons, which are between the space charge neutral ions. This high-frequency acceleration is basically different from the electron acceleration in the surface of a material or in the vacuum above a surface. The nonlinear force acts to accelerate the whole volume of electrons within the space charge neutralizing ions as a volume effect, and all our knowledge of the surface effects of electron emission are not relevant.

The generation of multifilaments by self-focusing in a material is very important, if the destruction of material is considered for laser of moderate intensity. Despite the belief that CO_2 laser beams have a very smooth lateral intensity profile, the fact that moderate laser beams at a very low aperture produce a granulated structure in irradiated solid hydrogen pellets, when used for filling magnetic confinement vessels [310], indicates the complexity of this process.

For laser compression of plasmas, the suppression of self-focusing is a very important condition, if a very homogeneous interaction with a spherical pellet surface is the aim. Therefore we can recommend, for various lower or medium scale experiments, the irradiation of plane surfaces with very large diameter laser beams of low aperture: these will also achieve laser beams with a very smooth and monotonic radial intensity profile. The comparison of these measurements with those of small diameters but the same intensities are then a next step towards clean experimental conditions for any further theoretical investigation. On the other hand, the generation of the filamentary behavior in the best possible way should be the tool for studying the destruction processes of solid targets by laser radiation. One of the aims is the largest possible amount of material destroyed, while other applications are the drilling of holes in materials with the best possible quality. One reason why the laser drilling of holes in ruby crystals for watch jewels did not succeed was because the holes were not smooth, but had very crazy surfaces [311].

12.6 Conclusions for Very High Laser Intensities

As soon as the conditions for relativistic self-focusing are reached, the effects of very high laser intensities in the focused filaments open the door to very interesting high-intensity effects. These effects are indeed not appropriate for laser fusion, and one has to known how to avoid the relativistic self-focusing in the case of laser fusion. For the physics of higher energies rather than for laser fusion, however, the fast shrinking of the laser beams to a diameter of a wavelength is very desirable.

One question is how the oscillation energy of the electrons can be increased in the relativistically self-focused filament, if the fact is taken into account, from Eq. (6.76), that the oscillation energy of the electrons increases by a square root law in the laser intensity only at superrelativistic intensities. This change in the exponent of I is a very important reason to seek to understand the laws of blackbody radiation and to discuss a derivation of the fine structure constant from basic physical laws [312]. This lower power increase of the electron energy, however, results in some disadvantages in reaching the highest possible electron oscillation energies. Another disadvantage is the fact that, if the focusing is performed in plasma densities closer to the cutoff density, the larger the effective wavelength the larger is the effective beam diameter at self-focusing. The relation of short self-focusing length is then influenced negatively by the necessary higher laser power. All these factors together have been evaluated [286] and the result is achieved in Fig. 12.15. The focusing of a neodymium glass laser beam in vacuum is assumed to be down to a diameter of $d_0=30$ wavelengths, which seems to be realistic. The resulting maximum oscillation energy of the electrons is then given for the various intensities I_v of the laser beams in such a vacuum focus of 30 wavelengths diameter, where the plasma density has been varied with $N=n_e/n_{ec}$ between 0.1 and 0.99. The resulting oscillation energies and self-focusing lengths are given in the diagram of Fig. 12.15.

It is remarkable that the oscillation energies of 3 MeV, which are necessary for a quantitative production of electron-positron pairs [313], are reached with neodymium glass laser radiation of only 5×10^{17} W/cm^2, corresponding to laser powers of 3×10^{11} W as an absolute minimum. The laser beam then has to be sufficiently smooth that only one filament is being produced. The laser powers, however, are within the state of the art.

A further result [314] is that the ion energy after nonlinear force acceleration from the extremely high intensity of laser beams with a diameter of one wavelength does not follow the relation of Z times the relativistic electron oscillation energy. The fact is that the ions have Z times the oscillation energy of the electrons as if these were following the subrelativistic law as long as the ion energies are subrelativistic. Using the relativistic threshold intensity I_{rel}, Eq. (6.73), the ion energies $\varepsilon_i^{\text{transl}}$ of translation after being accelerated by the nonlinear forces from the relativistic self-focused filament are [314]

$$\varepsilon_i^{\text{transl}}=\frac{Z_i m_e c^2 (I/I_{\text{rel}})}{4} \tag{12.54}$$

Depending on the laser power, the ion energy is independent of the wavelength. The result is given in Fig. 12.16. Historically, the measurement of the MeV ion energies [76] by Ehler, in full agreement with the later ones, Fig. 12.16, was very transparent, although several authors could not believe in the MeV ion energies. Hughes et al. [315] measured the MeV ions and

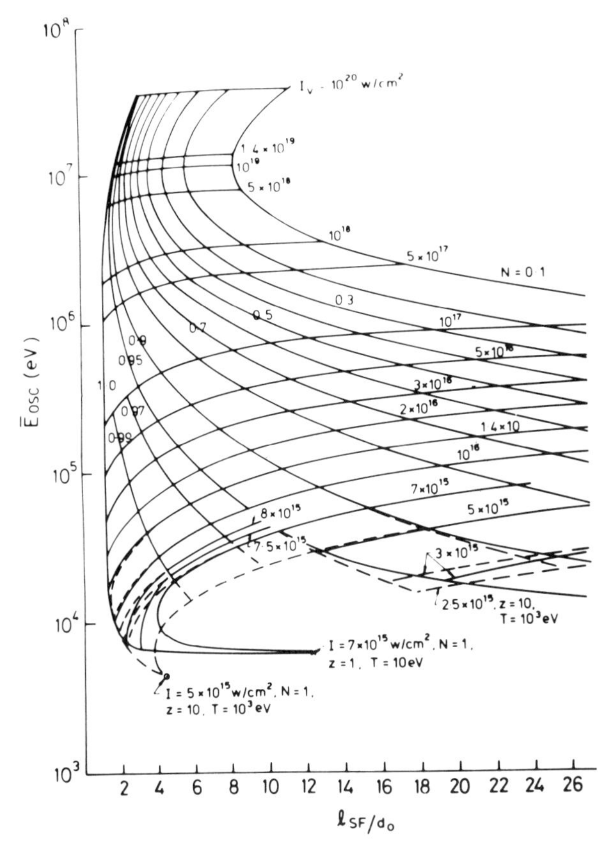

Figure 12.15 Maximum oscillation energy in the relativistically focused neodymium glass laser beam, where the vacuum focusing to 30 wave lengths diameter has reached. The maximum laser intensities I_v are shown, and the plasma densities are given by multiples N from 0.1 to 0.99 of the cut off density [286].

found agreement with the theory of relativistic self-focusing. The discussion of the Z-dependent peaks of the ion probe signals (Fig. 1.9) as obscure stray signals, or of real MeV ion signals, was immediately confirmed by the consideration of a kind of Eq. (12.25) or Fig. 12.16. When Hughes was submitting a paper with the first announcement ever about MeV ions at the Amsterdam Quantum Electronics Conference in 1976, it was rejected with the argument that everyone had seen these MeV ion signals (although nobody had really understood and interpreted them as MeV ions).

It is remarkable that the measured protons from a laser produced plasma of 15 MeV [316] would correspond to 450 MeV ions, which are 30 times

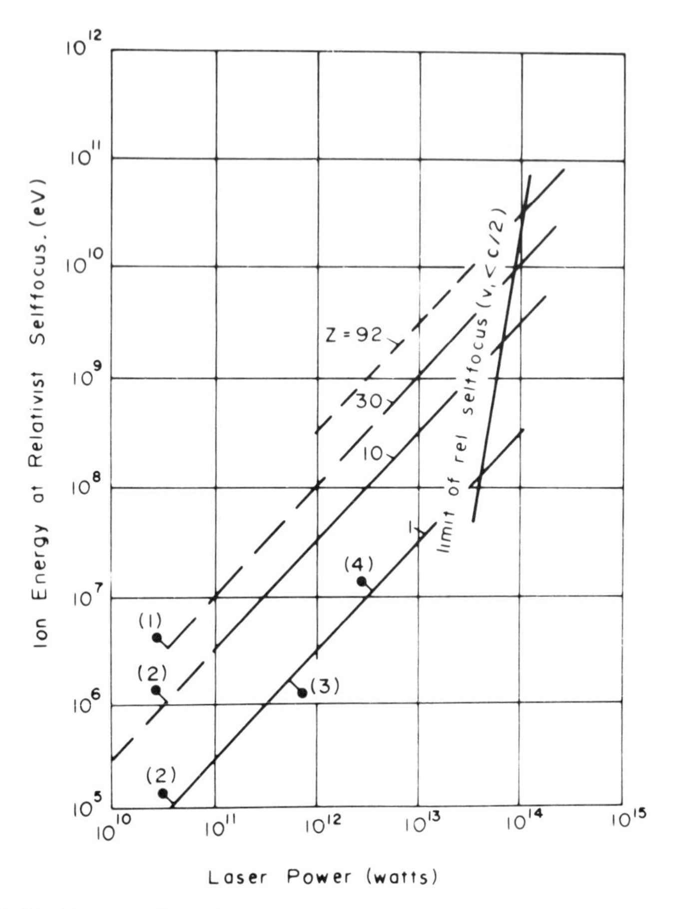

Figure 12.16 Energy of ions from a target with relativistic self-focusing according to Eq. (12.25)[15] depending on the laser power. There is no dependence on the wavelength. The dependence on the ion charge number is restricted by the expected degree of ionization. The measured ion energies correspond to (1) [314], (2) [51], (3) [316], and (4) [317].

ionized, if the same laser beam had been applied to a high-Z target. Ionizations of 40 and more are known from laser produced plasmas.

The production of high Z GeV ions should not be too far beyond the state of the art. Instead of promoting the next middle size accelerators in the range of $100 million of the conventional type for heavy ions, one should first discuss the prospects of producing a laser GeV heavy ion accelerator [318]. The properties of the laser produced multi-MeV ions produced now are quite different from the beams in conventional accelerators, but the disadvantages of a large energy spread and of short pulses should be compared with the advantage of producing short very high Z-ion bursts of some picosecond duration with an ion beam density, which is many orders of magnitude larger than any conventional ion beam. This advantage necessitates a detailed revision of the interesting experimental requirements of medium and high energy in nuclear physics.

THIRTEEN

Laser Compression of Plasma for Nuclear Fusion

The preceding sections developed the macroscopic nonlinear dynamics of laser-plasma interaction. One basic experience was the need for the most possible correctness of initial assumptions, for example, exact solutions of Maxwellian equations, if non-contradictory results will be produced. In strengthening these conditions, we can assume that our formulation of the nonlinear force Eq. (8.81) or Eq. (8.82) is complete according to all present knowledge. This resulted in an agreement of the nonlinear force acceleration of ions up to MeV energy proportional to Z after relativistic self-focusing in accordance with experiments. The momentum transfer (half Abraham and half Minkowski) is consistent with a basically new interpretation of the Abraham–Minkowski problem. In contrast to the thermokinetic interaction, the nonlinear force acceleration for quasi-plane fronts at sufficiently high intensities results in cavitons in full agreement with experiments. Extensive numerical calculations with plane waves indicated the highly efficient transfer of optical energy into fast motion of thick blocks of "cold" plasma. A later soliton process is one additional proof of the very general nonlinear code and explains a fast ionic thermalization of the corona at the later stages of interaction. Oblique incidence results in striated motion and (for p-polarization) in resonance absorption for which a nonlinear-force process of acceleration was developed indicating the electrostatic wave mechanism. For beams, after developing a quantitative model of nonlinear-force self-focusing and a new mechanism of relativistic self-focusing, the radial forces resulted in a measured electron acceleration (Boreham experiment) in agreement with the theory, *if* exact solutions of Maxwell's equations are used showing longitudinal components of electromagnetic waves in vacuum.

All these results indicate a completely new view of this field of physics with a nonclassical behavior, opening new fields of material treatment, pair production, or MeV to GeV ions pulses of very high flux densities, very short pulse and very intensive x-ray sources of much shorter wavelength than line radiation [313], or excitation of nuclei (isomeric ^{235}U) by laser irradiation [319]. The main effort in research, however, is at present to apply laser irradiation of plasma to ignite controlled thermonuclear fusion reactions for energy production. The need for this is underscored by the present energy crisis. The basic economic dilemma [320] was formulated [321] in the following way:

> The history of man is a record of progress directly tied to a continuing reduction in the cost of energy. From the discovery of fire and wheel, to the harnessing of coal, oil, and natural gas for the generation of electricity, man's material lot has improved in indirect proportion to the cost and availability of energy. Now that fossil fuels have reversed their curve of cost efficiency, and other sources of energy (winds, tides, solar energy, fission, geothermal and hydroelectric power) have become economically attractive *only* in comparison to the ever-increasing cost of fossil fuels, we must look to what is the cheapest source of future energy in our universe—fusion.

13.1 Energy from Nuclear Fusion

Energy production by thermonuclear reactions is the source of the immense energies emitted from stars. The only exothermal process of this kind produced hitherto on the earth is the explosion of H-bombs, where small fission reactions heat and compress solid-state material with light nuclei to temperatures above 10 million degrees (corresponding to 1 keV kinetic temperature), such that the subsequent expansion of the plasma still permits much more energy production by the fusion reaction than necessary for the ignition. This type of energy production is the aim of extensive research costing at present about $500 million per year studying the possibility of performing thermonuclear reactions in a controlled way. One concept is to generate and heat a stationary plasma confined by magnetic fields. The best known system of this kind is at present the tokamak torus [322], another is the use of mirror magnetic fields [323] or high-frequency fields (rotomac) [324] or wall confinement at reduced thermal contact to the walls by magnetic fields [325]. In contrast to this magnetic confinement, the advent of the laser gave the hope of simulating the macroscopic explosions, in a controlled way, by microscopic explosions. The task is to heat and to compress a plasma by the extremely brief laser irradiation at very high intensities to such high temperatures and densities that thermonuclear fusion reactions will produce more energy than would be necessary for producing the laser pulse. It must be mentioned that fast heating and compression may be possible in the future by using relativistic electron beams or ion beams [326–329].

The reactions of interest are deuterium with deuterium D and deuterium with tritium (D–T), as it was discovered by Oliphant, Harteck, and Rutherford 1933 [330].

$$D+D \rightarrow 50\% \begin{cases} {}^3He+n+3.27\ MeV \\ {}^3T+p+4.03\ MeV \end{cases} \tag{13.1}$$

$$D+T \rightarrow {}^4He+n+17.6\ MeV \tag{13.2}$$

The necessary tritium can be produced (bred) from ^7Li by the emitted neutrons from this reaction or by reaction (13.1).

The DT reaction (13.2) is the only one which can be used exothermally by the tokamak, as the other interesting reactions produce too high a loss by cyclotron radiation. The inertial fusion confinement is not restricted to reaction (13.2), and there is reasonable hope that clean nuclear reactions could be realized where no neutrons are produced, which damage the reactor or produce radioactive waste in the reactor material. Preferable are further those reactions where the reaction products are charged particles, whose kinetic energy can then be converted directly into electrical power. Heat pollution through thermomechanical conversion is avoided. Examples for the clean fuel fusion reaction are [331]

$$D+{}^3He \rightarrow {}^4He+p+18.3\ MeV \tag{13.3}$$

where the necessary ^3He can be bred by the fast protons of the reaction from ^6Li. Another example is the reaction of light hydrogen H with ^{11}B [332]

$$H+{}^{11}B \rightarrow 3\,{}^4He+8.9\ MeV \tag{13.4}$$

or the reaction of D with ^6Li, for which several branches exist. The evaluation of the cross sections of the 2 ^4He branch D ^6Li reaction and the comparison with D–D, D–T, and D–3He for use in fusion reactions has been done recently by Clark et al., Fig. 13.1 [333]. The average of the cross section σ for a thermalized plasma of temperature T for the spectrum of the thermal velocity v is for an averaged mass m_i of the nuclei

$$\langle \sigma v \rangle = \frac{\sqrt{m_i}}{2\sqrt{\pi}(KT)^{3/2}} \int_0^\infty \frac{m_i}{2} v^2 \sigma(v) \exp\left(-\frac{m_i v^2}{2KT}\right) dv^2$$

is given in Fig. 13.2.

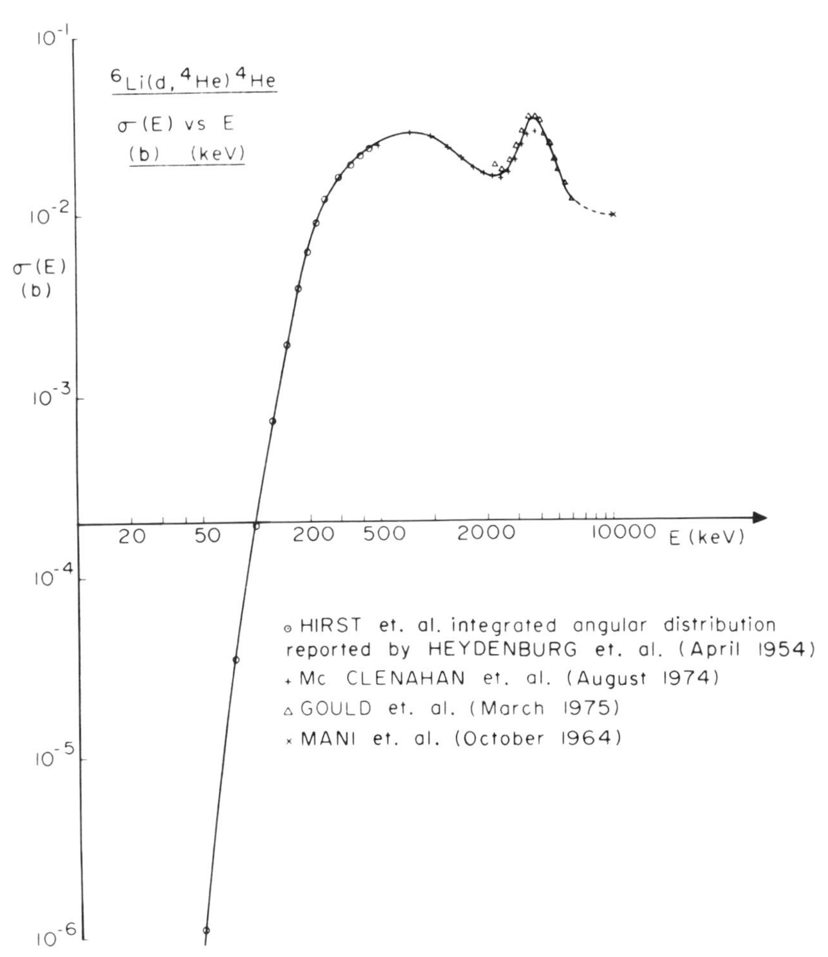

Figure 13.1 Measured and best fit cross sections for the $6Li(d, \alpha)\alpha$ reaction using the results of Hirst et al [334], McClehan et al. [335], Gould et al. [336] and Mani et al. [337] following Clark et al. [333].

13.2 Inertial Fusion Gain Calculations

There are two steps in a fusion gain calculation for all kinds of driving, either by lasers or particle beams. The first one describes the very primitive conditions of a spherical plasma of an initial volume V_0 and initial atomic density n_0 of fully ionized ions of charge Z, into which an energy E_0 had to be deposited in some undefined way to arrive at a homogeneous plasma

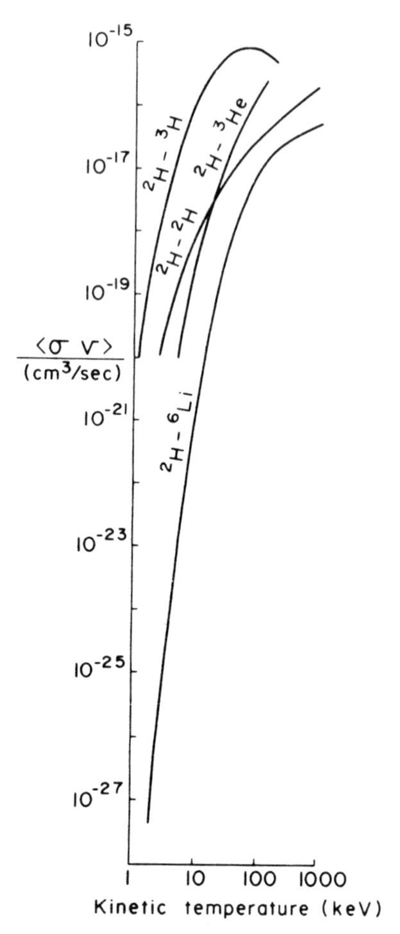

Figure 13.2 Velocity averaged cross section for plasma of various temperatures T for the reaction as in Fig. 12.1 compared with reactions of hydrogen isotopes and ^{3}He [333].

temperature T_0. The fusion reaction gain is defined by the ratio

$$G = \frac{\text{Reaction energy}}{\text{Input energy } E_0} = \frac{\varepsilon_R}{E_0} \int dt \int d^3r \frac{n_i^3}{A} \langle \sigma v \rangle \tag{13.5}$$

where ε_R is the energy per fusion reaction, n_i is the ion density, and $\langle \sigma v \rangle$ is the velocity averaged fusion cross section with a constant $A = 4$ for binary reactions. Gains depending on the initial volume, density, and input energy can be calculated, Fig. 13.3. The highest gains are those where the initial temperature for DT is 10.3 keV and for $H^{11}B$ is 98 keV. The optimum gains

[339] follow from the tangential line to the curves of Fig. 13.3 and result in the relation [340]

$$G=\left(\frac{E_0}{E_{\rm BE}}\right)^{1/3}\left(\frac{n_0}{n_{ec}}\right)^{2/3} \tag{13.6}$$

where n_s is the solid-state density of the fusion fuel, and $E_{\rm BE}$ is the break-even energy, which is 1.6 MJ for DT [339], [341–348] and nearly 1 TJ for $H^{11}B$. This formula has the ability to show immediately how an increase of the initial plasma density n_0 decreases the necessary input energy E_0 by the quadratic power, if the same gain G has to be produced, indicating the need for compression of the plasma. If one expresses the input energy

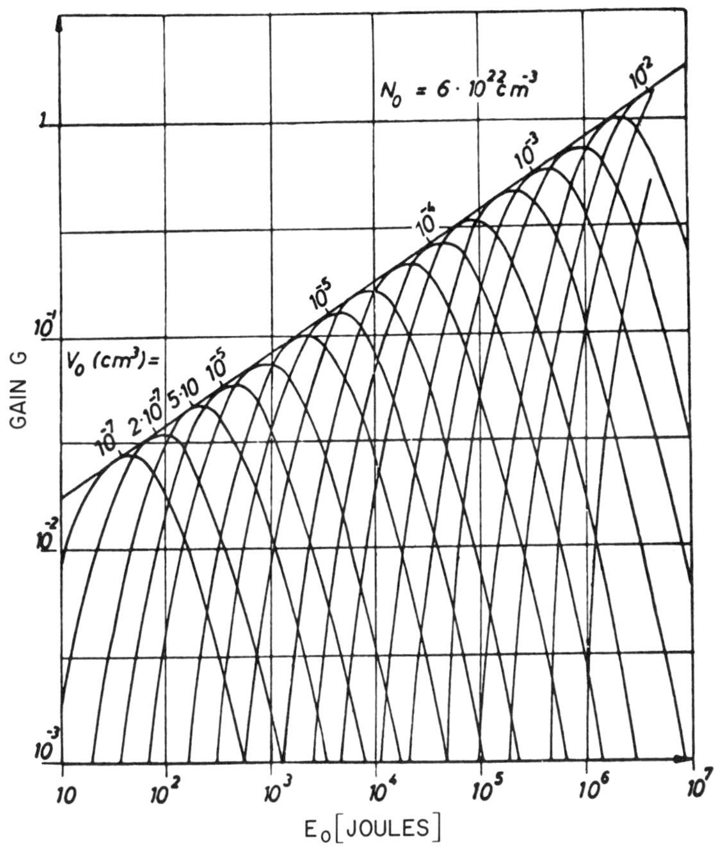

Figure 13.3 Fusion gains G from Eq. (13.5) using $\langle\sigma v\rangle$ values of J. Tuck [338] for DT depending on the input energy E_0 into the spherical pellet of solid state density of various initial volume V_0 [339].

$E_0 = 4R_0^3 n_{i0}(1+Z)kT_0/3\pi$ by the initial radius R_0 of the plasma and the initial temperature T_0 derived from the optimum calculation, formula (13.6) is then

$$G = \text{const } n_{i0}R_0; \qquad \text{const} = 1.66\times10^{-22}\ \text{cm}^2 \tag{13.7}$$

as first given by Kidder [349] with a constant for DT differing by a minor amount only from the derivation from Eq. (13.6).

The next step for a generalization of the gain calculations is to take into account the depletion of the fusion fuel and the losses by bremsstrahlung as long as the absorption length is larger than the plasma size. Furthermore, the reheat of the plasma by the generated alpha particles must be included. Because the alpha production is of a large intensity at a fusion reaction of the following ranges, the Fokker–Planck approximation for the stopping power cannot be used, since the approximation works with small perturbation only and with the first two Fokker–Planck coefficients only. Based on a concept that was successful for high-intensity electrons, and based on polarization effects, a collective model was used for calculating the stopping power [351].

It must be emphasized that Ray was first discussing the Fokker–Planck range, reproducing the Winterberg approximation of the range $R = \pi^{3/2}$ (T = plasma temperature) [350] but more generally showing a kink at temperatures near 1 keV, Fig. 13.4. There was no change due to a quantum

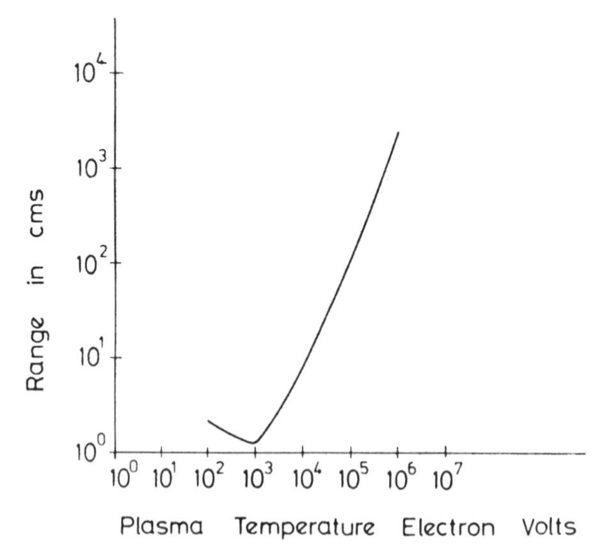

Figure 13.4 Fokker–Plank approximation of the range R of 14.7 MeV protons in a DT plasma of solid-state density depending on the plasma temperature T [351].

electrodynamic generalization [351]. The low-temperature part merges into the ranges calculated from the collective model [352] where the range R is

$$R = \frac{e^2}{2KT} \frac{m_H}{m} E_i[\ln(\lambda E_H)^2]; \qquad Ei(X) = \int_{-\infty} \frac{\exp(t)}{t} dt \tag{13.8}$$

where

$$\lambda = \left(\frac{m}{m_H}\right)^2 \frac{KT}{\pi n_e Z_H^2 e^6} \tag{13.9}$$

for an initial energy E_H and mass m_H of the high-energy particle of charge Z_H and the plasma temperature T. An example for the range of the alphas of the $H^{11}B$ reaction is given in Fig. 13.5, which always corresponds to the range left of the kink of the Fokker–Planck result. To the right side of the kink (for temperatures above 10^3 to 10^4 eV) the discrepancy between the collective model and the Fokker–Planck approximation could be large, as has been shown in details in 1977 [353]. It is evident that one has to use the shortest possible R if there were competitive models for R for calculating the reheat. With this stopping power (which was always that of the collective model), we arrived at fusion gains given in Figs. 13.6 and 13.7 for DT [354] and $H^{11}B$ [244].

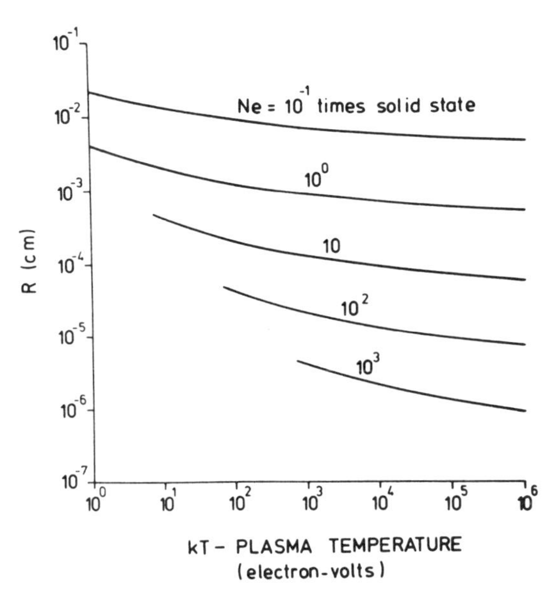

Figure 13.5 Range R of 2.89 MeV alphas from the $H^{11}B$ reaction on the temperature of plasmas of various electron densities n_e [352].

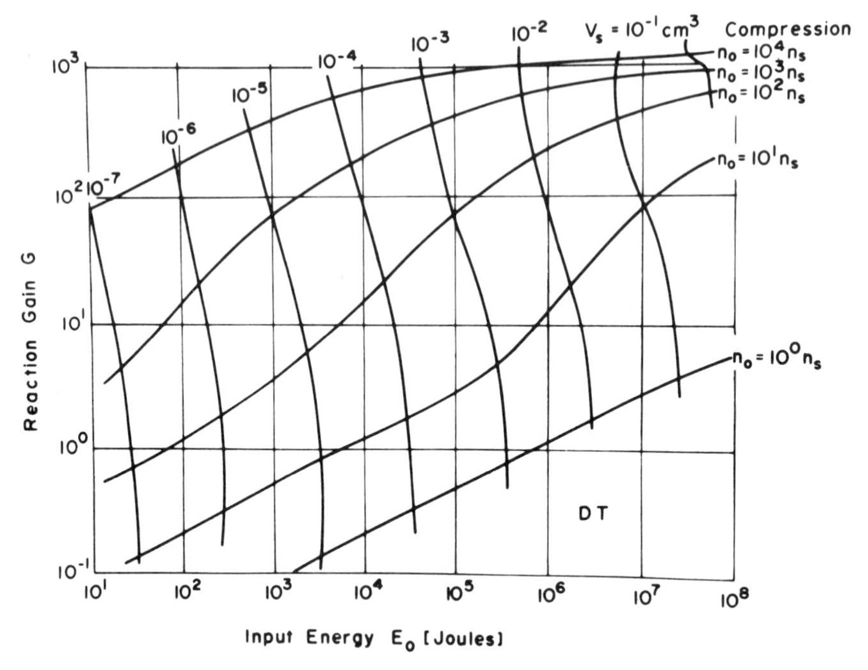

***Figure** 13.6* Fusion gains [Eq. (13.5)] for DT, with bremsstrahlung loss, reheat and fuel depletion, depending on the energy E_0 deposited in the plasma of an initial density n_0 n_s = solid-state density), v_s is the volume at solid state before compression to n_0 [244, 354].

Although relatively simple assumptions are still involved and although for final validity the reheat model may still need a further, more profound derivation, it is remarkable that the results for DT agree to within a few percent with the very extensive gain calculations of Nuckolls [355] for comparable parameters [356–358], where the reheat values were taken from the measurements of large-scale fusion reactions [358].

It is a significant result of the gain calculations of Figs. 13.6 and 13.7 that the optimum temperature drops to lower values than in the case without reheat calculations. This can be seen from the vertical lines in Figs. 13.6 and 13.7, which are giving the initial volume of the plasma at solid-state density before compression to the given densities. These lines are vertical if the reheat processes are not included. The drift of these lines to the left-hand side for increasing initial density is proportional to the decrease of the optimum initial temperature. The relatively low optimum temperatures for DT at compressions of 1000 times solid-state density for input energies E_0 near 10 kJ are about 2 keV only. The correctness of this surprising result

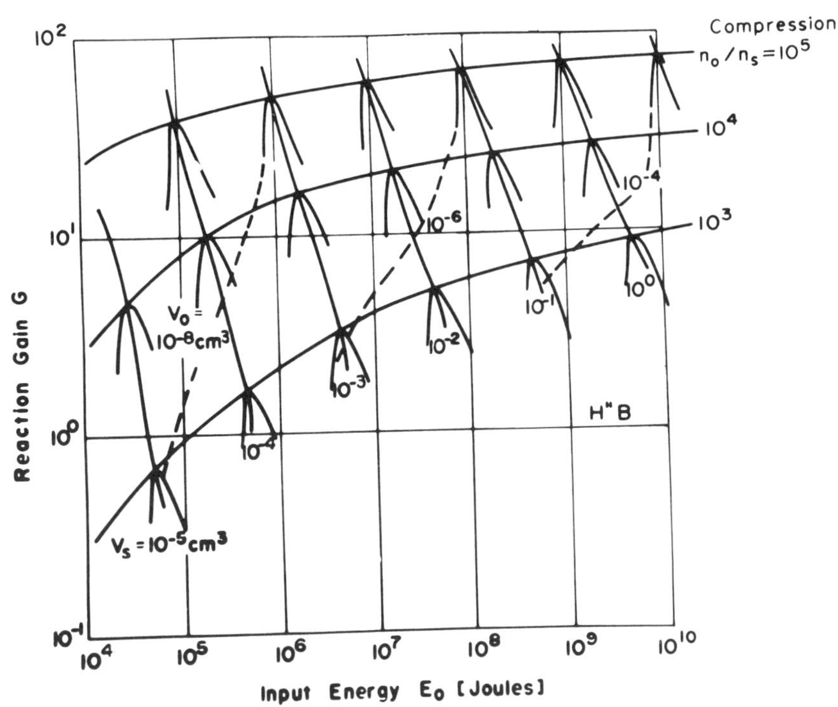

Figure 13.7 Fusion gains for the $H^{11}B$ reaction for the same conditions as in Fig. 13.6 [244].

can be seen when the time dependence of the plasma temperature is printed out [354]. After a first slow increase of the temperature in time, a strong rise is observed within a few picoseconds up to temperatures exceeding 50 keV, followed by a fast drop of the temperature due to the fast expansion, see Fig. 13.8. This process corresponds to a homogeneous ignition of the compressed plasma, which differs from the ignition process due to fusion combustion fronts described by Brueckner and Jorna [359].

The ignition process is the reason that the attractive clean fuel fusion process of $H^{11}B$ has considerable chances in the next possible developments in inertial fusion confinement. If the concepts of heavy ion beam fusion, or of the electron beam fusion, or if the development of new laser systems—such as nuclear pumped lasers [33] with favorable energy storage and pulse compression of several MJ energy—permit a high-efficiency transfer of energy E_0, in the MJ range into the plasma compressed up to 10,000 times the solid-state density, the hydrogen boron reaction could be used. It is important to consider how the scenario of the whole inertial fusion energy development will be changed, if the use of $H^{11}B$ could be done soon, for example,

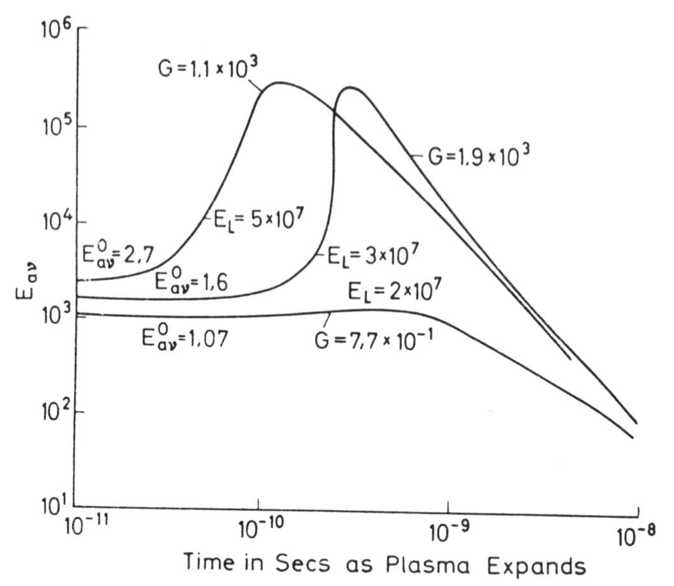

Figure 13.8 Time dependence of the temperature of a DT laser plasma for compression 10^3 times solid initial volume 10^{-3} cm^3 and varying initial energy E_0. A strong increase of the gain results by a small increase of E_0 when ignition and self-burning is happening [354].

within this century. The early achievement of this scientific goal could greatly simplify the technological concepts. As there is no primary production of neutrons, pellet preparation can be done at room temperature and does not need cryogenics, and the energy can be converted directly into electrical energy, thereby providing an efficiency exceeding 80% and reducing the heat pollution, which is involved in nearly all present power stations [360].

Our calculations of the gain G for the alpha branch of the D^6Li reaction did not show an ignition. Instead of the left-hand drift of the vertical lines (lower ignition temperature at higher densities), a right-hand drift was found. Gains did not exceed the number 2. The reason is that the losses by bremsstrahlung are stronger than any improvement by reheat. It should be mentioned that Dawson [361] reported that he had no ignition for $H^{11}B$ but ignition for D^6Li. For the most skeptical reader, this should indicate that clean fuel fusion does have a content of truth for pellet compression by lasers or charged beams. Our understanding of the preference of $H^{11}B$ is based on the shorter R of the 2.9 MeV alphas compared with the much larger R (and therefore less reheat) of the 11 MeV alphas in the case of D^6Li.

13.3 Results of Laser Fusion

The idea of the use of lasers for inertial fusion confinement was obvious after the discovery of the laser in 1960 [362]. The first publication appeared in 1963 by Basov and Krokhin [363] and by Kastler [364], followed by several authors. The report of the first detected fusion neutrons from a laser irradiated target containing deuterium was in 1968 [365], where the number of the fusion neutrons was just above the level of detection. Quantitative numbers of fusion neutrons were reported by Lubin in August 1969 [366] and in September 1969 by F. Floux [367]. The essential point of this experiment was the steep increase of the laser pulse and the suppression of the preceding laser intensities by at least 10^{-5} before the main pulse by electro-optical switches.

Following this knowledge, Sklizkov et al. [368] built a nine beam laser system, Kalmar, with which up to 10^8 fusion neutrons were produced from deuterated targets by 1973. A very strong stimulation came from the disclosure by E. Teller in 1972 [369] reporting on many years of unpublished computer work by J. Nuckolls [370] and many others, where the use of laser irradiation on a spherical target causes a strong compression and heating to achieve the appropriate fusion gains. The next important step was the use of DT gas filled glass balloons under spherical laser irradiation to demonstrate laser compression of the pellet using pinhole camera pictures of the emitted x-rays [371]. While the peripheral parts of the plasma were emitting x-rays due to the immediate laser interaction with the plasma corona, no fusion neutrons were produced there as only glass plasma was generated in the corona. Obscure neutron production in the outer plasma part by plasma jets, by striated motion, or by resonance absorption could then be excluded. The x-rays from the center of the plasma, which appeared about 100 psec later, correspond to the genuine CTR (controlled thermonuclear reaction) fusion neutrons.

The compression of plasma up to 100 times the solid-state density has been reached in 1980 (Fig. 13.9), where the initial density was much below the solid-state density and was that of a pressured gas. The number of 10^3 genuine CTR neutrons in 1974 [372] has been increased to 3×10^{10} neutrons per shot in 1979 [373] with irradiation by a 10 kJ laser pulse of the SHIVA laser system [374]. The development of the 10 kJ HELIOS laser system [375] achieved more than 10^8 neutrons. The use of very short laser pulses from neodymium glass laser pulses (30 psec pulses) seemed to be more favorable compared to the other cases of long pulses. M. Lubin reported [376] about 10^9 neutrons from DT filled glass ballons with an input energy of 100 J only. Similar excellent results with short pulses are known [377], where the small ARGUS laser produced at least the same number of fusion

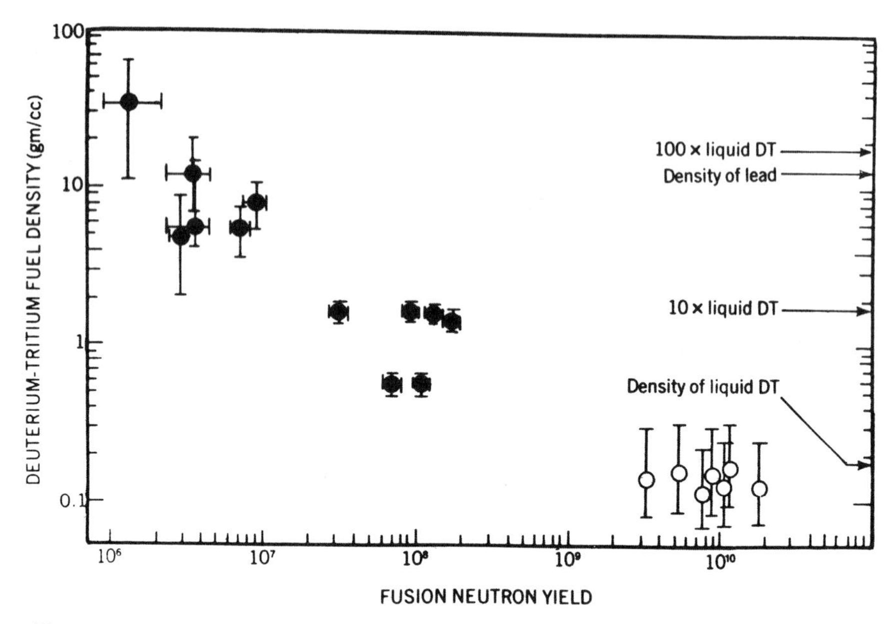

Figure 13.9 Compression and neutron gain of laser irradiated DT filled pellets from the Shiva experiment. After Lubkin [378].

neutrons, for pure deuterium target conditions, per unit input energy [374] as came from the most advanced Princeton tokamak experiment per input neutral beam energy. It must be emphasized that the laser compression experiments produced genuine CTR neutrons, while the aforementioned tokamak experiments [378] had a high ion energy maximum, which indicates that no thermal equilibrium was reached and the neutrons were not genuine CTR neutrons.

The details of the diagnostics of large-scale laser compression experiments have been developed to extremely refined methods [379], and techniques for producing pellets of very high-quality are available including hollow spheres of solid deuterium [380]. The development of large laser systems on the basis of free-electron lasers [303, 381], nuclear pumped lasers [382], or other concepts opens the possibility of designing the high-power lasers required for transferring MJ pulses to pellets.

In view of the preceding considerations of laser plasma interactions, there are some general requirements that have to be fulfilled for plasma compression by lasers for fusion. Apart from the numerous processes mentioned and other mechanisms, which will be discovered in future developments, and apart from the highly complex involvement of different nonlinear processes, one main postulate will be to provide a *symmetric irradiation* of the pellets.

Though the thermokinetic forces are determined only by the gradient of the density, and even the nonlinear net forces, which accelerate the plasma toward the lower density, do not depend on the direction of laser incidence or on the polarization (in lower order approximation), the nonlinear processes (striated motion and resonance absorption) depend on the *p*-polarization at oblique incidence and cause a very high asymmetry. The hope that the asymmetric irradiation can be compensated by thermal interaction processes can simply be neglected, when considering the x-ray pinhole pictures of pellet coronas, when perpendicular irradiation by highly focused beams is considered [383]. The interaction and pusher-type compression occurs at localized spots of irradiation only, and no exchange of the heating to less irradiated areas can be observed.

A further obvious aspect for symmetry [384] is the need to irradiate the plasma in such a way that self-focusing is avoided; otherwise one cannot expect highly symmetric spherical compression fronts. It is important to suppress both types of self-focusing: the nonlinear force self-focusing with its low-power threshold and long delay, and the relativistic self-focusing with its instantaneous action at high intensities.

The study of the processes at oblique laser incidence on a plasma is important for analysis of the mechanisms for perpendicular incidence also and should not be underestimated. However, for the final goal of laser fusion, the symmetric spherical illumination of the targets will be of central interest. If the results of the nonlinear force interaction lead to fast low-entropy producing generation of thick blocks for a pusher-type compression in the initial stages of interaction, then the sufficient conditions for the concept of a laser pellet compression scheme will be given. If a very fast thermalization of the plasma corona for driving an isentropic thermokinetic ablation scheme is necessary, the nonlinear force interaction also provides the necessary fast thermalization for perpendicular incidence of high intensity laser radiation, when the soliton decay of the plasma corona due to the nonlinear forces is taken into account [386].

13.4 Thermokinetic and Nonlinear-Force Compression

There are two essentially different schemes for compression of laser produced plasmas. The earlier developed scheme of Nuckolls [370] and several other authors starts from the irradiation of a spherical pellet by laser radiation, the intensity of which has to increase according to a certain time-dependent program. For neodymium glass laser pulses, the intensity is slowly increasing over many orders of magnitude during the first few nanoseconds, while 50% of the laser energy has to be deposited during the last 60 to 100 psec [370]. The laser radiation generates a plasma at the pellet surface, where the plasma expands into the vacuum due to the high temperatures generated in the plasma corona. As a reaction, the plasma below the cutoff density (where the electron density causes a plasma frequency higher than the laser frequency) causes a motion toward the plasma interior with a compression of plasma. It is remarkable that this kind of compression was the immediate result of the hydrodynamic calculations, as published first by Mulser [88] and by Rehm [215]. The Nuckolls-type increase of laser intensity causes an addition to the compression process, where finally a compressed core of 10,000 times the solid-state density will be achieved [370]. This result can be obtained automatically from the hydrodynamic calculation of the laser plasma interaction, where extensive computer capacity is necessary.

A similar kind of compression to very high densities can be calculated from the sequence of shock waves produced in the plasma pellet by a sequence of laser pulses with increasing intensity. This process follows a model of Guderley [220], where the timing of the pulses, of their intensities, and of the speed of propagation of the shocks to the center are scaled in such a way that the shocks are arriving at the pellet center at the same time. It must be noted that this shock wave model describes some essential properties of the Nuckolls scheme; however, the high entropy production in a purely shock wave type process is highly undesirable. The selection of the appropriate increase of the laser intensity is one of the parameters to be chosen, in an optimized way, for a minimum entropy production. The total efficiencies for the compression of plasma by neodymium glass laser pulses arrived at low values, since, for example, only 5% of the incident laser energy can be transferred into the compressed core. These efficiencies are high enough to achieve a nuclear fusion gain of 40 for deuterium tritium reactions if the incident laser energy is 200 kJ.

The hydrodynamic properties of the compression by the gasynamic ablation scheme of Nuckolls seems to follow automatically from the hydrodynamic codes. If one looks into the details of the process, there are several points where complications arise. The first point is the need to transfer the laser energy in the plasma corona sufficiently fast into heat of electrons and ions for their subsequent hydrodynamic motion. It has been pointed out that the usual thermalization by the Coulomb frequency is too slow, Eq. (10.2), Fig. 10.4. The collision time for 10^{16} W/cm^2 neodymium glass laser radiation exceeds the time of 60 psec of the main laser pulse in the Nuckolls scheme. It was therefore necessary to look for fast thermalization processes, if one would not step back to comparably low-intensity laser interaction of less than 10^{14} W/cm^2 for neodymium glass lasers, while 10^{19} W/cm^2 is needed at the compression core [387]. As was mentioned, the parametric decay instabilities could not provide the sufficient magnitude of energy absorption necessary for the dynamic processes.

Another possibility of a fast energy transfer is the resonance absorption for oblique incidence at p-polarization. However, an asymmetry of irradiation of the pellet will result. One way out for achieving fast thermalization in the plasma corona at very high laser intensities under symmetric (perpendicular) irradiation is the process of soliton decay in the corona due to the nonlinear force interaction, as described in connection with Fig. 10.22. These processes are fast for the scales of laser-plasma interaction, as the soliton decay may occur within some picoseconds of interaction or longer. Very probably, the interaction times up to nano seconds for neodymium glass laser radiation will be too long.

Another problem in the Nuckolls scheme comes from processes of the laser radiation in the interior of the plasma. The processes of electron heat waves for oblique incidence and p-polarization were studied extensively; the resonance absorption should result in high-energy electrons directed toward the interior of the plasma, supporting the heating of the compressed front of the plasma. These transport processes and the thermal conductivity in the compressed pellets are considered as crucial problems together with the absorption processes in the plasma corona for the Nuckolls scheme.

The use of very long neodymium glass laser pulses (a few nanoseconds) and moderate intensities below 10^{14} W/cm^2 should avoid the anomalies mentioned and achieve the ideal gas dynamic ablation mode [388]. The too low temperatures measured in this case with too low neutron gains [385] should be overcome when very large pellets and laser pulses are used. It should be emphasized that in these larger cases the ratio of the interaction interface to the volume to be heated is worse than in the measured cases of ablation modes [385] where high compression but too low temperatures were the result; Fig. 13.9.

The alternative second scheme of laser compression is that of producing a very quick pusher in the plasma, where the nonlinear forces could be used for a very efficient transfer of optical energy into kinetic energy of thick blocks of fast-moving plasma. This interaction process causes nearly no change in the compressed plasma temperature, Fig. 10.18*b*. As shown by the example in Fig. 10.20*a*, the generation of the fast-moving compressing block is during the first period of interaction before the soliton decay begins. As can be seen from Fig. 10.20*a*, the fast motion of the compressing block still holds at later times, but it is obvious that no further transfer of laser radiation to increase the kinetic energy of this block is possible, after the soliton decay of the surface absorbs the irradiated laser energy before permitting an interaction with the compression block.

The high efficiency of optical energy transfer into kinetic energy of compressed plasma can be seen by earlier calculations, result 2 on page 172, where 43% of the incident radiation was received in the compressing plasma block. The more recent calculations for interaction times of several picoseconds, [153], Fig. 10.18, resulted in 47% of the energy of incident laser radiation being transferred into the compressing block. The following motion of the compressing block can be described as an imploding spherical shell, which after collapsing will be compressed. The kinetic energy of the plasma will then be transferred into thermal energy. If the initial conditions at the time of collapsing correspond to a Gaussian density and linear velocity profile, the conditions of the self-similarity model are fulfilled, Section 5, and an ideal adiabatic or isentropic compression and expansion of the spherical plasma will occur. Including nuclear fusion reactions, a certain disturbance of the ideal adiabatic conditions will be unavoidable. Strong deviations from the ideal adiabatic conditions result in undesired shock processes and heating, which can cause decreases of the final nuclear fusion gains by up to a factor of 10. In these calculations with entropy production, however, the initial conditions have not been chosen in optimized way; therefore an improvement is possible even for nonideal initial conditions of the plasma at the time of compression.

The velocity of the compressing block of plasma has to be chosen of such a value that at the maximum compression, the optimized fusion temperature is produced. This is between 2 and 10 keV for DT reactions and between 30 and 100 keV for the $H^{11}B$ reaction for compressions up to 10^3 and 10^4 times the solid-state density, respectively. The optimized parameters for the fusion reaction gains are taken from the following.

The initial temperatures of the plasma, given as a hollow sphere of a certain density profile with a maximum density near the cutoff value, have to be very low for long wavelengths. These values were given first for the adiabatic compression case [255]. With respect to the low cutoff density for the long wavelength of the CO_2 laser, these initial conditions were very extreme, yet not unfeasible. Using the following improved nuclear fusion gains, the conditions for CO_2 are better. Even in this very unfavorable case, the following conditions [244] arrive at a solution. An incident laser pulse of this wavelength, of 0.3 nsec duration, and 400 kJ energy will transfer 50% of the laser energy into a plasma of 3.5 keV temperature of 100 times compression. The nuclear reaction energy is 20 MJ, arriving in a total reaction gain of 50 based on the incident laser energy. The target is a DT gas of 10^{19} atoms per cm^3 initially limited by a sphere of thin CD_2 of 2.62 cm radius and by a photoexplosive rigid shell of 1 cm radius. In the initial stages, a bi-Rayleigh-type profile with a parameter 1.9×10^3 cm^{-1} [244] is used.

The laser intensity at the shell center is 3.2×10^{13} W/cm^2, and the acceleration due to the nonlinear force for generating the block motion is 2.43×10^{17} cm/sec^2 at the initial point of maximum swelling of the laser intensity.

13.5 Conclusions for Laser Fusion

Despite of the great efforts with very large projects exceeding $100 million each, laser compression of plasma arrived at some critical point at the beginnings of the 1980s. If one takes the 10^8 DD neutrons produced in 1973 by the nine-beam Kalmar [368] corresponding to 10^{10} DT neutrons using an imput energy E_0 of about 0.5 kJ, the progress to 3×10^{10} DT neutrons using several kilojoules of Shiva in 1979 [385] seems to be small. However, in the first case, the neutrons may be due to obscure surface effects and in the second case there are really thermonuclear neutrons from a compressed pellet core avoiding surface effects. Nevertheless, the gain of Shiva is 10^4 times less than expected, according to Brueckner [389]. At present, the ion beam fusion is progressing quickly. Cooperstein et al. [329] reported the production of 10^{13} DD neutrons by focusing 1.4 MeV-deuterium ion pulses of 50 kJ energy of 4 mm diameter and some 10 nsec duration on a deuterated target. This is 20 times higher gain than from the best tokamak [390]. Heavy GeV ion beams [327] may even provide better conditions because of the well-defined beams in high vacuum and their cleaner focusing onto pellets.

There is agreement [391] that the optimum energy densities for the thermokinetic compression scheme of 10^{15} W/cm^2 will be the domain of the more efficient and successful ion beams, while the same intensities for laser beams led to the described difficulties.

One way out for lasers could be the use of the nonlinear force compression at 10^{17} W/cm^2, which resulted in a low-entropy (highly efficient) transfer of optical energy into kinetic energy of fast thick blocks of "cold" plasma for compression. This intensity cannot be reached by ion beams because of space charges. However, 10^{17} W/cm^2 is closer to the necessary 10^{19} W/cm^2 needed at the compressed core [388]. The fact is that the highest neutron gains published yet were achieved with the short pulses of 25 psec of the symmetrically irradiating Argus laser [392], which is closer to the short-pulse nonlinear force compression than the long pulse Shiva experiments, Fig. 13.9. The explanation of the different thermokinetic schemes for these "long pulses" are masterpieces of physics but they did not increase the gains. It has been found that at very long pulses (>nsec) the plasma reaches very high compression but not high neutron yields because of insufficient energy transfer into the compressed core (*ablation mode*). If the pulses are not so very long (*pusher mode*), the higher laser intensities produce hot electrons which preheat the pellet core too strongly preventing the necessary high densities at adiabatic compression. In Fig. 13.9, the lowest densities are due to pushers, and the others due to ablation modes. In contrast to this, the *very fast nonlinear force pusher* will accelerate the plasma before the hot electrons can be generated. This has been confirmed by extensive numerical studies and can be the way out of the present difficulties of laser fusion.

APPENDIX A

The Effective Mass

In Section 2 we used the effective mass m^* of electrons. Though this is a property of condensed matter, a marginal discussion for high-density plasmas is useful under the aspect of the quantum properties of high-density plasmas. As it is the endeavor of this book to derive the physics *ab initio*, as, for example, the hydrodynamics or electrodynamics, a similar derivation of the quantum mechanics is given.

Quantum physics was due to the discovery of the fact that all quanties with the dimension of an action can appear only in multiples of Planck's number $h = 6.67 \times 10^{-27}$ erg sec, or $\hbar = h/2\pi$. This observation was not as easy as the observation of the atomistic structure of the electric charge (given by the electron charge), which everyone could see immediately in the Millikan experiment. The history of the discovery of the atomistic structure of action was pronounced where energies E [of electrons at photoemission or in gas discharges (Franck–Hertz effect)] were related to (optical) frequencies ν, where

$$\frac{E}{\nu} = h \tag{A.1}$$

Since the product of a momentum p and a length x is of the dimension of an action, there were difficulties on how to keep the beautiful knowledge of mechanics, which was formulated so successfuly by Newton, d'Alembert, Lagrange, and Hamilton. If we write the Hamilton function of a simple system as the sum of kinetic energy $p^2/2m$ and potential energy $V(x)$

$$H = \frac{p^2}{2m} + V(x) = \mathrm{E} \tag{A.2}$$

as the total energy E, then we have to be aware of the background of the Lagrangian and aware that this is not simply the difference of kinetic and potential energy for nonconservative forces [115].

One way to quantize Eq. (A.2) is to not use the quantities of p and x directly, but to describe them by a distribution function as shown in Section 3. Instead of getting the average value of a set of quantities q_n one could use a distribution function f_n to arrive at the average value of (see Eq. 3.2)

$$q = \frac{\sum f_n q_n}{\sum f_n} \tag{A.3}$$

Knowing this, Eq. (A.2) can be written by differential operators, where, however, only a distribution function will be defined from which the physical quantities have to be derived similarly to Eq. (A.3). If the operators

$$p = -i\hbar \frac{\partial}{\partial x}; \qquad E = -\frac{\hbar}{i}\frac{\partial}{\partial t} \tag{A.4}$$

are used we are in agreement with quantization

$$p\, \partial x = \text{``}\hbar\text{''}; \qquad E\, \partial t = \text{``}\hbar\text{''} \tag{A.5}$$

where the quotes are a symbolism which nobody would have accepted if the following steps were not performed historically by wave equations (de Broglie, Schrödinger). Only in retrospect, the motivation for (A.4) from a quantization as in (A.1) or (A.5) should be understood. Using (A.4) in Eq. (A.2), the Hamilton function becomes a Hamiltonian operator for a differential equation for a distribution function Ψ (Schrödinger equation)

$$\left\{ -\frac{\hbar^2}{2m}\frac{\partial^2}{\partial x^2} + V(x) \right\} \Psi(x, t) = -\frac{\hbar}{i}\frac{\partial}{\partial t} \Psi(x, t) \tag{A.6}$$

Stationary (time-independent) solutions of this wave equation can be expressed by

$$\Psi = \psi(r) \exp\left(-\frac{i}{\hbar} Et \right) \tag{A.7}$$

where E is an eigenvalue representing an energy in the time-independent Schrödinger equation expressing the spatial dependence now by the coordinates of

$$\left(-\frac{\hbar^2}{2m} \nabla^2 + V(r) - E \right) \psi(r) = 0 \tag{A.8}$$

If the potential $V = 0$, electrons in vacuum can be described from (A.8) by plane waves

$$\psi(r) = A \exp(i\mathbf{k}\cdot\mathbf{r}); \qquad \Psi = A \exp\left(\mathbf{k}\cdot\mathbf{r} - \frac{i}{h} Et \right) \tag{A.9}$$

where $E/\hbar = \omega$ is a radian frequency, and the wave vector $\mathbf{k}$ is from Eq. (A.8)

$$|\mathbf{k}| = \frac{1}{\hbar}\sqrt{2mE} \tag{A.10}$$

In order to arrive at a physical quantity (expectation value) from the distribution functions, one has to proceed as in Eq. (A.3); however, as Ψ can be complex we have then to include the conjugate complex value Ψ^*

$$Q = \frac{\int \Psi^* q \Psi \, d^3\tau}{\int \Psi^* \Psi \, d^3\tau} \tag{A.10a}$$

integrating over the whole space. The normalization of the amplitude A in Eq. (A.9) is to fulfill $\int \Psi^* \Psi \, d^3\tau = 1$. For example, to arrive at the momentum of the electron, the quantity q in (A.10a) is the operator p of Eq. (A.4), and we find the momentum by spatial differentiation of (A.9)

$$\mathbf{p} = \int \psi^* \frac{\hbar}{i} \nabla \psi \, d^3\tau = \hbar \mathbf{k} \tag{A.11}$$

using the result (A.10), we find the point mechanical relation between momentum and energy of a free electron

$$p = \sqrt{2mE} \tag{A.12}$$

It is a classical example in this method of quantum mechanics to use the Coulomb potential of a proton for $V(r)$ in Eq. (A.8) to arrive at the stationary (bound) states of the electron. The solutions of the equation for the distribution functions ψ arrive by mathematical reasons at eigenvalues $E_n (n = 1, \ldots, \infty)$ which are the energy levels of the electrons in the atom. The spatial distribution of $\Psi^*\Psi$ corresponds to the electron density in the atom arriving at the diameter of twice the Bohr radius for $n = 1$, or eight Bohr radii at $n = 2$, and so on (see Section 2.3).

For the theory of condensed matter, the case of a periodic potential

$$V(\mathbf{r} + \mathbf{d}) = V(\mathbf{r}) \tag{A.13}$$

with a periodicity vector

$$\mathbf{d} = \mathbf{i}_1 d_1 a_1 + \mathbf{i}_2 d_2 a_2 + \mathbf{i}_3 d_3 a_3 \tag{A.14}$$

is of interest, where d_i are distances of atoms in the three crystal directions $i = 1, 2, 3$, and the a_i are integers. Bloch discovered that the solutions of Eq. (A.8) for the periodic potential (A.13) are of the form

$$\psi(\mathbf{r}) = u(\mathbf{k}, \mathbf{r}) \exp(i\mathbf{k} \cdot \mathbf{r}) \tag{A.15}$$

The Schrödinger equation has then the form

$$\left\{-\frac{\hbar^2}{2m}\nabla^2+V(\mathbf{r})-E(\mathbf{k})\right\} u(\mathbf{k},\mathbf{r})\exp(i\mathbf{k}\cdot\mathbf{r})=0 \tag{A.16}$$

Mathematically, the mechanical problem is determined, if $V(\mathbf{r})$ is given, resulting in a uniquely defined $\psi(\mathbf{r})$. This is uniquely related to the function $E(\mathbf{k})$. Instead of describing electrons in a crystal by $V(\mathbf{r})$, or $\psi(\mathbf{r})$, one can uniquely describe them by considering $E(\mathbf{k})$. $\mathbf{k}$ is defining the momentum of the electrons; therefore the energy momentum relation is unique for the description. It was the discovery of Bloch that instead of the parabolic relation $p^2/2m=E$ of Eq. (A.12) for free electrons, there are forbidden gaps for E. The $E(\mathbf{k})$ functions are periodic also, and for $k=0$, that parabolic relation for free electrons can be approximated by the $E(\mathbf{k})$ functions. The parabolas only are more or less curved. The dimensionless factor between the curvatures is simply given by

$$m^*=\frac{1}{\hbar^2}\frac{[\partial E/\partial(p^2)]_{\text{vacuum}}}{\partial E/\partial(\mathbf{k}^2)} \tag{A.17}$$

which is the effective mass.

The structure of energy bands for electrons can occur in very high-density low-temperature (degenerate) plasmas. The physics of laser compressed plasmas is now going into these conditions which will need more detailed study in the future.

To complete the conceptual framework of quantum mechanics drawn in this Appendix, it should be mentioned that the discussion of the quantum mechanical problems with distribution functions and expectation values to conserve the Newton–Hamiltonian mechanics by Schrödinger's differential equation has a mathematical equivalence with an integral equation problem (transformation theory by Weil). The eigenvalues of the differential equation correspond then to the elements of the infinite matrices of the integral equation problem. The matrices alone can be used in a Hamiltonian as it was described by Heisenberg (matrix mechanics).

APPENDIX B

The Maxwell–Boltzmann Distribution

The distribution function of the energy to the particles of a plasma or a gas at equilibrium that was used in Section 3 is derived now. There is a correlation between the entropy S_{12} and the probability W_{12} of the microscopic structure of the states of two thermodynamic systems with the respective values S_1, S_2, W_1, and W_2 [394, 395]

$$S_{12}=S_1+S_2 \tag{B.1}$$

$$W_{12}=W_1 W_2 \tag{B.2}$$

from their definition. The function that reproduced the correlation [396]

$$f(x_1x_2)=f(x_1)+f(x_2) \tag{B.3}$$

is given by Boltzmann's relation

$$S=K \ln W \tag{B.4}$$

using the Boltzmann constant K as the gas constant per particle.

The philosophy for describing a plasma by the probabilities of the distribution of energy to its individual particles is an extremistic picture and may not cover all facts of reality. It implies, for example, that the forces between the particles are small or negligible in first-order or only during negligible times, while the interactions are necessary on the other hand to achieve equilibrium. The other extreme with its insufficiencies is the description of phenomena by differentiable or by analytic (holomorphic) functions which may run into a superdeterminism (Laplace, Cauchy). This can even be a consequence of quantum mechanics (not only in the Schrödinger picture) if the correlation between object and measuring apparatus is considered [397].

In the Boltzmann statistics—in contrast to the quantum statistics—the possibility to distinguish between the particles of an ensemble is assumed. Using six-dimensional volume elements $\Delta\tau_i = \Delta x\ \Delta y\ \Delta z\ \Delta v_x\ \Delta v_y\ \Delta v_z$, the number N_i of particles in this element is given by a distribution function $f(i)$

$$N_i = f(i)\Delta\tau_i \tag{B.5}$$

The total number N of particles should be constant

$$N = \sum f(i)\Delta\tau_i; \qquad \delta N = \sum \delta f(i)\Delta\tau_i = 0 \tag{B.6}$$

where the constancy of $\Delta\tau_i$ at any variation (due to the Liouville theorem) has been used. The energy $U(\delta)$ of the particles in the ith cell results is an energy $N_i u(i)$ in the cell. The total energy U should be constant

$$U = \sum u(i) f(i)\Delta\tau_i; \qquad \delta U = \sum \delta f(i) u(i)\Delta\tau_i = 0 \tag{B.7}$$

The probability of the system whose states of each cell are weighted by

$$G_i = \Delta\tau_i \tag{B.8}$$

is given by the number of the combination of all cases $N!\Pi G_i$ that are possible by permutation without repetition if we use distinguishable particles:

$$W = \frac{N!\Pi G_i^{N_i}}{\Pi N_i!} \tag{B.9}$$

Using Eqs. (B.5), (B.8), and the Stirling formula (on approximation for large N)

$$N! = \frac{N^N}{e^N} \tag{B.10}$$

we find from (B.9)

$$W = \frac{N^N \Pi \Delta\tau_i^{\ f(i)\Delta\tau_i}}{\Pi(f(i)\Delta\tau_i)^{\ f(i)\Delta\tau_i}} \tag{B.11}$$

and the entropy from Eq. (B.4)

$$S = KN \ln N - K \sum f(i)\Delta\tau_i \ln f(i) \tag{B.12}$$

Equilibrium corresponds to the value of the highest probability W, or $\delta S = 0$ at the secondary conditions of constant total particle number N, Eq. (B.6), and total energy U, Eq. (B.7), from Eq. (B.12)

$$0 = \sum \delta_i f(i)\Delta\tau_i \ln f(i) + \sum \delta f(i)\Delta\tau_i \tag{B.13}$$

To include the secondary conditions, the method of multiplicators is used by adding conditions (B.6) after multiplying with α to Eq. (B.13) and to proceed with (B.7) after multiplying with β in the same way. The result is

$$\ln f(i) + 1 + \alpha + \beta u(i) = 0 \tag{B.14}$$

This leads immediately to the desired energy distribution function

$$f(i)=A\exp(-\beta u(i)) \tag{B.15}$$

where

$$A=\exp[-(1+\alpha)] \tag{B.16}$$

is given by the constant number N of all particles from Eqs. (B.6) and (B.15)

$$N=A\sum\exp(-\beta u(i))\Delta\tau_i \tag{B.17}$$

or

$$f(i)=\frac{N\exp(-\beta u(i))}{\sum\exp[-\beta u(i)]\Delta\tau_i} \tag{B.18}$$

The denominator is called the sum of states

$$\sigma=\sum\Delta\tau_i\exp(-\beta u(i)) \tag{B.19}$$

The physical interpretation of the multiplicator β is given from the definition of the entropy. Using Eqs. (B.18) and (B.19) in (B.12)

$$S=K\ln N-K\sum\frac{N}{\sigma}\Delta\tau_i\exp(-\beta u(i))[\ln N-\beta u(i)-\ln\sigma] \tag{B.20}$$

and Eq. (B.7)

$$\frac{N}{\sigma}\sum\Delta\tau_i\exp(-\beta u(i))=U \tag{B.21}$$

Eq. (B.20) reduces to

$$S=K\beta U+KN\ln\sigma \tag{B.22}$$

Thermodynamics defines the relation between S, U, and the temperature T for conditions of constant volume V

$$\frac{1}{T}=\left(\frac{\partial S}{\partial U}\right)_v \tag{B.23}$$

U is a function of β by Eq. (B.21), therefore

$$\left(\frac{\partial S}{\partial U}\right)_v=\frac{dS}{d\beta}\left(\frac{\partial\beta}{\partial U}\right)_v=\frac{dS}{d\beta}\frac{1}{(\partial U/\partial\beta)_v} \tag{B.24}$$

After differentiating Eq. (B.22) we find

$$\frac{dS}{d\beta}=KU+K\beta\frac{\partial U}{\partial\beta}+\frac{KN}{\sigma}\frac{\partial\sigma}{\partial\beta} \tag{B.25}$$

by substitution of the differential quantities of Eqs. (B.19) and (B.21)

$$\frac{\partial\sigma}{\partial\beta}=-\sum u(i)\Delta\tau_i\exp[-\beta u(i)]=-\frac{U\sigma}{N} \tag{B.26}$$

and from Eq. (B.25)

$$\frac{dS}{d\beta}=K\beta\frac{\partial U}{\partial\beta} \tag{B.27}$$

Taking the differential form and Eq. (B.24) we find

$$\left(\frac{\partial S}{\partial U}\right)_v=\beta K=\frac{1}{T} \tag{B.28}$$

and finally

$$\beta=\frac{1}{KT} \tag{B.29}$$

The distribution function (B.18) arrives then at the Maxwell–Boltzmann distribution

$$f(i)=\frac{N\exp(-u(i)/KT)}{\sum\Delta\tau_i\exp(-u(i)/KT)} \tag{B.30}$$

as used in Eq. (3.20).

It should be noted that the discussion of a correction of the dimension of the quantity σ in Eq. (B.22) which has to be of no dimension, while σ is of the dimension of an action to the cube, led Planck [398] to the addition of an arbitrary constant h in the entropy of Eq. (B.22)

$$S=\frac{U}{T}+KN\ln\frac{\sigma}{h^3} \tag{B.31}$$

arriving at a free energy

$$F=U-TS=-KNT\ln\frac{\sigma}{h^3} \tag{B.32}$$

This conclusion of the atomistic structure of action (quantization) from the classical statistics required that oscillators of a frequency ν can gain multiples $h\nu$ of energy only. This caused a reformulation of the sum of states σ (for the oscillator only)

$$\sigma=\frac{h}{1-\exp(-h\nu/KT)} \tag{B.33}$$

and in combination with the Rayleigh–Jeans density of low frequency modes of blackbody radiation led to Planck's radiation law from which the comparison with the fully fitting experiments resulted in the number of h. Remembering that Einstein discovered the stimulated emission (the laser) from his derivation of Planck's radiation law [399], these far ranging facts may be considered [400] as a consequence of Boltzmann's statistics.

APPENDIX C

Derivation of the General Two-Fluid Equations

The direct derivation of the two-fluid equation of motion, (8.3) or (8.6), from the Euler equation of electrons and that for ions is presented in this appendix even though it contains several trivial steps that are usually omitted in textbooks. The derivation is similar to Schlüter's original work [136]. We use the Euler equation with a general viscosity term (determined by the collision frequency ν) for ions [as Eq. (6.1)]

$$m_i n_i \left[\frac{\partial}{\partial t} \mathbf{v}_i + \mathbf{v}_i \cdot \nabla \mathbf{v}_i \right] = -\nabla n_i K T_i + Z n_i e \mathbf{E} + \frac{Z n_i e}{c} \mathbf{v}_i \times \mathbf{H} - \frac{m_i n_i m n_e}{m_i n_i + m n_e} \nu(\mathbf{v}_i - \mathbf{v}_e) + \mathbf{K}_i \tag{C.1}$$

and a Euler equation for electrons [as Eq. (6.2)]

$$m n_e \left[\frac{\partial}{\partial t} \mathbf{v}_e + \mathbf{v}_e \cdot \nabla \mathbf{v}_e \right] = -\nabla n_e K T_e - n_e e \mathbf{E} - \frac{n_e e}{c} \mathbf{v}_e \times \mathbf{H} + \frac{m_i n_i m n_e}{m_i n_i + m n_e} \nu(\mathbf{v}_i - \mathbf{v}_e) + \mathbf{K}_e \tag{C.2}$$

which both may be considered to be based on hydrodynamics, as derived from the kinetic theory in Section 3. The viscosity terms will be canceled when we add both equations later.

In this section we shall use the net pressure of the plasma

$$p = n_i K T_i + n_e K T_e \approx n_i (1 + Z) K T \tag{C.3}$$

with the (optional) simplification

$$T_i \approx T_e \approx T \tag{C.4}$$

of thermal equilibrium.

The discussion that follows basically involves the question of quasi-neutrality

$$n_e \approx Z n_i \tag{C.5}$$

which is used to such an extent that it usually is not perfect [sign of equality in Eq. (C.5)], but it is so well realized within all known properties of fluctuations that the net plasma velocity

$$\mathbf{v} = \frac{m_i n_i \mathbf{v}_i + m n_e \mathbf{v}_e}{m_i n_i + m n_e} \tag{C.6}$$

can be approximated by [see Eq. (6.3)]

$$\mathbf{v} \approx \frac{m_i \mathbf{v}_i + Z m \mathbf{v}_e}{m_i + Z m} \tag{C.7}$$

The velocity difference permits a similar sufficient approximation

$$\mathbf{v}_e - \mathbf{v}_i \approx \frac{n_e \mathbf{v}_e - Z n_i \mathbf{v}_i}{n_e} \frac{e}{e} = -\frac{\mathbf{j}}{e n_e} \tag{C.8}$$

where the definition of the electric current density [see Eq. (6.4)]

$$\mathbf{j} = e(Z n_i \mathbf{v}_i - n_e \mathbf{v}_e) \tag{C.9}$$

is included. A derivation without all these assumptions was performed by Lüst and a result is given in Ref. 401. Agreement with this result will not only justify the assumptions used here, but will also emphasize the essential properties of the relations.

In order to achieve the equation of motion of the plasma by adding Eqs. (C.1) and (C.2), we get for the right-hand side

$$-\nabla p + (Z n_i - n_e) e \mathbf{E} + (Z n_i e \mathbf{v}_i - n_e e \mathbf{v}_e) \times \frac{\mathbf{H}}{c} + \mathbf{K}_i + \mathbf{K}_e \tag{C.10}$$

or using Eq. (C.9),

$$-\nabla p + \mathbf{E} e (Z n_i - n_e) + \mathbf{j} \times \frac{\mathbf{H}}{c} + \mathbf{K}_i + \mathbf{K}_e \tag{C.10a}$$

where the viscosity terms have canceled.

By adding (C.1) and (C.2) and by using an identity for the left-hand side of Eq. (C.2), we get for the left-hand side

$$m_i n_e \left(\frac{\partial}{\partial t} \frac{m}{m_i} \mathbf{v}_e + \frac{m}{m_i} \mathbf{v}_e \cdot \nabla \mathbf{v}_e \right) \tag{C.10b}$$

We arrive at the following expression after adding several more terms and subtracting them again

$$
\begin{aligned}
& m_i n_i \frac{\partial}{\partial t}\mathbf{v}_i + m_i n_i \frac{\partial}{\partial t}\frac{m}{m_i} Z\mathbf{v}_e + m_i n_i \mathbf{v}_i\cdot\nabla\mathbf{v}_i + m_i n_i \frac{Zm}{m_i}\mathbf{v}_e\cdot\nabla\mathbf{v}_e \\
& + m_i N_i \frac{Zm}{m_i}\mathbf{v}_e\cdot\nabla\mathbf{v}_i - m_i n_i \frac{Zm}{m_i}\mathbf{v}_e\cdot\nabla\mathbf{v}_i + m_i n_i \frac{Zm}{m_i}\mathbf{v}_i\cdot\nabla\mathbf{v}_e \\
& - m_i n_i \frac{Zm}{m_i}\mathbf{v}_i\cdot\nabla\mathbf{v}_e - m_i n_i \left(\frac{Zm}{m_i}\right)^2 \mathbf{v}_e\cdot\nabla\mathbf{v}_e \\
& - m_i n_i \frac{Zm}{m_i}\mathbf{v}_i\cdot\nabla(\mathbf{v}_e-\mathbf{v}_i) + m_i n_i \frac{Zm}{m_i}\mathbf{v}_i\cdot\nabla(\mathbf{v}_e-\mathbf{v}_i)
\end{aligned} \tag{C.11}
$$

Here the term with the quadratic mass ratio m/m_i can be neglected as well as the very last term with $\mathbf{v}_i\cdot\nabla\mathbf{v}_i$ in the third term because of the mass ratio. All other terms of Eq. (C.11) can be combined to result in

$$
m_i n_i \left[\frac{\partial}{\partial t}\mathbf{v} + \mathbf{v}\cdot\nabla\mathbf{v}\right] + m_i n_i \frac{Zm}{m_i}(\mathbf{v}_e - \mathbf{v}_i)\cdot\nabla(\mathbf{v}_e - \mathbf{v}_i) \tag{C.12}
$$

The last term in Eq. (C.12) can be rewritten by using Eqs. (C.8) and (C.9)

$$
mZn_i(\mathbf{v}_e - \mathbf{v}_i)\cdot\nabla(\mathbf{v}_e - \mathbf{v}_i) = \frac{m\mathbf{j}}{e}\cdot\nabla\frac{\mathbf{j}}{en_e} = \frac{4\pi}{\omega_p^2}\mathbf{j}\cdot\nabla\mathbf{j} - \frac{m}{e^2 n_e^2}\mathbf{j}\mathbf{j}\cdot\nabla n_e \tag{C.13}
$$

From Eqs. (C.10), (C.12), and (C.13), the result of adding Eqs. (C.1) and (C.2) arrives at the net force density $\mathbf{f}$ in the plasma

$$
\begin{aligned}
\mathbf{f} = m_i n_i \left[\frac{\partial}{\partial t}\mathbf{v} + \mathbf{v}\cdot\nabla\mathbf{v}\right] &= -\nabla p + \mathbf{E}e(Zn_i - n_e) + \mathbf{j}\times\frac{\mathbf{H}}{c} + \mathbf{K}_i + \mathbf{K}_e \\
&\quad - \frac{4\pi}{\omega_p^2}\mathbf{j}\cdot\nabla\mathbf{j} + \frac{4\pi}{\omega_p^4}\mathbf{j}\mathbf{j}\cdot\nabla\omega_p^2
\end{aligned} \tag{C.14}
$$

This is the complete result of Schlüter at an appropriate interpretation of his velocities [136] where we have permitted ions of a general charge Z, as one point of generalization. The same result was reported as the outcome of an exact treatment by Lüst [401] without our assumptions of eqs. (C.5) to (C.9). Additional terms were negligible because of the space charge neutrality.

Although Schlüter [136] regarded forces by high-frequency electromagnetic fields such as radiation pressure to be included in the unspecified forces $\mathbf{K}_i + \mathbf{K}_e$, the treatment of the high-frequency fields is possible directly by Eq. (C.14). We neglect $\mathbf{K}_i$ and $\mathbf{K}_e$ and use then monochromatic time dependence given by a frequency ω for all quantities $\mathbf{E}$, $\mathbf{j}$, and $\mathbf{H}$ as in Eq.

(6.19). We further use Eq. (6.8)

$$\frac{\partial}{\partial t}\mathbf{j}+\nu\mathbf{j}=\frac{\omega_p^2}{4\pi}\mathbf{E} \tag{C.15}$$

which is the result of subtracting Eq. (C.1) from (C.2). Nonlinear terms have been neglected in Eq. (C.15) as is permissible for subrelativistic high-frequency fields.

The steps of subtracting Eq. (C.2) from (C.1) which led to Eqs. (C.15) and (6.7) should be shown. Equation (C.1) is multiplied by Zm and (C.2) by m_i. After subtraction, the left-hand side is

$$m_i m\left[Zm_i\left(\frac{d}{dt}\right)_i \mathbf{v}_i - n_e\left(\frac{d}{dt}\right)_e \mathbf{v}_e\right] \approx m_i m Z n_i \frac{d}{dt}(\mathbf{v}_i-\mathbf{v}_e) \tag{C.16}$$

where use was made of Eq. (C.5). This implies also approximate equality of $(d/dt)_{e,i}=\partial/\partial t+\mathbf{v}_{e,i}\cdot\nabla$ and $d/dt=\partial/\partial t+\mathbf{v}\cdot\nabla$ using $\mathbf{v}_e\approx\mathbf{v}_i\approx\mathbf{v}$. The necessary neglect of coherent quiver motion in $\mathbf{v}_e$ does not affect the following conclusions. Another limiting case is where the spatial derivations are less compared with the $\partial/\partial t$ terms and thus no restriction is given to the amplitudes [if not Eq. (C.5) is violated]. Using Eq. (C.8), the left-hand side of the result of subtraction is then

$$\frac{m_i m}{e}\left[\frac{d}{dt}\mathbf{j}-\frac{\mathbf{j}}{n_e}\frac{d}{dt}n_e\right] \tag{C.17}$$

The second term in brackets is usually neglected in order to achieve the well-known result that follows [Eq. (C.20)].

The right-hand side after adding and subtracting one more term is

$$\begin{aligned}&-m_e\nabla Zn_iKT_i+m_i\nabla n_eKT_e+\mathbf{E}e(mn_iZ^2+n_em_i)\\&+(Z^2mn_ie\mathbf{v}_i+m_in_ee\mathbf{v}_e)\times\frac{\mathbf{H}}{c}\\&+[Zm^2n_e\nu+m_imn_e\nu](\mathbf{v}_e-\mathbf{v}_i)\\&+m_iZn_ie\mathbf{v}_i\times\frac{\mathbf{H}}{c}-m_iZn_ie\mathbf{v}_i\times\frac{\mathbf{H}}{c}\end{aligned} \tag{C.18}$$

Dividing both sides by m_i and neglecting mn_iZ^2 as compared to n_em_i results in

$$\frac{m}{e}\left[\frac{d}{dt}\mathbf{j}+\nu\mathbf{j}\right]=\nabla n_eKT_e+\mathbf{E}en_e-\mathbf{j}\times\frac{\mathbf{H}}{c}+Zn_ie\mathbf{v}_i\times\frac{\mathbf{H}}{c}$$

Dividing by en_e, using $\mathbf{v}_i\approx\mathbf{v}$, and using $p_e=p/(1+1/Z)$ at thermal equilibrium, we arrive at the generalized Ohm's law (diffusion equation)

$$\frac{4\pi}{\omega_p^2}\left[\frac{d}{dt}\mathbf{j}+\nu\mathbf{j}\right]=\mathbf{E}-\frac{1}{en_e c}\mathbf{j}\times\mathbf{H}+\mathbf{v}\times\frac{\mathbf{H}}{c}+\frac{1}{en_e}\nabla\frac{p}{1+1/Z} \tag{C.20}$$

which is identical with Eq. (6.7) and the derivations of Schlüter [136] and Lüst [401].

If we had not neglected the second term in brackets of expression (C.17), the generalized Ohm's law would be

$$\frac{4\pi}{\omega^2}\left[\frac{d}{dt}\mathbf{j}+\mathbf{j}\left(\nu-\frac{1}{n_e}\frac{d}{dt}n_e\right)\right]=\mathbf{E}-\frac{1}{en_e c}\mathbf{j}\times\mathbf{H}+\mathbf{v}\times\frac{\mathbf{H}}{c}+\frac{1}{en_e}\nabla p_e \tag{C.21}$$

This shows that an additional damping mechanism appeared, given by an effective collision frequency

$$\nu_{\text{eff}}=-\frac{1}{n_e}\frac{d}{dt}n_e=-\frac{1}{n_e}\left[\frac{\partial}{\partial t}n_e+\mathbf{v}_e\cdot\nabla n_e\right] \tag{C.22}$$

This damping—which obviously has not been recognized before—will find a very special interpretation for the special case where one has a one-dimensional variability of n_e only (using the x-direction) and especially if

$$\frac{\partial}{\partial t}n_e+\mathbf{v}_e\frac{\partial}{\partial x}n_e=-\mu\frac{\partial^3}{\partial x^3}n_e \tag{C.23}$$

The damping is due to Langmuir solitons where Eq. (C.23) is the Korteweg–de Vries equation for the electron density n_e (Langmuir waves). The dispersion relation μ is not necessarily the usual dispersion but can be much more complex as the example in Section 10.6 has shown, where for the numerically derived dynamic solitons [$\mathbf{v}$ instead of n_e in Eq. (C.23)] a dispersion appeared that was not of the usual value but of the same formulation as Denisov derived for resonance absorption.

Ohm's law [Eq. (C.21)] determines the high-frequency electromagnetic waves in a plasma, and the new damping process, given by an effective collision frequency [Eq. (C.22)] or—in the special case of Eq. (C.23)

$$\nu_{\text{eff}}=-\frac{1}{n_e}\frac{d}{dt}n_e=\mu\frac{\partial^3}{\partial x^3}n_e \tag{C.24}$$

indicates a dissipation of transversal waves by longitudinal (Langmuir) waves. Remembering that these electrostatic oscillations are always damped by Landau damping (Section 3.4), a direct relation for the damping of transversal (electromagnetic) waves in a plasma by Landau damping has been achieved.

Let us return to our discussion of the equation of motion. Using the monochromatic time dependence [Eq. (6.20)], one arrives from Eq. (C.15) at

$$\mathbf{j}\left(1-i\frac{\nu}{\omega}\right)=\frac{\omega_p^2}{4\pi\omega^2}\frac{\partial E}{\partial t} \tag{C.25}$$

and

$$\frac{\partial \mathbf{E}}{\partial t}\frac{\partial \mathbf{E}}{\partial t}=-\omega^2\mathbf{E}\mathbf{E} \tag{C.26}$$

We assume $\nu \ll \omega$ which is always possible at very high amplitude fields, therefore we use

$$1-\frac{i\nu}{\omega}\approx 1 \tag{C.27}$$

The two last terms in Eq. (C.14) can then be written with Eqs. (C.25) and (C.26)

$$-\frac{4\pi}{\omega_p^2}\frac{\omega_p^2}{4\pi\omega^2}\frac{\partial \mathbf{E}}{\partial t}\cdot\nabla\frac{\omega_p^2}{4\pi\omega^2}\frac{\partial \mathbf{E}}{\partial t}+\frac{4\pi}{\omega_p^4}\frac{\omega_p^4}{(4\pi)^2\omega^4}\frac{\partial \mathbf{E}}{\partial t}\frac{\partial \mathbf{E}}{\partial t}\cdot\nabla\omega_p^2 \tag{C.28}$$

Using Eq. (C.26) and the refractive index [Eq. (6.28)]

$$\mathbf{n}^2=1-\frac{\omega_p^2}{\omega^2(1-i\nu/\omega)}$$

including Eq. (C.27), we find for the expression (C.28)

$$-\frac{1}{4\pi}\mathbf{E}\cdot\nabla\mathbf{E}(1-\mathbf{n}^2)+\frac{1}{4\pi}\mathbf{E}\mathbf{E}\cdot\nabla(1-\mathbf{n}^2)=-\frac{1}{4\pi}(1-\mathbf{n}^2)\mathbf{E}\cdot\nabla\mathbf{E} \tag{C.29}$$

This is the Schlüter term, Eq. (6.7), as the high-frequency result of the two last terms in the equation of motion (C.14).

Up to this stage, the discussion of the two-fluid theory of plasma followed the convenient derivations, with some modifications only with respect to Z-time ionized ions, or with respect to fields oscillating with a high radian frequency ω. The only operation we had to carry out for arriving at the most general equation of motion (8.3) or (8.6) was with respect to the second term on the right-hand side of Eq. (C.14), which describes the action of electric fields $\mathbf{E}$ due to space charges. In the cases preceding Ref. 138, these space charges in plasmas had been neglected because of space charge neutrality. We had to adhere strictly to this convention with regard to static or stationary distributions of space charges. A difference exists, however, with respect to (time-averaged vanishing) oscillating high-frequency space charges. This is due to the knowledge of the exact description of plane electromagnetic waves in stratified plasmas at oblique incidence and p-polarization, where longitudinal optical fields appear that drive oscillations of high-frequency space charges [138].

The important question was then [138], how should the second term on the right-hand side of the equation of motion (C.14) be interpreted? To avoid any difficulty with the convention of dielectric or diamagnetic properties in the description of plasma, a consequent Lorentz picture with $\varepsilon=\mu=1$ was used. Nobody, however, will doubt that the high-frequency oscillation of space charges are due to polarization currents, determined by a complex refractive index $n=\varepsilon^{1/2}$, as it determined the refraction of the propagation of the optical waves and their change of direction in inhomogeneous plasmas at oblique incidence. As it was the confirmation of the correctness of the subsequent treatment [138], the space charge term in Eq. (C.14) had to be formulated with inclusion of the dielectric displacement,

$$\frac{1}{4\pi}\mathbf{E}e(Zn_i-n_e)=\frac{1}{4\pi}\mathbf{E}\nabla\cdot n^2\mathbf{E}=+\frac{1}{4\pi}\mathbf{E}\nabla\left(1-\frac{\omega_p^2}{\omega^2}\right)\mathbf{E} \qquad \text{(C.30)}$$

$$=\frac{1}{4\pi}\mathbf{E}\nabla\cdot\mathbf{E}-\frac{1}{4\pi}\mathbf{E}\nabla\cdot\mathbf{E}\frac{\omega_p^2}{\omega^2}$$

$$\frac{1}{4\pi}\mathbf{E}e(Zn_i-n_e)=\frac{1}{4\pi}\mathbf{E}\nabla\cdot\mathbf{E}-\frac{1}{4\pi}(1-n^2)\mathbf{E}\nabla\cdot\mathbf{E}-\frac{1}{4\pi}\mathbf{E}\mathbf{E}\cdot\nabla(1-n^2) \qquad \text{(C.31)}$$

Using this result in the equation of motion (C.14) and the Schlüter term [Eq. (C.29)] instead of the last two terms in Eq. (C.14), we arrive at

$$f=m_in_i\left[\frac{\partial}{\partial t}\mathbf{v}+\mathbf{v}\cdot\nabla v\right]=-\nabla p+\mathbf{j}\times\frac{\mathbf{H}}{c}+\frac{1}{4\pi}\mathbf{E}\nabla\cdot\mathbf{E}$$

$$-\frac{1}{4\pi}(1-n^2)\mathbf{E}\nabla\cdot\mathbf{E}-\frac{1}{4\pi}\mathbf{E}\mathbf{E}\cdot\nabla(1-n^2)-\frac{1}{4\pi}(1-n^2)\mathbf{E}\cdot\nabla\mathbf{E} \qquad \text{(C.32)}$$

which is identical with Eq. (8.3) or (8.6) [138].

The final proof of the correctness of Eq. (C.32) was given (as shown in Section 8) by the correct result of the nonlinear forces at oblique incidence. Using $n\equiv 1$ in Eq. (C.26) gave wrong results. The unsolved question was, how the electrostatic field (or very low frequency fields) have to follow, which usually are described by

$$\mathbf{E}\nabla\cdot\mathbf{E}=\frac{1}{4\pi}\mathbf{E}e(Zn_i-n_e) \qquad \text{(C.33)}$$

where there is no refractive index n.

The solution of this question may be related to Novak's answer to experiments where the low-frequency description of the field is given by the

Abraham tensor, while the high-frequency case is given by the Minkowski tensor [179] as the then valid approximation of the Abraham tensor. The description of plasmas using quantities **E** and space charges is strange to the earlier concepts of plasma theory. There seems, however, a much more general aspect given by astrophysics than by our question of laser plasma dynamics in favor of this new development as Alfvén has derived [402].

LIST OF SYMBOLS

$\mathbf{a}$	acceleration of a particle (4.1)
a'	constant for absorption (6.48)
A	second order term (8.48)
c	vacuum velocity of light, see, for example, Eq. (6.16)
c_p	specific heat at constant pressure
c_s	velocity of sound (ion acoustic waves) (4.36)
c_v	specific heat at constant volume
c_ψ	phase velocity (6.26)
E	energy
$\mathbf{E}$	electrical field strength
E_F	Fermi energy (2.47)
E_{kin}	kinetic energy of plasma (10.17)
E_r	amplitude of laser field (6.20) depending on $\mathbf{r}$ only
E^r	relativistic oscillation energy of electron (threshold, 6.71)
E_V	Vacuum amplitude of E
E^*	threshold for predominance of nonlinear force (9.4)
f	distribution function (3.3)
$\hat{f}_M$	Maxwell–Boltzmann distribution (3.20)
F	exponent in WKB approximation (7.9)
F_0	real part of F (8.31)
$\mathbf{F}$	force (4.1)
G_0	length (5.23)
G	for WKB solution (7.29)
G'	Eq. (8.64)
$\mathbf{H}$	magnetic field strength
I	laser intensity (6.55)

I_r	relativistic threshold laser intensity (6.73)
k	wave number
$\mathbf{k}$	wave vector (7.19)
$\mathbf{k}_s$	wave vector for electrostatic waves (9.69)
$\bar{k}$	average absorption constant (7.13)
K	Boltzmann constant
$\bar{K}$	optical absorption constant
$\tilde{K}$	effective wave number (11.51)
$\mathbf{K}_e$	force density to electrons by gravitation
$\mathbf{K}_i$	force density to ions by gravitation
$\bar{K}_{NL}$	nonlinear absorption constant
l	mean radius of a Gaussian density profile (5.21)
$\bar{l}$	mean free path (3.69)
L	Lagrangian for parametric resonance (9.56)
$\mathscr{L}$	Lagrangian of Proca equation (9.50)
L'	Denisov length (11.62)
m	mass of electron
$\bar{m}$	mass of particle (4.1)
m_i	mass of ion
m_0	rest mass of electron
M	total mass of plasma from pellet
$\bar{M}$	radially averaged mass (5.14)
n	particle density
n_e	electron density
n_i	ion density
n	(optical) refrective index (6.27)
n'	real part of the refractive index
N_i	total number of ions in sphere (5.26)
p	pressure
$\bar{p}$	classical Coulomb cross section (2.32)
p_A	Abraham momentum of photon (9.31)
p_M	Minkowski momentum of photon (9.32)
p_0	initial pressure
p_ϕ	momentum per photon (9.29)
$p_{\phi,pl}$	p_ϕ in plasma (9.30)
P	momentum of electromagnetic energy

P_{inh}	P in inhomogeneous plasma
P_{int}	internal radiation pressure
P_0	P in vacuum (9.10)
P'	laser power (12.22)
r	radius
r_F	focus radius (5.41)
r_0	Coulomb impact parameter (2.29)
R	radius of spherical pellet (5.1)
R_0	initial radius of pellet
$\dot{R}$	expansion velocity of pellet radius
$\dot{R}_0$	initial expansion velocity of pellet radius
Re_{kr}	Reynolds number (11.42)
$\tilde{R}$	Fresnel reflection coefficient (9.34)
$\mathbf{S}$	Poynting vector (8.7)
t	time
t_0	initial time
t_{TP}	time at which laser irradiated pellet becomes transparent (5.31)
T	temperature
$\tilde{T}$	transmission coefficient (9.34)
T_{th}	thermokinetic temperature (6.57)
$\mathbf{T}$	Maxwellian stress tensor (8.26)
$T^{\mu\nu}$	canonical energy-momentum tensor (9.50)
u	angle of incidence
u_x, u_y, u_z	angles for the direction of plane waves (7.19)
U	blackbody radiation density (9.40)
U_p	Planck's value of U for vacuum (9.41)
$\mathbf{v}$	macroscopic (drift) velocity (3.34)
v_r	radial velocity
v_{r0}	initial radial velocity (5.13)
v_ϕ	phase velocity (9.70)
V	volume
V_0	initial volume
W	input power density (4.39)
W_1	special input power density (5.24)

x	Cartesian coordinate
y	Cartesian coordinate
z	Cartesian coordinate
Z	number charges of ions
α	parameter for Rayleigh profile (7.29)
α_0	angle of incidence in vacuum (8.43)
$\alpha(x)$	angle of incidence in plasma
$\bar{\alpha}_{(n)}$	Fokker–Planck coefficients (3.18)
β	angle of polarization (8.42)
γ	ratio of specific heats (4.23)
γ_e	Spitzer's correction for collisions (2.35)
ε	dielectric constant
$\bar{\varepsilon}$	complex dielectric constant (6.27)
ε_D	Debye energy (2.15)
ε_{osc}	oscillation energy of electrons (6.54)
ε_{osc}^{kin}	average kinetic energy of quivering electron (6.53)
ζ	spatial variable (7.83)
η	viscosity (4.4; 11.38)
κ	imaginary part of the refractive index
$\bar{\kappa}$	compressibility (4.21)
κ_T	thermal conductivity (4.39)
λ_D	Debye length (2.14)
μ	magnetic permeability
μ'	dispersion function (10.21)
ν	collision frequency
ν_e	collision frequency of electrons
ν_{ei}	electron ion collision frequency
ρ	density of plasma (4.2)
ρ_0	initial density
σ	electric conductivity (2.42)
σ_o	Stefan–Boltzmann constant (9,46)
σ_{op}	radiation constant in plasma (9.45)
τ_{col}	electron collision time (10.1)
τ_0	duration of laser pulse (half-width) (10.4)
τ_R	rise time of laser pulse (8.7)
τ^*	minimum rise time of laser pulse for thermalization (10.5)

ϕ	potential (4.1)
ω	laser radian frequency
ω_e	Bohm–Gross frequency (9.69)
ω_p	plasma frequency (2.6)
Ω_0	vacuum resistivity (6.55)

REFERENCES

1 H. Hora, *Laser Plasmas and Nuclear Energy* (Plenum, New York, 1975), 424 pages.

2 T. P. Hughes, *Plasmas and Laser Light* (Adam Hilger, Bristol, 1975).

3 H. Motz, *The Physics of Laser Fusion* (Academic Press, London, 1979), 290 pages.

4 H. Hora, "Nonlinear Plasma Dynamics at Laser Irradiation," Lecture Notes in Physics (University of Berne, Springer, Heidelberg, 1979).

5 *Laser Interaction and Related Plasma Phenomena*, H. J. Schwarz *et al.*, Eds. (Plenum, New York) Vols. 1 to 5 (1971–1980).

6 R. Castillo, H. Hora, E. L. Kane, G. W. Kentwell, P. Lalousis, V. F. Lawrence, R. Mavaddat, M. M. Novak, P. S. Ray, and A. Schwartz, see Ref. 5, Vol. 5 (1980) p. 399

7 H. Alfvén, *Phys. Today* **24** (February 1971), 29.

8 R. T. Young, C. W. White, G. J. Clark, J. Narayan, W. H. Christie, M. Murakami, P. W. King, and S. D. Kramer, *Appl. Phys. Lett.* **32** (1978), 139; H. Hora, *Naturwissenschaften* **48** (1961), 641; Z. Angew Phys. **14** (1962), 9; S. Hinckley, H. Hora, and J. C. Kelly, *Phys. Status Solidi* **51A** (1979), 523.

9 P. D. Maker, R. W. Terhune, and C. M. Savage, *Proc. 3rd Int. Quant. Electron. Conf., Paris*, February 1963, N. Bloembergen and M. Grivet, Eds. (Dunod, Paris, 1964), Vol. 2, p. 1559.

10 R. G. Meyerand and A. F. Haught, *Phys. Rev. Lett.* **11** (1963), 401; C. DeMichelis, *IEEE J. Quantum Electron.* **5** (1969), 181, **6** (1970), 630; E. Panarella and P. Savic, *Can. J. Phys.* **46** (1968), 183; S. A. Ramsden, *Physics of Hot Plasmas*, B. J. Rye et al., Eds. (Oliver & Boyd, Edinburgh, p. 346.

11 R. Papoular, see Ref. 5, (1972) Vol. 2, p. 79.

12 G. V. Ostrovskaya and A. N. Zaidel, *Usp. Fiz. Nauk* **111** (1973), 579; *Sov. Phys. Uspekhi* **16** (1974), 834.

13 P. Kolodner and E. Yablontovich, *Phys. Rev. Lett.* **37** (1976), 1754.

14 B. W. Boreham and H. Hora, *Phys. Rev. Lett.* **42** (1979), 776.

15 H. Hora, E. L. Kane, and J. L. Hughes, *J. Appl. Phys.* **49** (1978), 923.

16 H. Hora, *J. Opt. Soc. Am.* **65** (1975), 882.

17 D. J. Bradley, *Phys. Bull.* **29** (1978), 418; J. C. Diels, *Laser Weekly* **12**, no. 5 (1978), p. 1.

18 J. Trenholme, E. Bliss, J. Emmett, J. Glaze, T. Gilmartin, R. Godwin, W. Hagen, J. Holzrichter, G. Linford, W. Simmons, and R. Speck, see Ref. 5, 1977, Vol. 4A, 1.

19 J. L. Emmett, *Proc. IAEA Conf. Nuclear Fusion, Innsbruck*, 1978, paper B-1.

20 N. G. Basov, O. N. Krokhin, Yu. A. Mikhailov, G. V. Sklizkov, and S. I. Fedotov, see Ref. 5, 1977, Vol. 4A, 15.

21 V. V. Korobkin, V. M. Ovchinnikov, P. P. Pashinin, Yu. A. Pirogov, A. M. Prokhorov, R. V. Serov, *Digest 8th Nat Conf Laser and Nonlinear Optics, Tibilisi*, May 1976, p. 248.

22 W. Kroy, "CW pumped 1000 Hz Nd Glass Laser," (Messerschmitt-Bölkow-Blohm GmbH, Ottobrunn).

23 P. P. Pashinin, see Ref. 21, postdeadline paper.

24 R. B. Allen and S. J. Scalise, *Appl. Phys. Lett.* **14** (1969), 188; W. Koechner, *Solid-State Laser Engineering* (Springer, Heidelberg, 1976), p. 277.

25 G. V. Sklizkov, paper P1, presented at *13th Europ. Conf.* Laser Interaction with Matter, Leipzig, December 1979.

26 R. R. Jacobs and W. F. Krupke, *IEEE J. Quantum Electron.* **13** (1977), 103.

27 R. B. Perkins, *Plasma Physics and Controlled Nuclear Fusion* (IAEA, Vienna, Vol. 3 (1978), p. 41.

28 S. Singer, *Development in High Power Lasers*, C. Pellegrini ed. (E. Fermi Sch. Vol. 74, North Holland, Amsterdam, 1980).

29 H. S. Kwok and E. Yablontovitch, *Appl. Phys. Lett.* **30** (1978).

30 N. G. Basov, V. A. Boiko, V. A. Danylichev, V. D. Zvorykin, A. N. Lobanov, A. F. Suchkov, T. V. Holin, and A. Y. Chugunov, *Kvant. Elekt.* **4** (1977), 1761.

31 C. Yamanaka, "0.5 nsec Pulses from 50 Atm. Compact CO_2 Lasers," (Osaka 1978).

32 W. Kroy, J. Langhole, T. Halderson, *Appl. Opt.* **19** (1980), 6. H. K. Koebner, *Laser in Medicine*, (Wiley, New York, 1980).

33 G. H. Miley, see Ref. 5 (1977), Vol. 4A, p. 181.

34 K. Hohla, G. Brederlow, E. Fill, R. Volk, and K. J. Witte, see Ref. 5 (1977), Vol. 4A, p. 97.

35 K. Hohla and K. L. Kompa, *Chem. Phys. Lett.* **14** (1972), 445.

36 K. Witte, G. Brederlow, E. Fill, K. Hohla, and R. Volk, see Ref. 5 (1977) Vol. 4A, p. 155.

37 S. Witkowski, *Laser & Elektro-Optik*, 10, no. 3 (1978), 47.

38 R. Lüty and K. Witte, Dept. Laser Physics, Univ. Berne, Report (1978).

39 N. G. Basov and V. S. Zuev, *Nuovo Cim.* **31B** (1976), 129.

40 R. J. Jensen, *Laser Focus* **12** (May 1976), 51.

41 F. G. Houtermans, *Helv. Phys. Acta* **33** (1960), 933.

42 M. H. Hutchinson, C. C. Ling, and D. J. Bradley, *Opt. Com.* **26** (1978), 273.

43 D. Jacobi, G. J. Pert, S. A. Ramsden, L. D. Shorrock, and G. J. Tallents, *Phys. Rev. Lett* **45** (1980) 1826; P. Jaegle, G. Jamelot, A. Carillon, and A. Sureau, see Ref. 5, (1977), Vol. 4A, p. 229.

44 M. H. Key, M. J. Lamb, C. L. S. Lewis, J. G. Lunney and A. K. Roy, *Opt. Comm.* **18** (1976), 156.

45 H. Hora, G. V. H. Wilson, E. P. George, *Aust. J. Phys.* **31** (1978), 55.

46 K. Okamoto, see Ref. 5, (1977), Vol. 4A, p. 283; *J. Nucl. Sci. Tech.* **14** (1977), 762.

47 K. Okamoto, see Ref. 5 (1980), Vol. 5, p. 299.

48 Y. Izawa, H. Otari, and C. Yamanaka, see Ref. 5 (1980), Vol. 5, p. 289

49 D. A. G. Deacon, L. R. Elias, J. M. J. Madley, G. J. Ramian, H. A. Schwettman, and T. J. Smith, *Phys. Rev. Lett.* **18** (1977), 892.

50 H. Motz, *J. Appl. Phys.* **22** (1951), 527; **24** (1953), 826.

51 S. Pellgrini, presented at 4th General Conf. European Physical Soc., York, England, September 1978, invited paper.

52 S. B. Segall, *Laser and Elektro-Optik* **10**, no. 3 (1978), 27.

53 H. Hora, presented at 2nd Int. Conf. Energy Storage, Electron and Laser Beams, Venice, Dec. 1978, invited paper (in print); H. Hora, B. W. Boreham, and J. L. Hughes, *Sov. J. Quant. Elect.* **9** (1979), 464.

54 H. Hora, see Ref. 5 (1971), Vol. 1, p. 383.

55 H. J. Schwarz, *Phys. Rev. Lett.* **42** (1979), 1141.

56 H. Schwarz and H. Hora, *Appl. Phys. Lett.* **15** (1969), 349; H. Hora, *Nuovo Cim.* **26B** (1975), 295; E. T. Jaynes, in *Novel Sources of Coherent Radiation, Physics of Quantum Electronics* Vol. 5, S. F. Jacobs, M. Sargent and M. O. Scully, Eds. (Addison–Wesley, Reading, 1978), p. 1.

57 T. H. Maiman, *Nature* **187** (1960), 493.

58 K. H. Steigerwald, *Chem. Ing. Tech.* **33** (1961), 191; H. Hora, *Chem. Rundschau* **14** (1960), 395.

59 J. F. Ready, *Effects of High-Power Laser Radiation* (Academic Press, New York, 1971); R. E. Honig. *Appl. Phys. Lett.* **3** (1963), 8.

60 H. Zahn and H. J. Dietz, *Exp. Techn. Phys.* **20** (1972), 401.

61 M. von Allmen, W. Lüthy, and K. Affolter, *Appl. Phys. Lett.* **33** (1978), 824.

62 H. Schwarz and H. A. Tourtellotte, *J. Vac. Sci. Techn.* **6** (1969), 373.

63 F. J. McClung and R. W. Hellwarth, *Proc. IEEE* **51** (1963), 46.

64 W. I. Linlor, *Appl. Phys. Lett.* **3** (1963), 210.

65 N. R. Isenor, *Appl. Phys. Lett.* **4** (1964), 152.

66 H. Schwarz, *Laser Interact. Rel. Plasma Phenomena*, H. Schwarz and H. Hora, Eds. (Plenum, New York, 1971), Vol. 1, p. 207.

67 S. Namba, H. Schwarz and P. H. Kim, *Proc. IEEE Sympos. Electrons, Ion and Laser Beam Technol.*, Berkeley, May 1967, p. 861.

68 D. W. Gregg and S. J. Thomas, *J. Appl. Phys.* **37** (1966), 4313.

69 S. A. Metz, *Appl. Phys. Lett.* **22** (1973), 211; H. Hora, *Appl. Phys. Lett.* **23** (1973), 39; J. E. Lowder and L. C. Pettingill, *Appl. Phys. Lett.* **24** (1974), 204; P. T. Rumsby, M. M. Michaelis and M. Burgess, *Opt. Comm.* **15** (1975), 422; B. Steverding and A. H. Werkheiser, *J. Phys.* **D4** (1971), 545; J. E. Lowder, *Appl. Phys. Lett.* **24** (1974), 204; S. Zweigenbaum, Y. Gazit, and Y. Paiss, *J. Phys.* **E11** (1978), 830.–The driving of solid foils by laser irradiation up to velocities of 2×10^7 cm/sec resulted in the fastest moving objects produced by man corresponding to a speed of 600 Mach: B. H. Ripin, R. Decoste, S. P. Oberschain, S. E. Bodner, E. A. McLean, F. C. Young, R. R. Whitlock, C. M. Armstrong, J. Grun, J. A. Stamper, S. H. Gold, D. J. Nagel, R. H. Lehmberg and J. M. McMahon, *Phys. Fluids* **23** (1980) 1012. This can be used for impact fusion (F. Winterberg, *Z. Naturforsch.* **19A** (1964) 231) where the concept of laser driven foils was proposed by W. Kaiser, H. Opower, and B. H. Puell, German Patent No. 1 279 859 (1966) and the collapsing and adiabatic heating at compression will follow in the same way as achieved by the fast acceleration of thick blocks of plasma by nonlinear fores (see section 13.4). The use of the nonlinear force for propulsion by lasers was described by H. Hora, German Patent 1933 409 (1971).

70 S. Zweigenbaum, Y. Gazit, and Y. Komet, *Plasma Physics* **19** (1977), 1035; D. Salzmann, Y. Gazit, Y. Komet, A. D. Krumbein, H. M. Loebenstein, M. Oron, Y. Paiss, M. Rosenblum, H. Szichman, A. Zigler, H. Zmora, and S. Zweigenbaum, see Ref. 5 (1977), Vol. 4A, p. 407.

71 G. Siller, K. Büchl, and H. Hora, see Ref. 5 (1972), Vol. 2, p. 252.

72 A. G. Engelhardt, T. V. George, H. Hora, and J. L. Pack, *Phys. Fluids* **13** (1970), 212.

73 K. Eidman and R. Sigel, see Ref. 5 (1974), Vol. 3B, p. 667.

74 H. Hora, see Ref. 1, p. 4.

75 K. B. Büchl, K. Eidmann, P. Mulser, H. Salzmann, and R. Sigel, see Ref. 5, (1972), Vol. 2, p. 503.

76 A. W. Ehler, *J. Appl. Phys.* **46** (1975), 2464.

77 J. W. Shearer, J. Garrison, J. Wong, and J. E. Swain, *Phys. Rev.* **A8** (1973), 1582.

78 C. Yamanaka, T. Yamanaka, T. Sasaki, K. Yoshida, and M. Waki, *Phys. Rev.* **6A** (1972), 2342.

79 F. J. Mayer, R. K. Osborn, D. W. Daniels, and J. F. McGrath, *Phys. Rev. Lett.* **40** (1978), 30.

80 B. Luther–Davies and J. L. Hughes, *Opt. Com.* **18** (1976), 351; M. Siegrist, B. Luther–Davies, and J. L. Hughes, *Opt. Comm.* **18** (1976), 605.

81 H. Hora, E. L. Kane, and J. L. Hughes, *Nucl. Inst. Meth.* **150** (1978), 589.

82 P. Wägli and T. P. Donaldson, *Phys. Rev. Lett.* **40** (1978), 875.

83 T. P. Donaldson and I. J. Spalding, *Phys. Rev. Lett.* **36** (1976), 467.

84 R. A. Haas, W. C. Mead, W. L. Kruer, D. W. Phillion, H. N. Kornblum, J. D. Lindl, D. MacQuigg, V. C. Rupert, and K. G. Tirsell, *Phys. Fluids* **20** (1977), 322; E. B. Goldman, L. M. Goldman, J. Delettrez, J. Hoose, S. Jackel, G. W. Leppelmeier, M. J. Lubin, A. Nel, I. Pelak, E. Thorsos, D. Woodall, and B. Yaa kobi, see Ref. 5 (1977), vol. 4B, p. 535.

85 R. G. Evans, M. H. Key, D. J. Nicholas, F. O'Neil, A. Raven, P. T. Rumsby, I. N. Ross, W. T. Tower, P. R. Williams, M. S. White, C. L. S. Lewis, J. G. Lunney, A. Moore, J. M. Ward, T. A. Hall, J. Murdoch, D. J. Kilhenny, T. Goldsack, J. D. Hares, *Plasma Phys. Contr. Nucl. Fusion Res.*, 1978 (IAEA, Vienna, 1979) Vol. 3, p. 87.

86 C. Yamanaka, M. Yokoyama, S. Makar, T. Yamanaka, Y. Izawa, Y. Kato, T. Sasaki, T. Mochizuki, Y. Kitagawa, M. Matoba, and K. Yoshida, see Ref. 5 (1977), Vol. 4B, p. 577.

87 A. Bekiarian, E. Buresi, A. Coudeville, R. Dautray, F. Delobeau, P. Guillaneux, C. Patou, J. M. Reisse, B. Sitt, J. M. Vedel, and J. P. Watteau, *Plasma Physics Contr. Nucl. Fusion Res.*, 1978 (IAEA, Vienna, 1979), Vol. 3, p. 65.

88 P. Mulser, *Z. Naturforsch.* **25A** (1970), 282.

89 E. J. Valeo, W. L. Kruer, *Phys. Rev. Lett.* **33** (1974), 750.

90 D. Biskamp and H. Welter, *Plasma Phys. Contr. Nucl. Fusion Res.*, Tokyo, 1974 (IAEA, Vienna, 1975) Vol. 2, p. 507.

91 For the change from Gaussian cgs-units preferred in Plasma Physics, to MKQS units, see, for example, D. L. Book, *Formulas for Plasma* (Naval Res. Lab., Washington, 1975).

92 P. Debye and E. Hückel, *Phys. Zeitschr.* **24** (1923), 185.

93 S. R. Milner, *Phil. Mag.* **23** (1912), 551; **25** (1913), 743.

94 T. P. Donaldson, *J. Phys.* **D10** (1977), 1589; *Plasma Phys.* **20** (1978), 1279.

95 G. Ruthemann, *Naturwissensch.* **29** (1941), 648.

96 G. Möllenstedt, *Optik* **5** (1949), 499.

97 D. Bohm and D. Pines, *Phys. Rev.* **92** (1953), 609.

98 H. Ringler and R. A. Nodwell, *Phys. Lett.* **29A** (1969), 151; D. Ludwig and C. Mahn, *Phys. Lett.* **35A** (1971), 191.

99 L. A. Godfrey, R. A. Nodwell, and F. L. Curzon, *Phys. Rev.* **A20** (1979), 567.

100 H. Thomas, *Z. Phys.* **147** (1957), 395.

101 P. Görlich and H. Hora, *Optik* **15** (1958), 116.

102 H. Fröhlich, *Ann Phys.* **7** (1930), 103.

103 B. Yaakobi and S. Goldsmith, *Phys. Lett.* **37A** (1970), 408; M. Neiger and H. Griem, *Phys. Rev.* **A14** (1976), 289; D. R. Inglis and E. Teller, *Astrophys. J.* **90** (1939), 439; F. L. Mohler, *Astrophys. J.* **90**, (1939), 429; H. Margenau and M. Lewis, *Rev. Mod. Phys.* **31** (1959), 569; S. Volonté, *J. Phys. D* **11** (1978), 1615.

104 G. Ecker and W. Kröll, *Phys. Fluids* **6** (1963), 62; H. R. Griem, *Plasma Spectroscopy* (McGraw-Hill, 1964), p. 139.

105 D. Salzmann and A. Krumbein, *J. Appl. Phys.* **49** (1978), 3229.

106 B. I. Henry, "Polarization Shift," Honours Thesis, University of New South Wales (Sydney, 1980); B. I. Henry, and H. Hora (to be published).

107 L. Spitzer, Jr., *Physics of Fully Ionized Gases*, 2nd ed. (Wiley Interscience, New York, 1962).

108 L. Spitzer and R. Härm, *Phys. Rev.* **89** (1953), 977.

109 P. S. Ray and H. Hora, see Ref. 5 (1977), Vol. 4B, p. 1081.

110 J. M. Blatt and A. H. Opie, *J. Phys.* **A7** (1974), 1895.

111 J. M. Blatt, *Prog. Theor. Phys.* **22** (1959), 745; *About Theories of Superconductivity* (Academic Press, New York, 1959).

112 J. M. Dawson, *Phys. Fluids* **7** (1964), 981.

113 E. Hinnov and J. Hirschberg, *Phys. Rev.* **125** (1962), 795.

114 A. F. Haught and D. H. Polk, *Phys. Fluids* **13** (1970), 2825.

115 H. Hora, *Higher Mechanics*, Dept. Theor. Phys. Rept. No. 24 (Univ. New South Wales, Sydney, 1980).

116 L. D. Landau, *J. Phys. USSR* **10** (1946), 25.

117 J. Meixner, *Ann. Physik* **39** (1941), 39; about fluctuations in the electric field, see M. S. Sodha, and L. A. Patel, *J. Appl. Phys.* **51** (1980) 2381.

118 D. Enskog, *Svenska Akademia* (1928), 21.

119 R. Castillo, M.Sc. Thesis, University New South Wales, 1979; R. Castillo, H. Hora, E. L. Kane, V. F. Lawrence, M. B. Nicholson-Florence, M. M. Novak, P. S. Ray, J. R. Shepanski, R. Sutherland, A. I. Tsivinsky, and H. A. Ward, *Nucl. Inst. Meth.* **144** (1977), 27.

120 W. J. Fader, *Phys. Fluids* **11** (1968), 2200.

121 E. A. Milne, *Z. Astrophys.* **6** (1933), 1.

122 O. Heckmann, *Theorie der Kosmologie* (Springer, Heidelberg, 1965).

123 L. L. Lengyel and M. Salvat, *Z. Naturforsch.* **30A** (1975), 1577.

124 Ya. B. Zeldovich and Yu. P. Raizer, *Physics of Schock Waves and High Temperature Hydrodynamic Phenomena* (Academic Press, New York, 1966).

125 N. G. Basov and O. N. Krokhin, *3rd Int. Quantum Elect. Conf. Paris, 1963*, P. Grivet and N. Bloembergen, Eds. (Dunod, Paris 1964), Vol. 2, p. 1373; S. Kaliski, *Bull. de l'Academis Polonaise des Sciences-Serie des Sciences Technique* **20** (1972), 297; **23** (1975), 881; V. V. Demchenkov and N. M. El-Siragy, *Physica* **67** (1973), 336; T. P. Donaldson, J. E. Palmer, J. A. Zimmermann, *J. Phys.* **D13** (1980) 1221; K. E. Lonngren, *Plasma Phys.*

22 (1980) 511. Yu. V. Afanasyev, N. G. Basov, O. N. Krokhin, V. V. Pustovalov, V. P. Silin, G. V. Sklizkov, V. T. Tikhonchuk, and A. S. Shikanov, *Interaction of Strong Laser Light with Plasmas*, (Radiotekhnika Vol. 17, Moscow, 1978).

126 H. Hora, *Inst. Plasmaphys. Garching, Rept. 6/23*, 1964.

127 H. Hora, see Ref. 5 (1971), Vol. 1, p. 365; H. T. Suji, K. Sato, and T. Sekiguchi, *Jap. J. Appl. Phys.* **18** (1979), 1807; S. O. Dean, *Rept. NRL PRO* (1971); Y. Ohwadano and T. Sekiguchi, *Jap. J. Appl. Phys.* **16** (1977), 1025; G. J. Pert, *J. Phys.* **A5** (1972), 506; L. L. Lengyel, *Nucl. Fusion* **17** (1977), 805.

128 A. F. Haught and D. H. Polk, *Phys. Fluids* **9** (1966), 2047; R. G. Tuckfield and F. Schwrizke, *Plasma Phys.* **11** (1969), 11; H. J. Kunze, *Z. Naturforsch*, **A20** (1965), 801; A. Cavalieri, P. Guipponi, and R. Gratton, *Phys. Lett.* **A25** (1967), 636.

129 J. Jacquinot, C. Leloup, and F. Waelbrock, *Rapp. CEA No. 12.2617* (1964).

130 M. Mattioli, *Euratom-CEA-Fontenay-Report*, EUR-CEA-FC-523C (1969).

131 E. Fabre and P. Vasseur, *J. Physique* **29** (1968), 123.

132 T. V. George, A. G. Engelhardt, J. L. Pack, H. Hora, and G. Cox, *Bull. APS* **13** (1968), 1553; H. Hora, see Ref. 5, Vol. 1 (1971) p. 273.

133 R. Sigel, K. Büchl, P. Mulser, and S. Witkowski, *Phys. Lett.* **26A** (1968), 498; H. Puell, *Z. Naturforsch.* **A25** (1970), 1807.

134 P. Mulser and S. Witkowski, *Phys. Lett.* **28A** (1969), 703.

135 K. Hohla, *unpublished measurements* (1969); J. Tulip, K. Manes, and H. J. Seguin, *Appl. Phys. Lett.* **19** (1971), 433.

136 A. Schlüter, *Z. Naturforsch.* **5A** (1950), 72.

137 See for example, H. Alfvén and C. G. Fälthammer, *Cosmical Electrodynamics*, 2nd ed. (Oxford University Press, London, 1973).

138 H. Hora, *Phys. Fluids* **12** (1969), 182.

139 P. A. M. Dirac, *Directions of Physics*, H. Hora and J. R. Shepanski, Eds. (Wiley Interscience, New York, 1978).

140 H. Hora and H. Wilhelm, *Nucl. Fusion* **10** (1970), 111; J. L. Bobin, *Phys. Fluids* **14** (1971), 2341; N. H. Burnett, *Can. J. Phys.* **50** (1972), 3184.

141 C. W. Allen, *Astrophysical Quantities* (Athlon Press, London, 1955).

142 J. A. Gaunt, *Proc. Roy. Soc. (London)* **A126** (1930), 654.

143 S. F. Smard and K. C. Westfold, *Phil. Mag.* **40** (1949), 831.

144 J. M. Dawson and C. Oberman, *Phys. Fluid* **5** (1962), 517.

145 G. W. Spitzer and H. Y. Fan, *Phys. Rev.* **108** (1957), 268.

146 H. Hora, *Opto Electronics* **2** (1970), 202.

147 S. Rand, *Phys. Rev.* **B136** (1964), 231; T. P. Hughes and M. B. Nicholson–Florence, *J. Phys.* **A2** (1968), 588.

148 C. Max and F. Perkins, *Phys. Rev. Lett.* **27** (1971), 1342.

149 H. Schwarz and R. Tabenski, see Ref. 5 (1977), Vol. 4B, p. 961.

150 V. L. Ginzburg, *The Propagation of Electromagnetic Waves in Plasma* (Pergamon, Oxford 1964), p. 205.

151 H. Hora, *Jenaer Jahrbuch*, P. Görlich Ed. (Fischer, Jena, 1957), p. 131; *Inst. Plasmaphysik*, Garching, Rept. 6/5 (1963).

152 H. Osterberg, *J. Opt. Soc. Amer.* **48** (1950), 513.

153 H. Hora and V. F. Lawrence, see Ref. 5 (1977), Vol. 4B, p. 877; V. F. Lawrence and H. Hora, *Optik* **55** (1980), 291.

154 J. Lindl and P. Kaw, *Phys. Fluids* **14** (1971), 371.

155 N. G. Watson, *A Treatise on the Theory of Bessel Functions* (Cambridge University Press, 1922).

156 M. E. Marhic, *Phys. Fluids* **18** (1975), 837; F. F. Chen, *Plasma Physics* (Plenum, New York, 1975); R. Dragila, *J. Phys.* **D11** (1978), 683.

157 J. A. Stamper and D. A. Tidman, *Phys. Fluids* **16** (1975), 2024.

158 H. Hora, D. Pfirsch, and A. Schlüter, *Z. Naturforsch.* **22a** (1957), 278.

159 H. A. H. Boot, S. A. Self, and R. B. Shersby–Harvie, *J. Elect. Contr.* **22** (1959), 434.

160 V. A. Gapunov and M. A. Miller, *Sov. Phys. JETP* **7** (1958), 168.

161 S. Weibel, "TRW Report May 1957," *J. Electr. Contr.* **5** (1958), 435.

162 H. Hora, Second term in Eq. (25b) in Ref. 138. By this way the use of the nonlinear force has arrived in a correct formulation of the equation of motion even with absorption in agreement with the following described experiments. The very complex open problems with the ponderomotive force are described by V. I. Pavlov, *Sov. Phys. Uspekhi* **21** (1978), 171. See also Ref. 179; F. Panarella, *Can. J. Phys.* **46** (1969), 183; G. P. Banfi and P. G. Gobbi, *Plasma Phys.* **21** (1979), 845.

163 J. A. Stamper, see Ref. 5 (1977), Vol. 4B, p. 721.

164 R. D. C. Miller and H. Hora, *Plasma Phys.* **21** (1979), 183.

165 G. W. Kentwell and H. Hora, *Plasma Phys.* **22** (1980), 1051.

166 L. D. Landau and E. M. Lifshitz, *Electrodynamics of Continuous Media* (Pergamon, Oxford, 1966), p. 242.

167 L. P. Pitaevskii, *Sov. Phys. JETP* **12** (1961), 1008.

168 H. L. Berk, D. L. Book, and D. Pfirsch, *J. Math. Phys.* **8** (1967), 1611.

169 G. W. Kentwell, "Resonance Absorption and Striated Motion at Laser Plasma Interaction," *Honours Thesis*, University New South Wales (1979).

170 S. Hinckley, H. Hora, E. L. Kane, G. W. Kentwell, J. C. Kelly, P. Lalousis, V. F. Lawrence, R. Mavaddat, M. M. Novak, P. S. Ray, A. Schwartz, H. A. Ward, *Experim. Tech. Phys.* **28** (1980), 417.

171 J. W. Shearer, R. E. Kidder, and J. W. Zink, *Bull. Amer. Phys. Soc.* **15** (1970), 1483; J. W. Shearer, *LLL-Report*, UCID-15745 (December 1970).

172 R. B. White and F. F. Chen, *Plasma Phys.* **16** (1974), 565.

173 E. Valeo, *Phys. Fluids* **17** (1974), 1391; J. J. McClure, *Bull. Amer. Phys. Soc.* **19** (1974), 869.

174 J. A. Stamper and S. E. Bodner, *Phys. Rev. Lett.* **37** (1976), 435. R. G. Tuckfield, and F. Schwirtzke, *Plasma Phys.* **11** (1969) 11; E. Schmutzer, and B. Wilhelmi, *Plasma Phys.* **19** (1971) 799; P. Mulser, and C. van Kessel, *Phys. Rev. Lett.* **38** (1977) 902; *J. Phys.* **D11** (1978) 1085; P. Mulser, and H. Tasso, *Z. Naturforsch.* **33A** (1978) 85; A. Ng, L. Pitt, D. Salzmann, and A. A. Offenberger, *Phys. Rev. Lett.* **42** (1979) 703; P. Chandra, *J. Appl. Phys.* **47** (1976) 3447; J. R. Saraf, *Z. Naturforsch.* **31A** (1976) 1038; A. T. Lin, and J. M. Dawson, *Phys. Fluids* **18** (1975) 201; F. Winterberg, Z. Naturforsch. **30A** (1975) 976; N. E. Andreev, V. P. Silin, and G. L. Stenchik, *Zh. Ex. Teo. Fiz.* **78** (1980) 1396; J. L. Hughes, *Izvest. Akad. Nauk SSSR, Ser. Fiz.* **42** (1978) 2593; **43** (1979) 1523.

175 R. E. Kidder, *Nucl. Fusion* **14** (1974), 797.

176 K. A. Brueckner and S. Jorna, *Rev. Mod. Phys.* **46** (1974), 325.

177 C. E. Max, *Phys. Fluids* **19** (1976), 74.

178 R. S. Craxton and M. G. Haines, *Plasma Phys.* **20** (1978), 487.

179 M. M. Novak, "Interaction of Photons with Electrons in Dielectric Media," Ph.D. Thesis, University New South Wales (February 1979), *Forschr. Phys.* **28** (1980), 339.

180 L. C. Steinhauer and H. G. Ahlstrom, *Phys. Fluids* **13** (1970), 1103.

181 S. L. Shapiro, M. A. Duguay, and L. B. Kreuzer, *Appl. Phys. Lett.* **12** (1968), 36.

182 H. P. Weber, *Phys. Lett.* **27A** (1968), 321.

183 See Eq. (36) in Ref. 156, (Marhic).

184 R. Klima, *Plasma Phys.* **12** (1970), 123; R. Klima and V. A. Petrzilka, *Cz. J. Phys.* **B22** (1972), 896; *J. Phys.* **A11** (1978), 1687; V. A. Petrzilka, *Cz. J. Phys.* **B26** (1976), 115.

185 H. Hora, *Phys. Fluids* **17** (1974), 1042.

186 H. Bebié, Seminar Lecture, Deptartment of Laser Physics, University of Bern, November 1978.

187 G Bekefi. *Radiation Processes in Plasma* (John Wiley. New York. 1966)

188 J. M. Dawson, *Adv. Plasma Phys.* **1** (1968), 1.

189 A. Einstein, *Phys. Z.* **18** (1917), 121.

190 G. Badertscher (private communication).

191 F. J Belinfante, *Physica* **6** (1939), 887.

192 H. Hora, *Lett. Nuovo Cim.* **22** (1978), 55; *Atomkernenergie* **34** (1979), 297.

193 F. F. Chen, see Ref. 5, Vol. 3A, p. 291; J. A. Stamper, *Phys. Fluids* **18** (1975), 735; Y. Sakamoto, *Jap. J. Appl. Phys.* **16** (1977), 1015; M. S. Sodha, R. P. Sharma, and S. C. Kaushitz, *Plasma Phys.* **18** (1976), 879.

194 F. F. Chen, *Comments on Mod. Phys.*, Part E **1**, no. 3 (1972), 81.

195 V. N. Oraevski and R. Z. Sagdeev, *Sov. Phys.-Tech. Phys.* **7** (1963), 955.

196 V. P. Silin, *Sov. Phys. JETP* **21** (1965), 1127.

197 D. F. DuBois and M. V. Goldman, *Phys. Rev.* **164** (1967), 201.

198 K. Nishikawa, *J. Phys. Soc. Japan* **24** (1968), 1152.

199 F. Cap, *Plasma Instabilities* (Academic Press, New York, 1980).

200 L. D. Landau and E. M. Lifschitz, *Mechanics* (Pergamon, New York, 1969) p. 80.

201 W. Paul and M. Raether, *Z. Phys.* **140** (1955), 262; H. Hora, and H. J. Schwarz, *Jap. J. Appl Phys.* **12**, Suppl. 2 (1974) 69.

202 E. Kamke, *Differentialgleichungen, Lösungsmethoden und Lösungen* (Akademie Verlag Ges. Leipzig, 1943), Vol. 1, p. 397.

203 D. Biskamp and H. Welter, Inst. f. Plasmaphysik, Garching, Report (1972); see Ref. 90.

204 D. Bohm and E. P. Gross, *Phys. Rev.* **75** (1949), 1851.

205 G. A. Askaryan, *Sov. Phys. JETP* **15** (1962), 1088.

206 H. Hora, *Z. Phys.* **226** (1969), 156.

207 D. F. Dubois, see Ref. 5, Vol. 3A (1974), p. 267.

208 R. L. Dewar, *Phys. Fluids* **16** (1973), 431.

209 J. P. Watteau, see Ref. [87]. (IAEA, Vienna, 1979) *Vol. III, p.* **65**.

210 A. Y. Wong, see Ref. 5, Vol. 4B (1977), p. 783.

211 J. L. Bobin, W. Woo, and J. S. DeGroot, *J. Physique* **38** (1977), 769; J. Weiland and H. Wilhelmsson, *Coherent Non-linear Interaction of Waves in Plasma* (Pergamon, New York, 1977).

212 R. Balescu, *Developments in High Power Lasers*, C. Pellegrini, ed. (E. Fermi Sch. Vol. 74, North Holland, 1980).

213 H. H. Chen and C. S. Liu, *Phys. Rev. Lett.* **37** (1976), 693; **39** (1977), 881; C. S. Liu and M. N. Rosenbluth, *Phys. Fluids* **17** (1974), 778.

214 H. H. Chen, C. Grebogi, C. S. Liu, V. K. Tripathi, *Plasma Physics and Controlled Nuclear Fusion Research 1978* (IAEA, Vienna, 1979), Vol. 3, p. 181.

215 R. G. Rehm, *Phys. Fluids* **13** (1970), 282; F. Winterberg, *Z. Naturforsch*, **30A** (1975), 976; C. Yamanaka, T. Yamanaka, J. Mizui, and N. Yamaguchi, *Phys. Rev.* **A11** (1975), 2138; M. S. Sodha, S. Prasad, and V. K. Tripathi, *J. Appl. Phys.* **46** (1975), 637; V. A. Volkov, F. V. Grigorev, V. V. Kalinovski, S. B. Korner, L. M. Lavrov, Y. V. Maslov, V. D. Urkin, V. P. Chudinov, *Sov. Phys. JETP* **42** (1975) 58; E. B. Goldman, *J. Appl. Phys.* **45** (1974), 5211; V. D. Leuthauser, *Atomkernenergie* **24** (1974), 193; J. R. Saraf, *Naturforsch.* **31A** (1976), 1038; D. Baboneau, G. diBona, P. Chelle, M. Decroissette, and J. Martineau, *Phys. Lett.* **A57** (1976), 247; R. Dragila, *J. Phys.* **D11** (1978), 1683; V. H. Kulkarni, *Ind. J. Phys.* **A51** (1977), 356.

216 Yu. A. Afanasyev, O. N. Krokhin and G. V. Sklizkov, *IEEE J. Quant. Electron.* **2** (1966), 483.

217 A. Caruso and R. Gratton, *Plasma Phys.* **10** (1968), 867.

218 G. J. Pert, *J. Phys.* **A5** (1972), 506; **B12** (1979) 2067; G. J. Tallents, *J. Phys.* **B13** (1980) 3057; **B10** (1977), 1763.

219 J. Nuckolls, see Ref. 5 (1975), Vol. 38, p. 399.

220 G. Guderley, *Z. Luftfahrtforschung* **19** (1942), 302.

221 L. L. Lengyel, *AIAA J.* **11** (1973), 1347; *Nucl. Fusion* **17** (1977), 805

222 H. Hora, *Aust. J. Phys.* **29** (1976), 375.

223 H. Hora, see Ref. 5, Vol. 2 (1972), 341.

224 J. W. Shearer, see Figs. 6 to 8 in Ref. 171.

225 E. Goldman, H. Hora, and M. Lubin, presented at 7th Europ. Conf. Laser Interaction with Matter, Garching, April 1974.

226 H. Hora, *Atomkernenergie* **24** (1974), 187.

227 A. Y. Wong and R. L. Stenzel, *Phys. Rev. Lett.* **34** (1975), 727; R. L. Stenzel, *Phys. Fluids* **19** (1976), 865.

228 H. C. Kim, R. L. Stenzel, and A. Y. Wong, *Phys. Rev. Lett.* **33** (1974), 886; G. Farkas, *Opt. Comm.* **21** (1977), 408.

229 M. E. Marhic, see Fig. 10, Ref. 156. Indications of the Nonlinear Force were reported before by G. Beaudry and J. Martineau, *Phys. Lett.* **43A** (1973), 331.

230 Yu. A. Zakharenkov, N. N. Zorev, O. N. Krokhin, Yu. A. Mikhailov, A. A. Rupasov, G. V. Sklizkov, and A. S. Shikanov, *Sov. Phys. JETP* **43** (1976), 283.

231 R. Fedosejevs, I. V. Tomov, N. H. Burnett, G. F. Enright, and M. C. Richardson, *Phys. Rev. Lett.* **39** (1977), 932.

232 H. Azechi, S. Oda, K. Tanaka, T. Norimatsu, T. Sasaki, T. Yamanaka, and C. Yamanaka, *Phys. Rev. Lett* **39** (1977), 1144.

233 J. F. Lam, B. Lippman, and F. Tappert, *Phys. Fluids* **20** (1977), 1176.

234 W. Gekelman and R. L. Stenzel, *Phys. Fluids* **20** (1977), 1316.

235 B. Luther-Davies, *Opt. Comm.* **23** (1977) 98.

236 T. P. Donaldson, J. E. Balmer, P. Wägli, and P. Lädrach, *Plasma Physics and Controlled Nuclear Fusion Research 1978* (IAEA, Vienna 1979) Vol. 3, p. 157; see also Ref. 82.

237 D. C. Slater, *Appl. Phys. Lett.* **31** (1977), 196; see also Ref. 79.

238 B. Luther-Davies, *Appl. Phys. Lett.* **32** (1978), 209; J. C. Samson, and A. J. Alcock, *Phys. Rev Lett.* **A51** (1975), 315.

239 K. R. Manes, H. G. Ahlstrom, R. A. Haas, and J. F. Holzrichter, *J. Opt. Soc. Amer.* **67** (1977), 717; see also Ref. 84.

240 E. B. Goldman, W. Leising, A. Brauer, and M. Lubin, *J. Appl. Phys.* **45** (1975), 1158.

241 D. J. Bradley, presented at 4th General Conf. Europ. Phys. Soc. York, 1978, invited paper.

242 R. Castillo, H. Hora, E. L. Kane, V. F. Lawrence, M. B. Nicholson-Florence, M. M. Novak, P. S. Ray, J. R. Shepanski, R. Sutherland, A. I. Tsivinsky, and H. A. Ward, *Nucl. Inst. Meth.* **144** (1977), 27.

243 V. F. Lawrence, Ph.D. Thesis, Univ. New South Wales, March 1978.

244 H. Hora, R. Castillo, R. G. Clark, E. L. Kane, V. F. Lawrence, R. D. C. Miller, M. F. Nicholson-Florence, M. M. Novak, P. S. Ray, J. R. Shepanski, and A. I. Tsivinsky, *Plasma Physics and Controlled Nuclear Fusion Research 1978* (IAEA, Vienna 1979) Vol. 3, p. 237.

245 G. B. Lubkin, *Phys. Today* **30**, No. 9 (1977) p. 19.

246 H. Izeki, invited lecture, *Proc. 13th Int. Conf. Phen. Ionized Gases* (Berlin, 1977), p. 139

247 R. Godfrey, Honours Thesis, Theoret. Phys., University New South Wales, Sydney 1978.

248 N. G. Denisov, *Sov. Phys. JETP* **4** (1957), 544.

249 See Fig. 7.8b of Ref. 1.

250 C. Yamanaka, *Progress in Inert. Conf. Fusion*, IAEA, San Francisco, Tech. Comm., February 1978 (Science Application, McLean Va, 1978), p. 29; A. L. Peratt and R. L. Watterson, *Phys. Fluids* **20** (1977), 1911; J. L. Bocher, J. P. Flie, J. Martineau, M. Rabeau, and C. Patou, see Ref. 5 (1977), Vol. 4B, p. 657; V. V. Blazhenkov, *Zh. E. TF.* **78** (1980), 1386.

251 K. Sauer, N. E. Andreev, and K. Baumgärtel, *Plasma Physics and Controlled Nuclear Fusion Research 1978* (IAEA, Vienna, 1979) Vol. 3, p. 187.

252 H. Hora, in *Advances in Inertial Confinement Systems*, Chiyoe Yamanaka, Ed. (Institute of Laser Engineering, Osaka University, 1980) p. 263.

253 P. Lalousis, see Section 5) of Ref. 6.

254 See H. H. Chen, and Y. C. Lee, *Phys. Rev. Lett.* **43** (1979), 264; R. Klima and V. A. Petrzilka, *Cz. J. Phys.* **29** (1979), 863.

255 H. Hora, *Sov. J. Quant. Elect.* **6** (1976), 154.

256 H. Hora, *Phys. Fluids* **17** (1974) 939.

257 J. L. Carter, and H. Hora, *J. Opt. Soc. Am.* **61** (1971) 1640.

258 V. G. Borodin, A. V. Cjarukchev, V. K. Chevokin, A. A. Girokhov, M. F. Danilov, V. D. Dyatlov, V. M. Komarov, A. A. Mak, V. A. Malinov, R. N. Medvedev, G. V. Obraztsov, P. P. Pashinin, A. M. Prokhorov, M. Ya. Schelev, and A. D. Starikov, *Summaries of the 8th Natl. Conf. Laser and Nonlin. Optics* (Tbilisi, 1976), p. 245.

259 R. B. White and F. F. Chen, Eq. (31) of Ref. 172.

260 See p. 119, Ref. 202.

261 P. Lädrach (private communication 1979).

262 H. Kogelnik and H. P. Weber, *J. Opt. Soc. Amer.* **64** (1974), 174.

263 I. P. Kaminov, W. M. Mammel, and H. P. Weber, *Appl. Opt.* **13** (1976), 396; R. Renard,

J. Opt. Soc. Amer. **54** (1964), 1190; see also Ref. 257.

264 S. Eliezer and Z. Schuss, *Phys. Lett.* **A70** (1979), 307.

265 P. Lädrach and J. E. Balmer, *Opt. Comm.* **31** (1979), 350.

266 See Ref. 4, p. 182.

267 H. Maki and K. Niu, *J. Phys. Soc. Japan* **45** (1978), 269.

268 J. P. Freidberg, R. W. Mitchell, R. L. Morse, and L. J. Rudsinksi, *Phys. Rev. Lett.* **28** (1972), 795; D. W. Forslund, J. M. Kindel, K. Lee, E. L. Lindman, and R. L. Morse, *Phys. Rev.* **A11** (1975), 679; K. G. Estabrook, E. J. Valeo, and W. L. Kruer, *Phys. Fluids* **18** (1975), 1151.

269 H. Maki, *J. Phys. Soc. Japan* **46** (1979), 653; Y. Sakamoto, *Jap. J. Appl. Phys.* **16** (1977), 1015

270 P. Koch and J. Albritton, *Phys. Rev. Lett.* **32** (1976), 1420.

271 J. L. Bobin, see Ref. 5 (1980), Vol. 5, p. 000.

272 H. Hora, *Ann. Physik* **22** (1969), 402.

273 H. Hora, see Ref. 5 (1977), Vol. 4B, p. 841.

274 M. V. Goldman, and D. R. Nicholson, *Phys. Rev. Lett.* **41** (1978) 406.

275 P. Debye, *Ann. Physik* **30** (1909), 755.

276 H. Hora, *Optik* **17** (1960), 409.

277 The difference between optical waves and matter waves in the second order is seen in the long beating wavelength of modulated electron waves: H. Hora, *Light Scattering in Solids*, M. Balkanski, Ed. (Flamarion, Paris, 1972), p. 128.

278 R. Y. Chiao, E. Garmire, and C. H. Townes, *Phys. Rev. Lett.* **13** (1964), 479; A. G. Litvak, *Sov. Phys. JETP* **30** (1970), 364; S. A. Akhmanov, D. P. Krindakh, A. P. Sukhorokov, and R. V. Khokhlov, *JETP Lett.* **6** (1977) 38.

279 H. A. Ward, MSc Thesis, Univ. New South Wales, 1980.

280 A. Schlüter, *Plasma Phys.* **10** (1968), 471.

281 M. S. Sodha, R. S. Mittal, *Opto-Electron.* **6** (1974), 167; M. S. Sodha, A. K. Ghatak, and V. K. Tripathi, *Progress in Optics*, E. Wolf, Ed. (Academic Press, New York, 1976), Vol. 13, p. 171; W. Engelhardt, *Appl. Phys. Lett.* **15** (1969), 216; D. P. Tewari and A. Kamar, *Plasma Phys.* **17** (1975), 133; M. Y. Yu, K. H. Spatschek, and P. K. Shukla, *Z. Natf.* **A29** (1974), 1736; M. S. Sodha, A. K. Chakravarti, U. P. Phadke, G. D. Gautama, and I. Rattan, *Appl. Phys. Lett.* **22** (1973), 121; M. S. Sodha, L. A. Patel, and R. P. Sharma, *J. Appl. Phys.* **49** (1978), 3707.

282 A. J. Palmer, *Phys. Fluids* **14** (1971), 2714; J. W. Shearer and J. L. Eddleman, *Phys. Fluids* **16** (1974), 1753; E. Valeo, *Phys. Fluids* **17** (1974), 1391; G. J. Tallents, *J. Phys.* **B10** (1977), 796; B. Bhat and V. K. Tripathi, *J. Appl. Phys.* **46** (1975), 1141; W. M. Mannheimer, *Phys. Fluids* **17** (1974), 1413; V. del Pizzo, B. Luther–Davies, and M. R. Siegrist, *Appl. Phys.* **14** (1977), 381; Yu. My, and K. H. Spatschek, *Z. Naturforsch*, **39A** (1974), 1736; R. Dragila and J. Krepelka, *J. Physique* **39** (1978), 617; *J. Phys.* **D11** (1975), 217; A. J. Alcock, C. De Michelis, and M. C. Richardson, *IEEE J. Quant. Electron.* **6** (1970), 622; N. A. Amherd and G. C. Vlases, *Appl. Phys. Lett.* **24** (1974), 93; M. Hugenschmidt, K. Vollrath, and A. Hirth, *Appl. Opt.* **11** (1972), 339; L. C. Steinhauer and H. G. Ahlstrom, *Phys. Fluids* **14** (1971), 1109; V. S. Soni and V. P. Nayyar, *J. Phys.* **D13** (1980), 361; S. K. Sinha and M. S. Sodha, *Phys. Rev.* **A21** (1980), 633; W. M. Mannheimer and E. Ott, *Phys. Fluids* **17** (1974), 1413.

283 V. V. Korobkin and A. J. Alcock, *Phys. Rev. Lett.* **21** (1968), 1433; M. C. Richardson and A. J. Alcock, *Appl. Phys. Lett.* **18** (1971), 357.

284 N. Ahmad, B. C. Gale, and M. H. Key, *J. Phys.* **B2** (1969), 403; R. G. Tomlinson, *IEEE J. Quant. Electron.* **5** (1969), 591; F. V. Grigoev, V. V. Kalinovski, S. B. Kormer, L. M. Lavrov, Yu. V. Maslov, V. D. Orlin, and V. P. Chudinov, *Sov. Phys. JETP* **42** (1976), 58.

285 N. L. Tsintsatse and E. G. *Tsikarishvili, Astrophys, Space Sci.* **39** (1976), 191.

286 H. Hora and E. L. Kane, *Appl. Phys.* **13** (1977), 165.

287 K. H. Spatschek, *J. Plasma Phys.* **18** (1978) 293.

288 H. Hora, E. L. Kane, and J. L. Hughes, see Ref. 81, Fig. 1.

289 M. R. Siegrist, *J. Appl. Phys.* **48** (1977), 1378.

290 E. L. Kane, Ph.D. Thesis, University of New South Wales, April 1979.

291 D. A. Jones, E. L. Kane, P. Lalousis, P. R. Wiles, and H. Hora (unpublished); see Ref. 170

292 M. J. Hollis, *Opt. Comm.* **25** (1978), 395.

293 See Ref. 14 and B. W. Boreham, R. Mavaddat, J. L. Hughes, and H. Hora, *Proc. 11th Symp. Rarefied Gas Dyn.*, R. Campargue ed. (CEA, Paris 1979) Vol. 1, p. 505.

294 B. W. Boreham and B. Luther–Davies, *J. Appl. Phys.* **50** (1979), 2533; B. Luther–Davies, B. W. Boreham, and V. E. del Pizzo, see Ref. 5, Vol. 5 (1980), p. 000; M. S. Sodha, and D. Subbarao, *Appl. Phys. Lett.* **35** (1979), 851; R. Mavaddat, and K. Ghatak, *J. Appl. Phys.* **31** (1980) 3501.

295 B. W. Boreham (in print); B. W. Boreham, *J. Opt. Soc. Amer.* **68** (1978), 698.

296 V. L. Keldysh, *Sov. Phys. JETP* **20** (1965), 1307; G. V. Ostrovski, *Usp. Fi Nauk* **111** (1973), 579; S. A. Ramsden, *Physics of Hot Plasma*, B. J. Tye, et al., Eds. (Oliver & Boyd, Edinburgh 1970).

297 N. K. Bereshotshaya, G. S. Voronov, G. A. Delone, and G. K. Piskova, *Sov. Phys. JETP* **31** (1970), 403; A. V. Phelps in *Physics of Quantum Electronics* (McGraw–Hill, 1966) p. 538.

298 P. A. M. Dirac, *The Principles of Quantum Mechanics*, 3rd ed. (Oxford University Press, 1949) p. 284.

299 R. Castillo (unpublished, 1977).

300 See pp. 386–390 of Ref. 54.

301 H. Hora and J. L. Hughes, German Pat. 2 832 100; about a forward momentum of the electrons to take account for the increase of the momentum of the laser pulse, see G. Viera, and H. Hora (to be published).

302 D. A. G. Deacon, L. R. Elias, J. M. J. Madey, G. J. Ramian, H. A. Schwettman, and T. I. Smith, see Ref. 49; J. M. J. Madey and D. A. G. Deacon, Stanford Univ. Rept. HEPL-797 (1976).

303 H. Hora, J. L. Hughes, and B. W. Boreham, *Sov. J. Quant. Elect.* 9 (1979), 464.

304 J. A. Stamper, K. Papadopoulos, R. N. Sudan, S. O. Dean, E. A. McLean, and J. M. Dawson, *Phys. Rev. Lett.* **26** (1972), 1012; F. Schwirzke, see Ref. 5 (1974), Vol. 3A, p. 234.

305 J. A. Stamper, NRL-Report 7411 (May 1972); see Ref. 158; D. A. Tidman, *Phys. Rev. Lett.* **32** (1974), 1179

306 J. A. Stamper, E. A. McLean, and B. H. Ripin, *Phys. Rev. Lett.* **40** (1978), 1177; J. J.

Thomson, C. E. Max, and K. Estabrook, *Phys. Rev. Lett.* **35** (1975), 663; S. P. Obenschain and N. C. Luhmann, *J. Phys. Rev. Lett.* **42** (1979), 311; A. Hasegawa, M. Y. Yu, P. K. Shukla, K. H. Spatschek, *Phys. Rev. Lett.* **24** (1978), 1656.

307 M. H. Key, presented at 12th Europ. Conf. Laser Interact. Matter, Moscow, December 1978

308 T. Sakagenu, *Advances Inertial Conf. Fusion*, C. Yamanaka Ed. (Institute of Laser Engineering, Osaka 1980) p. 122.

309 H. A. Ahlstrom, "Inertial Confiement Fusion," *Opt. Soc. Amer.*, San Diego (February 1978), paper ThA2-1.

310 A. C. Walker, S. Kogoski, T. Samatak, I. Spalding *Opt. Comm.* **27** (1978) 247.

311 H. P. Weber (private communication 1979).

312 H. Hora and M. M. Novak, see Ref. 5, 1977, Vol. 4B, p. 999.

313 J. W. Sheaerer, J. Garrison, J. Wong and J. E. Swain, see Ref. 5 (1974), Vol. 3B, p. 803; H. Hora, *Opto-Electron* **5** (1973), 491; A. I. Tsivinsky, M.Sc. Thesis, University of New South Wales, 1977; F. V. Bunkin and A. M. Prokhorov, in *Polarization, Matiere et Rayonnement* (Soc. France de Physique, Paris, 1969).

314 H. Hora, E. L. Kane, and J. L. Hughes, Fig. 1 in Ref. 15.

315 See Ref. 80; B. Luther–Davies, *Opt. Comm.* **23** (1977), 98.

316 R. A. Haas, J. F. Holzrichter, H. G. Ahlstrom, E. Storm, and K. R. Manes, *Opt. Comm.* **18** (1976), 105.

317 Los Alamos, press release, 4th March 1977; R. P. Godwin and F. Engelmann, *Europhys. News* **8**, No. 5 (1977), 11.

318 W. Willis, see Ref. 5, 1977, Vol. 4B, p. 991.

319 C. Yamanaka, *Laser Focus*, No. 2 (1980), p. 40; see Refs. 47 and 48.

320 L. La Mer Slaner, *Prospectus*, Soc. to Advance Fusion Energy, 10 Normandy Lane, Scarsdale, New York 10583.

321 A. Grey, Jr., G. Miley, and G. Brumlik, *National Rev.* 150E 95th Street, New York, February 2, 1979.

322 S. J. Zweben, *Bull. Amer. Phys. Soc.* **24** (1979), 1072.

323 F. H. Coensgen, *Bull. Amer. Phys. Soc.* **84** (1979), 966.

324 W. N. Hugrass, I. R. Jones, K. F. McKenna, M. G. R. Phillips, R. G. Storer, and H. Tuczek, *Phys. Rev. Lett.* **44** (1980), 1676.

325 R. A. Gross and K. N. Kmetyk, *Bull. Amer. Phys. Soc.* **23** (1978), 884.

326 G. Yonas, *Scientif. Amer.* **239** (1978), 40.

327 T. F. Godlove and D. F. Sutter, *Plasma Physics and Controlled Nuclear Fusion Research*, 1978 (IAEA, Vienna, 1979) Vol. III, p. 211.

328 R. O. Bangerter, presented at 13th Europ. Conf. Laser Matter Interaction, Leipzig, December 1979.

329 G. Cooperstein et al., see Ref. 5 (1980), Vol. 5, p. 105.

330 M. L. E. Oliphant, P. Harteck, and Lord Rutherford, *Proc. Roy. Soc.* **A144** (1934), 692.

331 R. W. B. Best, *Nucl. Inst. Meth.* **144** (1977), 1.

332 M. L. E. Oliphant and Lord Rutherford, *Proc. Roy. Soc.* **A141** (1933), 259; R. W. Best, *Nucl. Inst. Meth.* **144** (1978) 1; K. Okamoto, *J. Nucl. Sci. Tech.* **14** (1977), 762.

333 R. G. Clark, H. Hora, P. S. Ray, and E. W. Titterton, *Phys. Rev.* **C18** (1978), 1127.

334 N. P. Heydenburg, C. M. Hudson, D. K. Inglis, and W. D. Whitehead, *Phys. Rev.* **74** (1948), 405.

335 C. R. McClenahan and R. E. Segel, *Phys. Rev.* **C11** (1975), 370.

336 C. R. Gould and J. M. Joyce, NBS Rept. 425 (1975), Vol. II, p. 697.

337 G. S. Mani, R. M. Freeman, F. Picard, D. Redon, and A. Sadeghi, *Proc. Phys. Soc.* **85** (1965), 281.

338 J. Tuck, *Nucl. Fusion* **1** (1961), 201.

339 H. Hora, Inst. f. Plasma Physik, Garching, Rep. 6/23 (1964); U.S. Govt. Res. Rep. NRC-TT-1193 (1965); see Ref. 5 (1971) Vol. 1, p. 427. F. Flona, *Nucl. Fusion* **11** (1971), 635; L. Rothardt, *Kernenergie* **13** (1970), 269.

340 H. Hora and D. Pfirsch, presented at 6th Int. Conf. Quantum Electron., Kyoto, Sept. 1970, *Conf. Digest* p. 10; J. P. Somon, *Nucl. Fusion* **12** (1972), 461; J. L. Bobin, *Phys. Fluids* **14** (1971), 2341; S. Martelluci, *Energia Nucleare* **18** (1971), 541; N. H. Burnett, *Can. J. Phys.* **50** (1972), 3184.

341 H. Opower and W. Press, *Z. Naturforsch.* **21A** (1966), 344.

342 K. H. Sun, J. M. Hicks, L. M. Epstein, E. W. Sucov, *J. Appl. Phys.* **38** (1967), 3402.

343 D. K. Bhadhra, *Phys. Fluids* **11** (1968), 234.

344 H. Opower, H. Puell, W. Heinicke, and W. Kaiser, *Z. Naturforsch.* **22A** (1967), 1392.

345 A. G. Engelhardt, *Westinghouse Res. Rept.* WERL-3472-5, March 27, 1967.

346 V. Ionescu, presented at Laser Applications in Plasma Physics, IAEA, Vienna 1969; D. L. Nguyen and K. J. Parbhakar, *J. Appl. Phys.* **45** (1974), 2089; O. R. Wood, *Proc. IEEE* **62** (1974), 355.

347 E. Hantzsche, *Physik und Technik. des Plasmas*, G. Wallis, Ed. (Phys. Ges. DDR, 1969) p. 326.

348 F. Schwirzke and A. W. Cooper, Report NPS-61SW0031A, April 15, 1970, p. 51.

349 R. E. Kidder, *Nucl. Fusion* **14** (1974), 797.

350 F. Winterberg, Desert Res. Inst. Rept. 64, March 1969.

351 P. S. Ray and H. Hora, *Z. Natuforsch.* **32A** (1977), 538.

352 P. S. Ray and H. Hora, *Nucl. Fusion* **16** (1976), 535; see Ref. 109.

353 P. S. Ray, Ph.D. Thesis, University of New South Wales, August 1977; UNSW-Dept. Theoret. Phys., Rept. No. 11.

354 H. Hora and P. S. Ray, *Z. Naturforsch.* **33A** (1978), 890.

355 J. H. Nuckolls, see Ref. 5 (1974), Vol. 3B, p. 403.

356 K. Boyer, *Astronautics and Aeronautics 11* (1973), 28.

357 H. Wobig, *Naturwiss.* **61** (1974), 97.

358 J. H. Nuckolls, presented at IAEA Inertial Fusion Committee Meeting, Livermore, February 6, 1978.

359 K. A. Brueckner and S. Jorna, p. 358, Ref. 136; P. Belland, C. DeMichelis, M. Mattioli, and R. Papoular, *Appl. Phys. Lett.* **18** (1971), 542.

360 H. Hora, presented at Colloquium Nuclear Club of Wall Street, Dreyfuss Foundation, New York, July 31, 1978.

361 J. M. Dawson, presented at Clean Fusion Fuel Meeting, Oak Ridge, October 1978.

362 T. H. Maiman, see Ref. 44; R. J. Collins, D. F. Nelson, A. L. Schawlow, W. Bond, C. G. S. Garret, and W. Kaiser, *Phys. Rev. Lett.* **5** (1960), 303.

363 N. G. Basov and O. N. Krokhin, *Sov. Phys. JETP* **19** (1964), 123; see Ref. 125.

364 A. Kastler, Compt. Rend., *Ac. Sc. Paris*, **258** (1964), 489.

365 N. G. Basov, P. G. Kriukov, S. D. Zakharov, Yu. V. Senatski, and S. V. Tchekalin, *IEEE J. Quant. Electron.* **4** (1968), 864.

366 M. J. Lubin, presented at Internal Conf. Laser Plasma Interact., August 1969.

367 F. Floux, presented at Belfast Conf., September 1969; F. Floux, D. Cognard, L. G. Denoeud, G. Piar, D. Parisot, J. L. Bobin, F. Delobeau, and C. Fauquignon, *Phys. Rev.* **A1** (1970), 821.

368 N. G. Basov, O. N. Krokhin, G. V. Sklizkov, S. J. Fedotov, and A. S. Shikanov, *Sov. Phys. JETP* **35** (1972), 109; L. Rothard, *Kernenergie* **13** (1970), 269.

369 E. Teller, *IEEE J. Quant. Electron.* **8** (1972), 564; *Bull. Amer. Phys. Soc.* **17** (1972), 1034.

370 J. H. Nuckolls, see Ref. 5 (1974), Vol. 3B, p. 399.

371 K. A. Brueckner, see Ref. 5 (1974), Vol. 3B, p. 427.

372 H. Gomberg, presented at 12th Europ. Conf. Laser Interact. Matter, Moscow, December 1978; R. Hofstadter, ibid.; Address to Panel Discussion.

373 J. F. Holzrichter, *Adv. Inertial Conf. Fusion*, C. Yamanaka, Ed. (Inst. Laser Eng., Osaka, 1980) p. 141.

374 G. H. Canavan, ibid., p. 129.

375 R. B. Godwin, ibid., p. 157.

376 M. J. Lubin, ibid., p. 165.

377 J. L. Emmet, presented at Conf. Laser Appl., Florida, December 1978.

378 G. B. Lubkin, *Phys. Today* **31**, No. 11 (1978), p. 20, 3rd col., 2nd paragraph.

379 H. G. Ahlstrom, J. F. Holzrichter, K. R. Manes, E. K. Storm, M. J. Boyle, K. M. Brooks, R. A. Haas, D. W. Philon, and V. C. Rupert, see Ref. 5 (1977), Vol. 4B, p. 437.

380 B. W. Weinstein, D. L. Willenborg, J. T. Weir, and C. D. Hendricks, Inertial Fusion Conf., Opt. Soc. Amer., San Diego, February 1978, paper Tu E9-1; R. J. Turnbull, Nucl. Eng. Univ. Ill, Rept. 1979.

381 R. Eckersley, *The Sydney Morning Herald* **148**, March 15 (1979), p. 1; see Refs. 14 and 301; H. Hora, *2nd Int. Conf. Energy Storage, Electron and Laser Beams*, H. Sahlin and O. Zucker, Eds. (Plenum, New York, 1979), in print.

382 R. T. Schneider, see Ref. 5 (1977), Vol. 3A, p. 85; G. H. Miley, ibid. Vol. 4A, p. 181; D. A. McArthur and P. B. Tollefsrud, *Trans. Amer. Nuc. Soc.* **19** (1974), 356; H. H. Helmik, J. L. Fuller, and R. T. Schneider, *Appl. Phys. Lett.* **26** (1975), 327; R. DeYoung, Ph.D. Thesis, *Nucl. Eng., Univ.* Illinois (1975); R. DeYoung, W. E. Wells, G. H. Miley, and J. T. Verdeyen, *Appl. Phys. Lett.* **28** (1976), 519.

383 J. P. Watteau, presented at 12th European Conf. Laser Interact. with Matter, Moscow, December 1978; S. A. Ramsden, *Physics of Hot Plasma*, B. J. Rye, Ed. (Oliver & Boyd, Edinburgh, 1970).

384 Yu. V. Raizer, *Laser Induced Discharge Phenomena* (Plenum, New York, 1977).

385 G. B. Lubkin, "The Shiva Experiment," *Physics Today* **33** (1979), No. 11, p. 21.

386 A. L. Peratt, Lawrence Livermore Lab. Rept. UCLR-80899 (1978) Phys. Rev. **A20** (1979) 2555; I. D. Larsen and J. Harte, Lawrence Livermore Lab. Rept. UCLR-79757-1 (1977).

387 J. L. Emmett, J. H. Nuckolls, and L. Wood, *Scientif. Amer.* **230** (1974), 24; H. Meldner, Lawrence Livermore Lab. Rept. UCLR 79380 (1979); "Laser Fusion", in Encyclopedia of Physics.

388 Yu. V. Afanasyev, N. G. Basov, P. P. Volosebni, E. G. Gamali, O. N. Krokhin, S. P. Kurdeomov, E. T. Levanov, A. A. Samarski, A. N. Tikhonov, *Plasma Physics and Thermonuclear Research 1974* (IAEA, Vienna, 1975) Vol. 2, p. 559.

389 K. A. Brueckner, see Walter Sullivan, *New York Times*, Jan. 13 (1980).

390 M. H. Brennan, T. S. Brown, H. Hora, C. Yamanaka, K. Yatsui, and M. Yokoyama, *Aust. Phys.* **17** (1980), 71.

391 R. E. Kidder, see [5] 1980, Vol. 5, p. 303.

392 E. Teller, presented at Laser Conference Miami, December 1978.

393 H. Hora, see Ref. 115, Chapter 3C.

394 G. Joos, *Theoretical Physics* (Blackie & Son, London, 1949), p. 554.

395 W. Weizel, *Lehrbuch der Theoretischen Physik* (Springer, Heidelberg, 1950) Vol. 2, p. 1172.

396 W. Weizel, *Z. Phys.* **135** (1953) 270.

397 R. Sutherland, Ph.D. Thesis, University of New South Wales, 1980.

398 M. Planck, *Ann. Physik* (*4*) **1** (1900), 719.

399 A. Einstein, *Mitt. Phys. Ges. Zürich*, No. 18, 1916; *Physikalische Zeitschr.* **18** (1917), 121.

400 H. Hora, *General Relativity and Gravitation One Hundred Years After the Birth of Albert Einstein*, A. Held, Ed. (Plenum, New York, 1980) Vol. 1, p. 17.

401 R. Lüst, *Fortschr. Phys.* **7** (1959) 503.

402 H. Alfvén, *Cosmic Plasma* (Reidel, Doordrecht, 1981).

Author Index

Numbers within brackets [ ] are reference numbers.

Subject Index